Cognitive Science and Technology

Series editor

David M. W. Powers, Adelaide, Australia

More information about this series at http://www.springer.com/series/11554

V. Srinivasa Chakravarthy
Ahmed A. Moustafa

Computational Neuroscience Models of the Basal Ganglia

Springer

V. Srinivasa Chakravarthy
Department of Biotechnology
Indian Institute of Technology Madras
Chennai, Tamil Nadu
India

Ahmed A. Moustafa
Cognitive and Behavioural Neuroscience
Western Sydney University
Sydney, NSW
Australia

ISSN 2195-3988 ISSN 2195-3996 (electronic)
Cognitive Science and Technology
ISBN 978-981-10-8493-5 ISBN 978-981-10-8494-2 (eBook)
https://doi.org/10.1007/978-981-10-8494-2

Library of Congress Control Number: 2018932536

Printed on acid-free paper

This Springer imprint is published by Springer Nature
The registered company is Springer Nature Singapore Pte Ltd.
The registered company address is: 152 Beach Road, #21-01/04 Gateway East, Singapore 189721, Singapore

Sri Aurobindo and The Mother
V. Srinivasa Chakravarthy

To my family: Marwa, Amr, Rasha, Mohamed, Kristina, Angelina, Hasan, and Haneen.
Ahmed A. Moustafa

Acknowledgements

Thanks to Prof. Joydeep Ghosh, who gave me the first taste of doing research under conditions of total freedom, necessary for any creative endeavour.

I am grateful to Prof. Read Montague who had inspired me to take up a career in computational neuroscience. Thanks to his mentorship, I learnt that in neuroscience, even in open and shut cases, there are always radically new ways of looking at a problem.

I am thankful to my friend and colleague, Prof. S. Bapiraju; the many hours of inspiring and insightful discussions with him had shaped my understanding of neuroscience. Thanks are also due to Prof. Srinivasa Babu for our collaborations on modelling the precision grip.

Thanks to my friends and colleagues in the basal ganglia community whose excellent work had paved the way to the efforts that led to this book: Dr. Michael Frank, Dr. Jeanette Kotaleski, Dr. Atsushi Nambu, Dr. Arvind Kumar, to mention a few.

I would like to thank my students who worked with me, over the years, on various aspects of modelling the basal ganglia to Dr. Sridharan Devarajan who contributed to our earliest efforts in this area; to Dr Gangadhar Garipelli and Denny Joseph who were part of our baby steps to create a synthesis of existing computational ideas of the basal ganglia; to Dr. Maitreye Rengaswamy and Deepika Sukumar who worked on aspects of modelling spatial navigation; to Ravi Krishnan who worked on modelling saccade generation; to Magdoom Mohammed, Dr. Shivkesavan Ratnadurai and Dr. Deepak Subramanian, who worked on an early version of the reaching model; to Sanjeev Kalva who had worked on the significance of the STN-GPe dynamics poised on the edge of chaos.

Last but not the least, I would like to thank my wife Indira and daughter Dyuti for their constant love and support, in matters neural and other.

V. Srinivasa Chakravarthy

I would like to thank Prof. Srinivasa Chakravarthy for always welcoming me at his laboratory in Chennai, and also for introducing me to his amazing students, from whom I have learnt so much about the basal ganglia.

I am very grateful to my friend Rosalyn Karin-D'arcy for helping me get started in my PhD upon arrival to America.

I am very thankful to my former supervisor and mentor Prof. Michael Frank, for teaching me the programming of reinforcement learning models and conducting experimental research with Parkinson's disease patients. Michael has opened my horizon to the field of deep brain stimulation. The time I spent at Michael's laboratory in Arizona is highly memorable.

Ahmed A. Moustafa

Contents

Chapter 1
Introduction

V. Srinivasa Chakravarthy and Ahmed A. Moustafa

Abstract The area of computational modeling of basal ganglia has seen an explosive growth in the last couple of decades. In this area, there is currently a multitude of modeling approaches, each approaching the functions of basal ganglia in a unique fashion, pursuing a specialized line of investigation. Existing models fall under certain prominent schools of thought, each successfully explaining a subset of basal ganglia functions that are amenable to that specific approach, while ignoring a host of other functions. The aim of this book is to describe a class of the basal ganglia models that comprehensively accommodates a wide range of the basal ganglia functions within a single modeling framework. This class of models is essentially based on reinforcement learning, a currently dominant paradigm for describing the basal ganglia function. However, the class of computational models described herein deviate significantly from some of the classical approaches like, for example, the Go-NoGo interpretation of the functional pathways of the basal ganglia. This class of models successfully explains a wide variety of motor functions, and some cognitive functions of the basal ganglia, in healthy and pathological conditions like the Parkinson's disease and other disorders associated with the basal ganglia.

It has been more than 20 years since James Houk, Joel Davis, and David Beiser published their superb book on Basal Ganglia models (MIT Press; ASIN: B010BF4U9K). Their very well-cited book (Houk, Davis, & Beiser, 1995) covered a variety of computational approaches to basal ganglia function. For example, Houk et al. (1995) proposed models that hypothesized that the matrisomes and striosomes within the basal ganglia subserve different functions. We discuss these in detail in Chap. 4. Some aspects of this hypothesis were confirmed in subsequent experimental studies (Brown et al., 2002; Wilson, 2004). Houk et al. (1995) have based their models on the Actor–Critic architecture, which has been repeatedly used in various experimental and computational studies of the basal ganglia and cortex (Atallah, Lopez-Paniagua, Rudy, & O'Reilly, 2007; Colas, Pauli, Larsen, Tyszka, & O'Doherty, 2017; Li, McClure, King-Casas, & Montague, 2006; Moustafa & Maida, 2007; Moustafa, Cohen, Sherman, & Frank, 2008; O'Doherty et al., 2004;

V. S. Chakravarthy and A. Moustafa, *Computational Neuroscience Models of the Basal Ganglia*, Cognitive Science and Technology,
https://doi.org/10.1007/978-981-10-8494-2_1

Piray et al., 2014). There are, however, some limitations of the collection of models described in Houk et al. (1995). The models in their book are quite diverse and do not present a coherent and unified picture of basal ganglia function. Second, the Houk et al. (1995) book mostly provides conceptual models without simulation studies to test the plausibility of model assumptions.

In addition, our knowledge on the basal ganglia has changed dramatically over the last couple of decades. Such diversity and multipolarity in approaches to basal ganglia function continue to date, though a large number of models are aligning themselves more and more with the reinforcement learning-based modeling framework. Within this framework, there is an emerging subclass of basal ganglia models that highlight the significance of complex dynamics of the indirect pathway in basal ganglia, and its contributions to exploration, an important ingredient of reinforcement learning (Balasubramani, Chakravarthy, Ravindran, & Moustafa, 2014, 2015; Mandali, Rengaswamy, Chakravarthy, & Moustafa, 2015; Muralidharan, Balasubramani, Chakravarthy, Lewis, & Moustafa, 2014). This class of models of basal ganglia has proven themselves to be capable of explaining a wide range of basal ganglia functions including action selection, spatial navigation, gait control, reaching and handwriting, precision grip control, saccade generation etc.

The purpose of our book is to amend and expand on James Houk's book by providing a comprehensive book on computational models of the basal ganglia. Our book provides a compendium of the aforementioned subclass of models of basal ganglia, which are partially based on our previously published studies on basal ganglia modeling (Balasubramani et al., 2014, 2015; Gangadhar, Joseph, & Chakravarthy, 2008; Gangadhar et al., 2009; Gupta, Balasubramani, & Chakravarthy, 2013; Helie, Chakravarthy, & Moustafa, 2013; Krishnan, Ratnadurai, Subramanian, Chakravarthy, & Rengaswamy (2011); Magdoom et al., 2011; Muralidharan et al., 2014, 2017; Sridharan, Prashanth, & Chakravarthy, 2006; Sukumar, Rengaswamy, & Chakravarthy, 2012). In addition, the models contained in the book present a coherent picture of basal ganglia. The book presents a long-awaited synthesis of some the key existent theories of basal ganglia function. In addition, our book presents computational models of basal ganglia-related disorders, including Parkinson's disease. For an integrative review on how the basal ganglia plays a key role in several motor processes, see Moustafa et al. (2016). In the last chapter, we will highlight the applications of understanding the role of the basal ganglia to treat neurological and psychiatric disorders. We also provide a roadmap for future work on basal ganglia modeling, including the simulation of the action of various neuromodulators in basal ganglia, as well as psychiatric disorders such schizophrenia. The MATLAB code for some of the simulation studies presented here is available upon request from the authors. These can be used and amended to simulate other functions of the basal ganglia, such as working memory, attention, as well as other basal ganglia-related disorders, such as ADHD.

References

Atallah, H. E., Lopez-Paniagua, D., Rudy, J. W., & O'Reilly, R. C. (2007). Separate neural substrates for skill learning and performance in the ventral and dorsal striatum. *Nature Neuroscience, 10*(1), 126–131. https://doi.org/10.1038/nn1817.

Balasubramani, P. P., Chakravarthy, S., Ravindran, B., & Moustafa, A. A. (2014). An extended reinforcement learning model of basal ganglia to understand the contributions of serotonin and dopamine in risk-based decision making, reward prediction, and punishment learning. *Frontiers in Computational Neuroscience, 8,* 47.

Balasubramani, P. P., Chakravarthy, S., Ravindran, B., & Moustafa, A. A. (2015). A network model of basal ganglia for understanding the roles of dopamine and serotonin in reward-punishment-risk based decision making. *Frontiers in Computational Neuroscience, 9,* 76.

Brown, L. L., Feldman, S. M., Smith, D. M., Cavanaugh, J. R., Ackermann, R. F., & Graybiel, A. M. (2002). Differential metabolic activity in the striosome and matrix compartments of the rat striatum during natural behaviors. *Journal of Neuroscience, 22*(1), 305–314.

Colas, J. T., Pauli, W. M., Larsen, T., Tyszka, J. M., & O'Doherty, J. P. (2017). Distinct prediction errors in mesostriatal circuits of the human brain mediate learning about the values of both states and actions: evidence from high-resolution fMRI. *PLoS Computational Biology, 13*(10), e1005810. https://doi.org/10.1371/journal.pcbi.1005810.

Gangadhar, G., Joseph, D., & Chakravarthy, V. S. (2008). Understanding Parkinsonian handwriting through a computational model of basal ganglia. *Neural Computation, 20*(10), 2491–2525.

Gangadhar, G., Joseph, D., Srinivasan, A. V., Subramanian, D., Shivakeshavan, R. G., Shobana, N., & Chakravarthy, V. S. (2009). A computational model of Parkinsonian handwriting that highlights the role of the indirect pathway in the basal ganglia. *Human Movement Science, 28* (5), 602–618.

Gupta, A., Balasubramani, P. P., & Chakravarthy, V. S. (2013). Computational model of precision grip in Parkinson's disease: A utility based approach. *Frontiers in Computational Neuroscience, 7.* https://doi.org/10.3389/fncom.2013.00172.

Helie, S., Chakravarthy, S., & Moustafa, A. A. (2013). Exploring the cognitive and motor functions of the basal ganglia: an integrative review of computational cognitive neuroscience models. *Frontiers in Computational Neuroscience, 7,* 174. https://doi.org/10.3389/fncom.2013.00174.

Houk, J. C., Davis, J. L., & Beiser, D. G. (1995). *Models of information processing in the basal ganglia.* Cambridge: The MIT press.

Krishnan, R., Ratnadurai, S., Subramanian, D., Chakravarthy, V. S., & Rengaswamy, M. (2011). Modeling the role of basal ganglia in saccade generation: is the indirect pathway the explorer? *Neural Networks, 24*(8), 801–813.

Li, J., McClure, S. M., King-Casas, B., & Montague, P. R. (2006). Policy adjustment in a dynamic economic game. *PLoS ONE, 1,* e103. https://doi.org/10.1371/journal.pone.0000103.

Magdoom, K., Subramanian, D., Chakravarthy, V. S., Ravindran, B., Amari, S.-I., & Meenakshisundaram, N. (2011). Modeling basal ganglia for understanding parkinsonian reaching movements. *Neural Computation, 23*(2), 477–516.

Mandali, A., Rengaswamy, M., Chakravarthy, S., & Moustafa, A. A. (2015). A spiking Basal Ganglia model of synchrony, exploration and decision making. *Frontiers in Neuroscience, 9,* 191.

Moustafa, A. A. & Maida, A. S. (2007). Using TD learning to simulate working memory performance in a model of the prefrontal cortex and basal ganglia. *Cognitive Systems Research, 8,* 262–281.

Moustafa, A. A., Chakravarthy, S., Phillips, J. R., Gupta, A., Keri, S., Polner, B., … Jahanshahi, M. (2016). Motor symptoms in Parkinson's disease: A unified framework. *Neuroscience & Biobehavioral Reviews, 68*, 727–740.

Moustafa, A. A., Cohen, M. X., Sherman, S. J., & Frank, M. J. (2008). A role for dopamine in temporal decision making and reward maximization in parkinsonism. *Journal of Neuroscience, 28*(47), 12294–12304. https://doi.org/10.1523/JNEUROSCI.3116-08.2008.

Muralidharan, V., Balasubramani, P. P., Chakravarthy, V. S., Gilat, M., Lewis, S. J., & Moustafa, A. A. (2017). A Neurocomputational Model of the Effect of Cognitive Load on Freezing of Gait in Parkinson's Disease. *Frontiers in Human Neuroscience, 10,* 649. https://doi.org/10.3389/fnhum.2016.00649.

Muralidharan, V., Balasubramani, P. P., Chakravarthy, V. S., Lewis, S. J., & Moustafa, A. A. (2014). A computational model of altered gait patterns in parkinson's disease patients negotiating narrow doorways. *Frontiers in Human Neuroscience, 7,* 190. https://doi.org/10.3389/fncom.2013.00190.

O'Doherty, J., Dayan, P., Schultz, J., Deichmann, R., Friston, K., & Dolan, R. J. (2004). Dissociable roles of ventral and dorsal striatum in instrumental conditioning. *Science, 304* (5669), 452–454.

Piray, P., Zeighami, Y., Bahrami, F., Eissa, A. M., Hewedi, D. H., & Moustafa, A. A. (2014). Impulse control disorders in Parkinson's disease are associated with dysfunction in stimulus valuation but not action valuation. *The Journal of Neuroscience, 34*(23), 7814–7824.

Sridharan, D., Prashanth, P., & Chakravarthy, V. (2006). The role of the basal ganglia in exploration in a neural model based on reinforcement learning. *International Journal of Neural Systems, 16*(02), 111–124.

Sukumar, D., Rengaswamy, M., & Chakravarthy, V. S. (2012). Modeling the contributions of Basal ganglia and Hippocampus to spatial navigation using reinforcement learning. *PLoS ONE, 7*(10), e47467.

Wilson, C. J. (2004). Basal ganglia. In G. M. Shepherd (Ed.), *The synaptic organization of the brain* (pp. 361–413). New York: Oxford University Press.

Chapter 2
The Molecular, Cellular, and Systems-Level Structure of the Basal Ganglia

Alekhya Mandali, V. Srinivasa Chakravarthy and Ahmed A. Moustafa

Abstract This chapter provides a brief overview of the systems, cellular, and molecular structure of the various nuclei of basal ganglia (BG) such as striatum, STN, GPe, GPi, and the SNr including the various neurotransmitters impacting its function. We start with the system-level connection between cortex and BG and then cover the various cell types, receptors (such as dopaminergic, acetylcholine) present on each of the BG nuclei. The effect of Parkinson's disease on their dynamics especially the STN–GPe oscillatory network is then discussed. The dopaminergic systems SNc and VTA are also covered in terms of their architecture and input–output synaptic projection patterns. Finally, a short intro to the multiple cortico-BG loops and their functional relevance is discussed. This brief overview helps provide background on BG structure, which is the basis of several models we present in this book.

2.1 Anatomical Structure of Basal Ganglia

The Basal Ganglia (BG) are a group of seven subcortical nuclei, involved in various important functions ranging from motor control to cognitive functions such as decision making, working memory, and action selection (Chakravarthy, Joseph, & Bapi, 2010; Chersi, Mirolli, Pezzulo, & Baldassarre, 2013; Gurney, Prescott, & Redgrave, 2001a, 2001b; Humphries & Gurney, 2002; Schroll, Vitay, & Hamker, 2012; Yucelgen, Denizdurduran, Metin, Elibol, & Sengor, 2012) (Fig. 2.1).

2.1.1 Systems-Level

The anatomical components of BG include the neo-striatum (caudate, putamen, and nucleus accumbens), Globus Pallidus externa, GPe, and Globus Pallidus interna, GPi, subthalamic nucleus (STN), and substantia nigra (pars compacta, SNc, and pars reticulata, SNr). The BG receive inputs from the cortex through the striatum and STN (Aravamuthan, Muthusamy, Stein, Aziz, & Johansen-Berg, 2007;

V. S. Chakravarthy and A. Moustafa, *Computational Neuroscience Models of the Basal Ganglia*, Cognitive Science and Technology,
https://doi.org/10.1007/978-981-10-8494-2_2

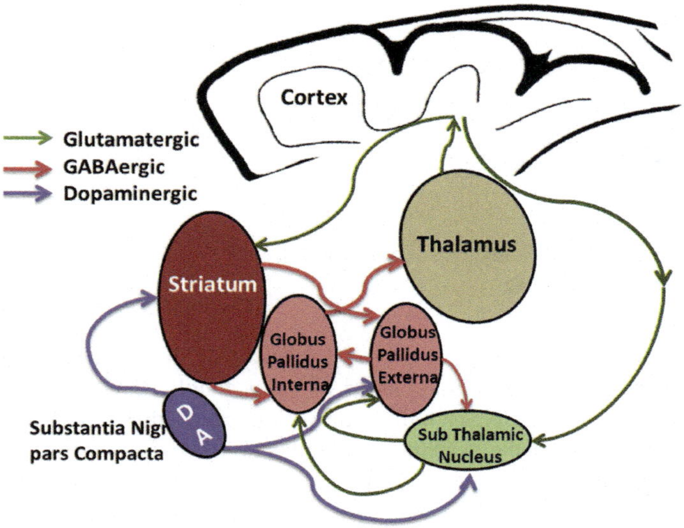

Fig. 2.1 This figure shows the Basal Ganglia with its major nuclei and their synaptic connections. The glutamatergic input from motor and executive cortices enters both striatum and STN and leaves through GPi or SNr (not shown in figure) via the thalamus. The dopaminergic input from SNc modulates the activity of neurons in striatal, STN, GPe nuclei

Maurice, Deniau, Glowinski, & Thierry, 1998) and projects through SNr and GPi, the output nuclei of BG, via thalamus (Albin, Young, & Penney, 1989) to motor and executive areas of the cortex (Steiner & Tseng, 2010) as well as sensory association cortex and temporal lobe (Middleton 1996). Classically, the BG includes two pathways: the indirect pathway (IP) constituting a part of the striatum, GPe and STN finally projecting to GPi (Gerfen & Surmeier, 2011), and the direct pathway (DP) constituting the direct projection from the striatum to GPi (Gerfen & Surmeier, 2011). A third pathway, dubbed the hyperdirect pathway from cortex to STN, has been added subsequently (Nambu, Tokuno, & Takada, 2002).

2.1.1.1 Multiple Cortico-BG Loops

Earlier studies argued for the presence of a single cortico-BG loop where all the cortical areas projected to BG. On further investigation using various anatomical and tracing studies, it has been observed that cortico-BG system indeed consists of multiple parallel loops, where cortical areas project to distinct and mostly non-overlapping areas of BG (Alexander, DeLong, & Strick, 1986; DeLong & Wichmann, 2010; Nakano, 2000). The parallel loops have been primarily segregated into motor (motor and oculomotor), associative (dorsolateral and orbitofrontal prefrontal cortex), and limbic loops. The projections from each of the cortical areas to the various BG nuclei in each of the loop are given in Fig. 2.2.

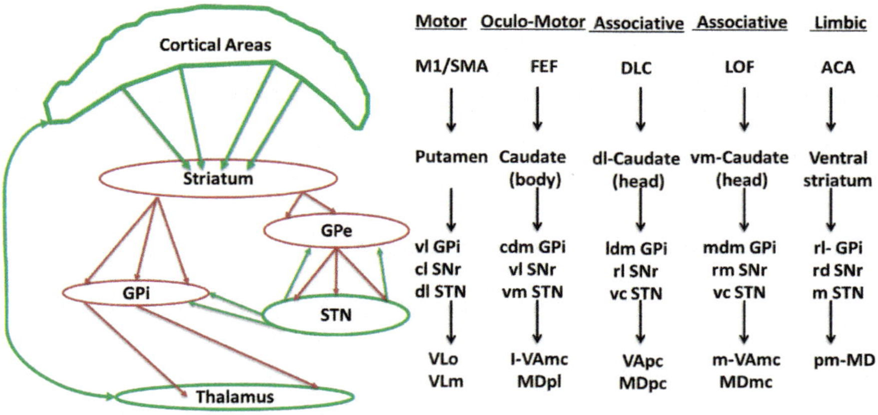

Fig. 2.2 This figure shows the parallel BG-cortico loops of motor, oculomotor, associative, and limbic areas with the specific areas M1: primary; SMA: supplementary motor area; FEF: frontal eye fields; DLC/DLPFC: dorsolateral prefrontal cortex; OFC/LOF: lateral orbitofrontal cortex; ACA: anterior cingulate area; Gpi: Globus Pallidus internus; SNr: substantia nigra pars reticulata; STN: subthalamic nucleus; MDpl: medialis dorsalis pars paralamellaris; MDmc: medialis dorsalis pars magnocellularis; MDpc: medialis dorsalis pars parvocellularis; VAmc: ventralis anterior pars magnocellularis; VApc: ventralis anterior pars parvocellularis; VLm: ventralis lateralis pars medialis; VLo: ventralis lateralis pars oralis; VP: ventral pallidum; VS: ventral striatum; c1: caudolateral; cdm: caudal dorsomedial; dl: dorsolateral; 1: lateral; Idm: lateral dorsomedial; m: medial; mdm: medialdorsomedial; pm: posteromedial; rd: rostrodorsal; r1: rostrolateral; rm: rostromedial; vm: ventromedial; vl: ventrolateral

Although the BG loops are functionally/anatomically segregated, it is important for these individual loops to interact with one another to ensure learning and information transfer occurs across motor, cognitive, and emotional domains. Newer anatomical evidence indeed suggests that there is an interaction among these closed loops (Haber & Calzavara, 2009).

Various theories proposed to explain the functionality of the multi-BG-cortex closed loops are explained briefly below. The first hypothesis suggests that information transfer across the loops is via the crossing of dendritic arbors from loop to another. The second is based on the overlap in the smaller BG structures which also have collaterals between them. This configuration creates an 'edge' where the neurons respond to more than one modality (motor/cognitive) (Yelnik, 2002). The third is the complex non-reciprocal connections providing directional flow of information. For example, it has been observed that limbic striatum can influence the motor output in rodents via striato-nigral pathway where the ventral striatum influences the dorsal via dopaminergic system (Haber, Fudge, & McFarland, 2000). The final one is based on 'hot spots' where an anatomical region within a structure receives input from multiple functional areas leading to integrative connectivity (Haber & Calzavara, 2009).

2.1.2 Cellular Level

In this section, we explain the features of the individual BG nuclei, the types of cells that constitute the nuclei, and the chemical messengers used by the cells for signaling.

2.1.2.1 Striatum—The Major Input Gateway

Based on the microanatomical studies, striatal neurons have been classified into two categories: spiny and aspiny neurons (Kreitzer, 2009). Spiny neurons in striatum, called the medium spiny neurons (MSNs), which constitute >95% of the population (Gerfen & Surmeier, 2011) receive projections from layer 5 of all neocortices onto their spines (Plenz & Kitai, 1998; Reig & Silberberg, 2014). The MSNs, whose major neurotransmitter is gamma-aminobutyric acid (GABA), have striato-nigral (direct, i.e., projections to GPi neurons) and striato-pallidal (indirect, i.e., projections to GPe then to STN then finally to GPi) projections (Gerfen & Surmeier, 2011). MSNs present a distinct compartmental pattern in terms of patch and matrix when viewed under the microscope. These compartments also have well-defined projection patterns (Bolam et al., 2006; Kreitzer, 2009). Although both striato-nigral and striato-pallidal neurons have patch–matrix compartments, striato-nigral patches predominantly project to SNc instead of SNr (Gerfen, 1984; Gerfen & Young, 1988).

Electrophysiologically, MSNs are characterized by their hyperpolarized resting membrane potential, lower input resistance, and bi-stable behavior (Kreitzer, 2009). This bistability is observed in terms of membrane potential, that is hyperpolarized (−90 to −70 mv, DOWN state) and depolarized (−60 to −40 mv, UP state) arising from the intrinsic membrane properties as well as from the glutamatergic input from cortex and thalamus (Kreitzer, 2009). The DOWN state has been mostly mediated by AMPA synaptic input, whereas the UP state is also modulated by the slow NMDA current. The main neuromodulator that affects the MSN's activity is DA (Kreitzer, 2009). Anatomically, the dorsal part of striatum receives dopaminergic projections from SNc (Gerfen & Surmeier, 2011) and ventral from Ventral Tegmental Area (Nicola, Surmeier, & Malenka, 2000). In the dorsal striatum, MSNs are the major targets for the dopaminergic projections arising from SNc targeting the spines and the axons (Surmeier, Ding, Day, Wang, & Shen, 2007; Surmeier, Song, & Yan, 1996). Using histochemical studies, the striato-nigral MSNs express D1 class dopaminergic receptors (D1 and D5) whereas striato-pallidal MSNs express D2 class (subdivided into D2, D3, and D4) (Kreitzer, 2009; Seeman, 1980; Surmeier et al., 2007). Recently, the presence of heterogeneous D1/D2 receptor which is a complex of D1 and D2 protomers was also observed in striatum (Rashid et al., 2007). Physiologically, the effect of dopamine on D1 and D2 expressing MSNs is opposite in nature when quantified in terms of firing rate. D1 (D2) receptors enhance (inhibit) the L-type calcium currents, thereby increasing (decreasing) the membrane potential eventually increasing (decreasing) the spiking rate of the MSNs (Kreitzer, 2009). In addition to DA, cholinergic modulation in MSNs is through the muscarinic ACh receptors for both D1 and D2 type MSNs. This modulation is through the activation of the A-type potassium currents and makes the neurons more hyperpolarized. However, in the presence of excitatory drive this A-type potassium current inactivates readily and decreases the delay in spiking. Experimental recordings show the activity of MSNs to be irregular and reach a maximum firing rate of 30 Hz (Kreitzer, 2009).

The second category of neurons in striatum is aspiny interneurons, further classified into fast spiking (FS), low threshold spiking (LTS), and cholinergic (TAN) neurons (Kawaguchi, 1993). The FS neurons though small in number help in regulating striatal activity and also receive input from cortex and thalamus. FS neurons also receive input from cortex and thalamus, not only to regulate their firing rate but also the cortical inputs received by the MSNs (Mallet, Le Moine, Charpier, & Gonon, 2005). Anatomical tracing studies show that a single MSN receives inhibitory synapses from 4 to 27 FS neurons and a single interneuron projects to 130 MSNs (Koós & Tepper, 1999). The dopaminergic modulation of FS neurons activity is mediated by D2 (D5) receptors which excite (inhibit) them. An increase in ACh levels also increases the firing rate of FS through direct depolarization of nicotinic receptors. The other kinds of interneurons are LTS, characterized by plateau potentials, and low threshold spikes also receive dopaminergic (through D5 receptors) and glutamatergic input from SNc and cortical areas (Kawaguchi, 1993) (Fig. 2.3).

The last ones are the TANs known to be large and constitute 1–2% of striatal neurons. These neurons are intrinsically active due to sodium and hyperpolarization-activated cation currents (Bennett, Callaway, & Wilson, 2000). They primarily receive input from MSNs and sparsely from thalamus and cortex.

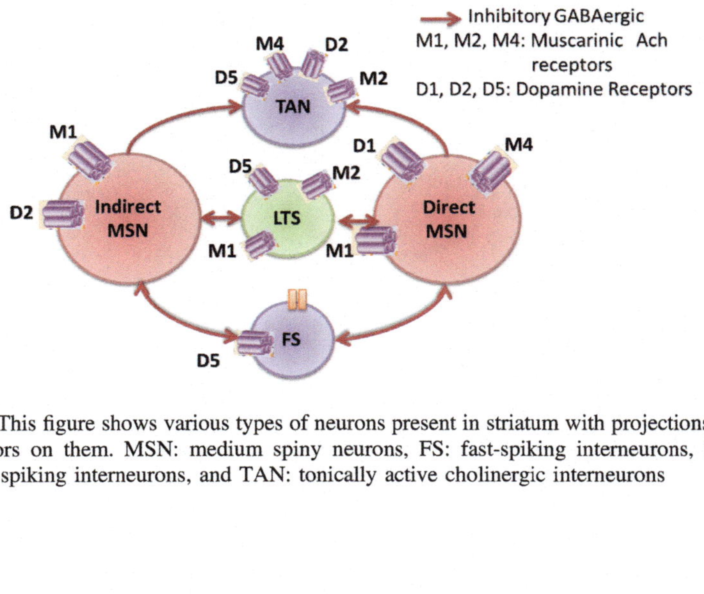

Fig. 2.3 This figure shows various types of neurons present in striatum with projections and type of receptors on them. MSN: medium spiny neurons, FS: fast-spiking interneurons, LTS: low threshold spiking interneurons, and TAN: tonically active cholinergic interneurons

They display a unique ability to pause their firing during salient cues including reward and reward prediction (Graybiel, Aosaki, Flaherty, & Kimura, 1994). TANs express both D5 and D2 receptors which control the spiking rate in similar way as in MSNs.

2.1.2.2 The Oscillator Network of BG

The reciprocally connected excitatory–inhibitory, i.e., STN–GPe network, is known for its active role in cognitive/motor process of healthy controls to pathological oscillations observed in PD patients (Baunez et al., 2001; Bergman, Wichmann, Karmon, & DeLong, 1994; Bevan, Magill, Terman, Bolam, & Wilson, 2002; Brown, 2003; Brown et al., 2001; Chakravarthy et al., 2010; Hammond, Bergman, & Brown, 2007; Heida, Marani, & Usunoff, 2008; Holgado, Terry, & Bogacz, 2010; Park, Worth, & Rubchinsky, 2010, 2011; Plenz & Kital, 1999). STN is unique among other nuclei of BG because it is the sole excitatory nucleus among BG nuclei (Charpier, Beurrier, & Paz, 2010). It is therefore named the 'driving force of BG.' It also receives direct input from cortex forming the hyperdirect pathway (Nambu et al., 2002), making it the fastest route for the cortico-thalamic influences to act on BG. STN neuronal activity is generally classified into three patterns: rhythmic, irregular, and bursting with average firing rate between 18 and 28 spikes/s in awake monkeys (Heida, Marani, et al., 2008). The irregular spontaneous spiking pattern is the most commonly observed which is due to large inward Na^+ currents independent of the GABAergic input (Bevan & Wilson, 1999) (Fig. 2.4).

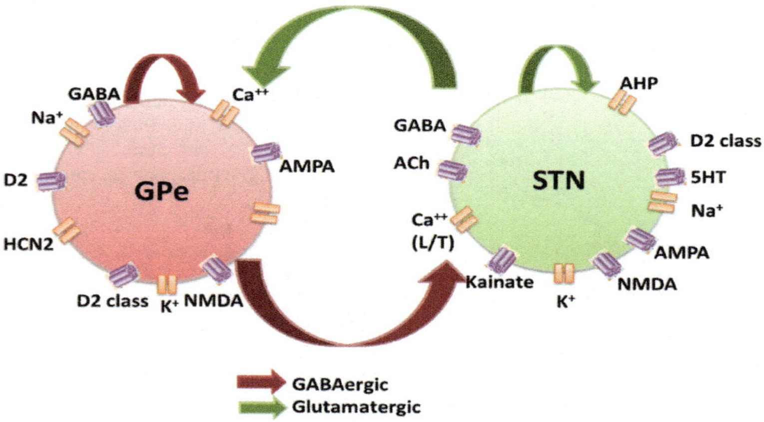

Fig. 2.4 This figure shows various receptors and channels with lateral connections in STN–GPe network. Receptors: gamma-aminobutyric acid (GABA), serotonin (5HT), dopamine (D2), kainate (KAR), α-amino-3-hydroxy-5-methyl-4-isoxazolepropionic acid (AMPA), N-methyl-D-aspartate (NMDA), cholinergic (ACh), Channels: calcium-L/T type (Ca^{++}), sodium (Na^+), potassium (K^+), hyperpolarized cation inward channel (HCN2), slow Ca^{++}-activated K^+ channel (AHP)

The ability of STN neurons to produce rebound and burst potentials comes from the presence of low threshold class-3 or T-type Ca^{2+} calcium channels which are active only during inhibitory GABAergic and do not participate in spontaneous behavior (Wilson & Bevan, 2011). STN neurons also display plateau potentials in response to depolarizing or hyperpolarizing current with sustained activity up to 500 ms (Beurrier, Congar, Bioulac, & Hammond, 1999). To obtain such a neural response, the ion channel responsible should be inactivated at resting membrane potential (RMP) and re-inactivated at hyperpolarization which was found to be due to the L-type calcium currents (Heida, Marani, et al.,). STN neurons also have dopamine (D2 class), serotonin (5HT), opioid, and cholinergic receptors which modulate their spiking pattern (Heida, Lakke, 2008& Usunoff, 2008). Glutamatergic input from the cortex (Kita, Chang, & Kitai, 1983) acts on the STN through both ionotropic (AMPA, NMDA, and kainite) and G-protein coupled metabotropic receptors. The main inhibitory influence from GPe (around 30%) is mediated through GABA (ionotropic and metabotropic) currents.

The GPe is an inhibitory nucleus that receives GABAergic projections from D2-expressing class of MSNs, through the indirect pathway. The majority of neurons are aspiny neurons and project to neurons of STN, GPi, and SNr (Kita & Kita, 1994; Sato, Lavallée, Lévesque, & Parent, 2000). Recent experimental studies show that around 2% of GPe neurons project to STN creating a strong inhibitory stimulus which aids in desynchronizing the activity of STN (Baufreton et al., 2009; Steiner & Tseng, 2010). Single unit recordings in awake monkeys show irregular firing (with few bursting neurons) around 20–100 Hz with average frequency around 60 Hz. Due to the presence of hyperpolarized-activated inward current (HCN2) and Nav1.6 currents, GPe neurons show autonomous firing property and are also capable of producing long pauses which is estimated to be due to low threshold Ca^{2+} or early K^{+} currents. The major source of excitatory input to GPe comes from STN (around 20%). Apart from inhibitory striatal current, GPe neurons also receive collateral GABAergic (mainly GABA-A) inputs from 15% of total neurons (Steiner & Tseng, 2010).

The STN–GPe loop together has a higher impact on the output of BG than individually. Experimental recordings of STN–GPe system under physiological condition show desynchronized activity, whereas under dopamine-deficient conditions, either in MPTP monkeys or PD patients, exhibits synchronized bursts within and between STN and GPe neurons (Bergman et al., 1994; Gillies, Willshaw, Gillies, & Willshaw, 1998; Park et al., 2011) (Bergman et al., 1994; Bevan et al., 2002; Hammond et al., 2007; Tachibana, Iwamuro, Kita, Takada, & Nambu, 2011; Weinberger & Dostrovsky, 2011) (Bergman et al., 1994, 1998). Plenz and Kitai (1998) worked on in vitro STN–GPe slices and proposed that they act as a pacemaker (Plenz & Kital, 1999), a source for generating oscillations in pathological conditions such as Parkinson's disease. This oscillatory activity was found to be present in two frequency bands, one around the tremor frequency [2–4 Hz] and another in beta [10–30 Hz] frequency range (Weinberger & Dostrovsky, 2011). Also, an increase of correlations in firing patterns of STN neurons was observed in PD state (Benazzouz et al., 2002; Brown, 2003; Brown et al., 2001;

Foffani, Bianchi, Baselli, & Priori, 2005; Levy et al., 2002; Willshaw & Li, 2002). Park et al. (2011) report the presence of intermittent synchrony between STN neurons and its local field potentials (LFP), recorded using multiunit activity electrodes from PD patients undergoing DBS surgery (Park et al., 2011). They also calculated the duration of synchronized and desynchronized events in neuronal activity by estimating transition rates, which were obtained with the help of first return maps plotted using phase of neurons (Park et al., 2010, 2011).

2.1.2.3 The Output Ports of BG (GPi and SNr)

GPi is the most common output port of BG which receives inhibitory GABAergic input from D1-expressing striatal MSNs (Gerfen, 1984; Gerfen & Surmeier, 2011; Surmeier et al., 2007) and GPe (Sato et al., 2000), and excitatory glutamatergic input from STN (Heida, Marani, et al., 2008); GPi integrates these inputs and influences the final selection of an action. The physiological characteristics of GPi neurons are very similar to GPe neurons, but GPi neurons fire at a much higher rate ($\sim$60–70 Hz). It has been observed that irregular spiking activity (of GPi) in physiological condition changes into bursting, synchronized pulses in PD/MPTP condition (Bergman et al., 1994; Raz, Vaadia, & Bergman, 2000). It has been hypothesized that it is due to this bursting activity that there is corruption in the transmission of information back to the cortex via thalamus (Rubin & Terman, 2004).

SNr, a nucleus in the ventrolateral part of the substantia nigra system, fires at a much higher rate (20–40 Hz) compared to its counterpart (SNc). It receives input from D1-expressing striatal MSNs, GPe, STN (Nakanishi, Kita, & Kitai, 1987; Robledo & Féger, 1990), and dopaminergic cells of SNc (Björklund & Dunnett, 2007) and projects to Superior Colliculus (Deniau, Hammond, Riszk, & Feger, 1978), ventral part of thalamus and SNc (Marsden, 1986; Tepper, Martin, & Anderson, 1995) via inhibitory GABAergic projections. Electrophysiological recordings from SNr slices reveal the presence of two types of neurons: type I and II (Nakanishi et al., 1987). Type I are spontaneous with short action potential intervals and a strong delayed rectification. Type II are not spontaneous and have large action potential durations and relatively large post-active hyperpolarization and less prominent delayed rectification (Nakanishi et al., 1987). At system level, SNr has been mainly involved during saccadic eye movements due to its projections to Superior Colliculus (Basso, Powers, & Evinger, 1996).

2.1.2.4 Dopaminergic System (SNc)

SNc, a part of the nigrostriatal pathway, is one of the clusters of dopaminergic cells in the midbrain. The SNc dopaminergic neurons display different types of activity ranging from regular/pacemaker (6.5 Hz), irregular/random (4 Hz) to bursting (4.25 Hz) and are classified based on the synchronous activity within themselves (Lee & Tepper, 2009). The dopaminergic neuron action potential can be divided into four components: (1) a slow depolarization, (2) an initial segment spike, (3) a somato-dendritic spike, and (4) an after hyperpolarization (Grace & Bunney, 1983). Irregular and bursting activity is often followed by slow after-depolarizations, and a second short latency action potential was seen riding on the depolarizing after potential following this first spike. Each of these firing patterns modulates the amount of dopamine released at their target locations. Inputs from other BG nuclei such as striatum (D1 receptors MSN), SNr (Björklund & Dunnett, 2007), is mostly GABAergic except the glutamatergic one from STN (Lee & Tepper, 2009). Apart from these nuclei, the interneurons within SNc and SNr also modulate the neural patterns in SNc. Cholinergic projections from pendenculopontine nucleus influenced through nicotinic and muscarinic receptors, as well as metabotropic glutamate receptor are believed to be another source of excitation apart from STN. Dopaminergic projections from SNc are targeted to multiple areas of brain including striatum (dorsal), GPe, and STN, specifically in modulating the activity patterns in GPe and STN. As stated in the earlier sections, the dopaminergic receptors predominantly come under either D1 family (D1, D5) or D2 family (D2, D3, D4) (Beaulieu & Gainetdinov, 2011). But recently neurons that express both D1/D2 receptors (heteromers) have been discovered in striatum, hippocampus, and cortex (Hasbi, O'Dowd, & George, 2011).

The death of SNc neurons is believed to be the primary cause for Parkinson's disease symptoms though the etiology of their death has been debated by multiple mechanisms (Blandini, 2010; Rodriguez-Oroz et al., 2010; Singleton et al., 2003; Wood-Kaczmar, Gandhi, & Wood, 2006). Due to the dopaminergic control on the activity of several BG nuclei and its pathways, multiple abnormalities have been reported in PD patients in both cognitive (Chaudhuri, Healy, & Schapira, 2006; Chaudhuri, Odin, Antonini, & Martinez-Martin, 2011; Merello, 2007) and motor (Brown, 2007; Schrag & Quinn, 2000; Xia & Mao, 2012) domains. A decrease in the activity of striatal neurons with D1 family receptors and an increase in the activity of D2-expressing striatal neurons have been observed (Gerfen et al., 1990; Gerfen & Surmeier, 2011; Gerfen & Young, 1988). Along with this, the excitatory–inhibitory circuit of BG, the STN–GPe network transits its activity from chaotic irregularity to synchronous bursting behavior (Brown, 2003, 2007; Fan, Baufreton, Surmeier, Chan, & Bevan, 2012; Holgado et al., 2010; Plenz & Kital, 1999). This is believed that loss of DA causes the observed molecular/cellular level changes at STN–GPe neurons (Brown, 2003, 2007; Fan et al., 2012).

2.1.2.5 Ventral Tegmental Area (VTA)

Apart from SNc, another major source of dopaminergic release is from Ventral Tegmental Area (VTA) which is located around the midline and the floor of midbrain and constitutes the mesocorticolimbic system (Yamaguchi et al., 2011). Due to its heterogeneous cytoarchitecture, VTA is named as the A10 area which is further divided into four regions, the paranigral nucleus (PN), the parabrachial pigmented area (PBP), the parafasciculus retroflexus area (PFR), and the rostromedial tegmental nucleus (RMTg) (Morales & Margolis, 2017). The PN and PBP are rich in dopaminergic cells compared to the other regions. Within the VTA neurons, part of the A10 cells consists of dopaminergic cells which express the dopamine-producing enzyme tyrosine hydroxylase (TH) and release dopamine. VTA also has VTA-GABA and VTA-glutamate neurons which not only regulate the local neuronal activity but also send long-range projections to areas that are innervated by the dopaminergic neurons (Morales & Margolis, 2017). It is reported that VTA has combinatorial neurons which co-express dopamine and GABA or DA and glutamate with the mechanism of the individual neurotransmitter being a hot topic currently (Morales & Margolis, 2017).

VTA receives inhibitory input from nucleus accumbens (nAcc) on both of its dopaminergic and GABA neurons. VTA dopaminergic neurons receive projections from the anterior cortex including the mesial prefrontal cortex (mPFC) and have reciprocal connectivity to mPFC (Han et al., 2017). Apart from the receiving projections from external structures, local synaptic projections from VTA-GABA and VTA-glutamate are also observed. VTA-DA neurons have divergent projections to various cortical areas, amygdala, nAcc, hippocampus, raphe nucleus locus coeruleus, mammillary body, lateral habenula (LHb), and the pallidum (Han et al., 2017; Morales & Margolis, 2017; Swanson, 1982).

Similar to SNc neurons, VTA neurons also respond to reward and shift their activation to cues that predict reward (Schultz, 1998). The role of DA in the aspect of motivation has been extensively studied especially with respect to modulating nAcc activity. It has also been observed that optogenetic stimulation of dopamine transporter expressing neurons in the dorsal hippocampus improved the recall accuracy in a complex spatial navigation task (Morales & Margolis, 2017). The projections from LHb on to the VTA-DA neurons have been implicated in aversive learning (Stamatakis et al., 2013) which is also being studied extensively in the area of depression (Lawson et al., 2016). Similarly, projections from raphe nucleus on to nAcc via VTA also play a major role in reward conditioning by increasing the dopamine release into nAcc (Morales & Margolis, 2017).

Pathophysiologically, in combination with meso-striatal (SNc) network, VTA has been involved in a variety of disorders from Parkinson's disease (Alberico, Cassell, & Narayanan, 2015), addiction, schizophrenia (Knable & Weinberger, 1997), and attention deficit hyperactivity disorder (Viggiano & Sadile, 2000). One of the major psychiatric problems that involve nAcc and VTA is addiction (Oliva & Wanat, 2016) as they are primary targets where the addictive drugs such as opioids, amphetamine act on.

References

Alberico, S. L., Cassell, M. D., & Narayanan, N. S. (2015). The vulnerable ventral tegmental area in Parkinson's disease. *Basal ganglia, 5*(2), 51–55.

Albin, R. L., Young, A. B., & Penney, J. B. (1989). The functional anatomy of basal ganglia disorders. *Trends in Neurosciences, 12*(10), 366–375.

Alexander, G. E., DeLong, M. R., & Strick, P. L. (1986). Parallel organization of functionally segregated circuits linking basal ganglia and cortex. *Annual Review of Neuroscience, 9*(1), 357–381.

Aravamuthan, B., Muthusamy, K., Stein, J., Aziz, T., & Johansen-Berg, H. (2007). Topography of cortical and subcortical connections of the human pedunculopontine and subthalamic nuclei. *Neuroimage, 37*(3), 694–705.

Basso, M. A., Powers, A. S., & Evinger, C. (1996). An explanation for reflex blink hyperexcitability in Parkinson's disease. I. Superior colliculus. *The Journal of Neuroscience, 16*(22), 7308–7317.

Baufreton, J., Kirkham, E., Atherton, J. F., Menard, A., Magill, P. J., Bolam, J. P., et al. (2009). Sparse but selective and potent synaptic transmission from the globus pallidus to the subthalamic nucleus. *Journal of Neurophysiology, 102*(1), 532–545.

Baunez, C., Humby, T., Eagle, D. M., Ryan, L. J., Dunnett, S. B., & Robbins, T. W. (2001). Effects of STN lesions on simple vs choice reaction time tasks in the rat: preserved motor readiness, but impaired response selection. *European Journal of Neuroscience, 13*(8), 1609–1616.

Beaulieu, J. M., & Gainetdinov, R. R. (2011). The physiology, signaling, and pharmacology of dopamine receptors. *Pharmacological Reviews, 63*(1), 182–217.

Benazzouz, A., Breit, S., Koudsie, A., Pollak, P., Krack, P., & Benabid, A. L. (2002). Intraoperative microrecordings of the subthalamic nucleus in Parkinson's disease. *Movement Disorders, 17*(S3), S145–S149.

Bennett, B. D., Callaway, J. C., & Wilson, C. J. (2000). Intrinsic membrane properties underlying spontaneous tonic firing in neostriatal cholinergic interneurons. *The Journal of Neuroscience, 20*(22), 8493–8503.

Bergman, H., Feingold, A., Nini, A., Raz, A., Slovin, H., Abeles, M., & Vaadia, E. (1998). Physiological aspects of information processing in the basal ganglia of normal and Parkinsonian primates. *Trends in Neurosciences, 21*(1), 32–38.

Bergman, H., Wichmann, T., Karmon, B., & DeLong, M. (1994). The primate subthalamic nucleus. II. Neuronal activity in the MPTP model of Parkinsonism. *Journal of Neurophysiology, 72*(2), 507–520.

Beurrier, C., Congar, P., Bioulac, B., & Hammond, C. (1999). Subthalamic nucleus neurons switch from single-spike activity to burst-firing mode. *The Journal of Neuroscience, 19*(2), 599–609.

Bevan, M. D., Magill, P. J., Terman, D., Bolam, J. P., & Wilson, C. J. (2002). Move to the rhythm: Oscillations in the subthalamic nucleus–external globus pallidus network. *Trends in Neurosciences, 25*(10), 525–531.

Bevan, M. D., & Wilson, C. J. (1999). Mechanisms underlying spontaneous oscillation and rhythmic firing in rat subthalamic neurons. *The Journal of Neuroscience, 19*(17), 7617–7628.

Björklund, A., & Dunnett, S. B. (2007). Dopamine neuron systems in the brain: An update. *Trends in Neurosciences, 30*(5), 194–202.

Blandini, F. (2010). An update on the potential role of excitotoxicity in the pathogenesis of Parkinson's disease. *Functional Neurology, 25*(2), 65.

Bolam, J., Bergman, H., Graybiel, A., Kimura, M., Plenz, D., Seung, H., … Wickens, J. (2006). Microcircuits, molecules and motivated behaviour: Microcircuits in the striatum. Paper presented at the Microcircuits: The Interface Between Neurons and Global Brain Function, Dahlem Workshop Report.

Brown, P. (2003). Oscillatory nature of human basal ganglia activity: relationship to the pathophysiology of Parkinson's disease. *Movement Disorders, 18*(4), 357–363.

Brown, P. (2007). Abnormal oscillatory synchronisation in the motor system leads to impaired movement. *Current Opinion in Neurobiology, 17*(6), 656–664.

Brown, P., Oliviero, A., Mazzone, P., Insola, A., Tonali, P., & Di Lazzaro, V. (2001). Dopamine dependency of oscillations between subthalamic nucleus and pallidum in Parkinson's disease. *The Journal of Neuroscience, 21*(3), 1033–1038.

Chakravarthy, V., Joseph, D., & Bapi, R. S. (2010). What do the basal ganglia do? A modeling perspective. *Biological Cybernetics, 103*(3), 237–253.

Charpier, S., Beurrier, C., & Paz, J. (2010). The subthalamic nucleus: from in vitro to in vivo mechanisms. *Handbook of Basal Ganglia Structure and Function*, 259–273.

Chaudhuri, K. R., Healy, D. G., & Schapira, A. H. (2006). Non-motor symptoms of Parkinson's disease: diagnosis and management. *The Lancet Neurology, 5*(3), 235–245.

Chaudhuri, K. R., Odin, P., Antonini, A., & Martinez-Martin, P. (2011). Parkinson's disease: The non-motor issues. *Parkinsonism & Related Disorders, 17*(10), 717–723.

Chersi, F., Mirolli, M., Pezzulo, G., & Baldassarre, G. (2013). A spiking neuron model of the cortico-basal ganglia circuits for goal-directed and habitual action learning. *Neural Networks, 41,* 212–224.

DeLong, M., & Wichmann, T. (2010). Changing views of basal ganglia circuits and circuit disorders. *Clinical EEG and Neuroscience, 41*(2), 61–67.

Deniau, J., Hammond, C., Riszk, A., & Feger, J. (1978). Electrophysiological properties of identified output neurons of the rat substantia nigra (pars compacta and pars reticulata): Evidences for the existence of branched neurons. *Experimental Brain Research, 32*(3), 409–422.

Fan, K. Y., Baufreton, J., Surmeier, D. J., Chan, C. S., & Bevan, M. D. (2012). Proliferation of external globus pallidus-subthalamic nucleus synapses following degeneration of midbrain dopamine neurons. *The Journal of Neuroscience, 32*(40), 13718–13728.

Foffani, G., Bianchi, A., Baselli, G., & Priori, A. (2005). Movement-related frequency modulation of beta oscillatory activity in the human subthalamic nucleus. *The Journal of Physiology, 568* (2), 699–711.

Gerfen, C. R. (1984). The neostriatal mosaic: compartmentalization of corticostriatal input and striatonigral output systems. *Nature, 311*(5985), 461.

Gerfen, C. R., Engber, T. M., Mahan, L. C., Susel, Z., Chase, T. N., Monsma, F., & Sibley, D. R. (1990). D1 and D2 dopamine receptor-regulated gene expression of striatonigral and striatopallidal neurons. *Science, 250*(4986), 1429–1432.

Gerfen, C. R., & Surmeier, D. J. (2011). Modulation of striatal projection systems by dopamine. *Annual Review of Neuroscience, 34,* 441.

Gerfen, C. R., & Young, W. S. (1988). Distribution of striatonigral and striatopallidal peptidergic neurons in both patch and matrix compartments: An in situ hybridization histochemistry and fluorescent retrograde tracing study. *Brain Research, 460*(1), 161–167.

Gillies, A., Willshaw, D., Gillies, A., & Willshaw, D. (1998). A massively connected subthalamic nucleus leads to the generation of widespread pulses. *Proceedings of the Royal Society of London, Series B: Biological Sciences, 265*(1410), 2101–2109.

Grace, A., & Bunney, B. (1983). Intracellular and extracellular electrophysiology of nigral dopaminergic neurons—2. Action potential generating mechanisms and morphological correlates. *Neuroscience, 10*(2), 317–331.

Graybiel, A. M., Aosaki, T., Flaherty, A. W., & Kimura, M. (1994). The basal ganglia and adaptive motor control. *Science, 265*(5180), 1826–1831.

Gurney, K., Prescott, T. J., & Redgrave, P. (2001a). A computational model of action selection in the basal ganglia. I. A new functional anatomy. *Biological Cybernetics, 84*(6), 401–410.

Gurney, K., Prescott, T. J., & Redgrave, P. (2001b). A computational model of action selection in the basal ganglia. II. Analysis and simulation of behaviour. *Biological Cybernetics, 84*(6), 411–423.

Haber, S. N., & Calzavara, R. (2009). The cortico-basal ganglia integrative network: The role of the thalamus. *Brain Research Bulletin, 78*(2), 69–74.

Haber, S. N., Fudge, J. L., & McFarland, N. R. (2000). Striatonigrostriatal pathways in primates form an ascending spiral from the shell to the dorsolateral striatum. *The Journal of Neuroscience, 20*(6), 2369–2382.

Hammond, C., Bergman, H., & Brown, P. (2007). Pathological synchronization in Parkinson's disease: Networks, models and treatments. *Trends in Neurosciences, 30*(7), 357–364.

Han, X., Jing, M.-y., Zhao, T.-y., Wu, N., Song, R., & Li, J. (2017). Role of dopamine projections from ventral tegmental area to nucleus accumbens and medial prefrontal cortex in reinforcement behaviors assessed using optogenetic manipulation. *Metabolic Brain Disease*, 1–12.

Hasbi, A., O'Dowd, B. F., & George, S. R. (2011). Dopamine D1-D2 receptor heteromer signaling pathway in the brain: emerging physiological relevance. *Molecular Brain, 4*(1), 26.

Heida, T., Lakke, E. A., & Usunoff, K. G. (2008a). *Subthalamic nucleus Part I: Development, cytology, topography and connections, the advances in anatomy, embryology and cell biology*. Berlin: Springer.

Heida, T., Marani, E., & Usunoff, K. G. (2008b). *The subthalamic nucleus: Part II: Modelling and simulation of activity*. Berlin: Springer.

Holgado, A. J. N., Terry, J. R., & Bogacz, R. (2010). Conditions for the generation of beta oscillations in the subthalamic nucleus–globus pallidus network. *The Journal of Neuroscience, 30*(37), 12340–12352.

Humphries, M., & Gurney, K. (2002). The role of intra-thalamic and thalamocortical circuits in action selection. *Network: Computation in Neural Systems, 13*(1), 131–156.

Kawaguchi, Y. (1993). Physiological, morphological, and histochemical characterization of three classes of interneurons in rat neostriatum. *The Journal of Neuroscience, 13*(11), 4908–4923.

Kita, H., Chang, H., & Kitai, S. (1983). The morphology of intracellularly labeled rat subthalamic neurons: A light microscopic analysis. *Journal of Comparative Neurology, 215*(3), 245–257.

Kita, H., & Kita, S. (1994). The morphology of globus pallidus projection neurons in the rat: An intracellular staining study. *Brain Research, 636*(2), 308–319.

Knable, M. B., & Weinberger, D. R. (1997). Dopamine, the prefrontal cortex and schizophrenia. *Journal of psychopharmacology, 11*(2), 123–131.

Koós, T., & Tepper, J. M. (1999). Inhibitory control of neostriatal projection neurons by GABAergic interneurons. *Nature Neuroscience, 2*(5), 467–472.

Kreitzer, A. C. (2009). Physiology and pharmacology of striatal neurons. *Annual Review of Neuroscience, 32,* 127–147.

Lawson, R., Seymour, B., Nord, C., Thomas, D., Roiser, J., Dayan, P., & Pilling, S. (2016). Disrupted habenula function in major depression. *Molecular psychiatry, 22*(2), 202.

Lee, C. R., & Tepper, J. M. (2009). Basal ganglia control of substantia nigra dopaminergic neurons. In *Birth, life and death of dopaminergic neurons in the substantia nigra* (pp. 71–90), Berlin: Springer.

Levy, R., Ashby, P., Hutchison, W. D., Lang, A. E., Lozano, A. M., & Dostrovsky, J. O. (2002). Dependence of subthalamic nucleus oscillations on movement and dopamine in Parkinson's disease. *Brain, 125*(6), 1196–1209.

Mallet, N., Le Moine, C., Charpier, S., & Gonon, F. (2005). Feedforward inhibition of projection neurons by fast-spiking GABA interneurons in the rat striatum in vivo. *The Journal of Neuroscience, 25*(15), 3857–3869.

Marsden, C. (1986). Movement disorders and the basal ganglia. *Trends in neurosciences, 9*, 512–515.

Maurice, N., Deniau, J.-M., Glowinski, J., & Thierry, A.-M. (1998). Relationships between the prefrontal cortex and the basal ganglia in the rat: physiology of the corticosubthalamic circuits. *The Journal of Neuroscience, 18*(22), 9539–9546.

Merello, M. (2007). Non-motor disorders in Parkinson's disease. *Revista de neurologia, 47*(5), 261–270.

Middleton, F. A., & Strick, P. L. (1996). The temporal lobe is a target of output from the basal ganglia. *Proceedings of the national academy of sciences, 93*(16), 8683–8687.

Morales, M., & Margolis, E. B. (2017). Ventral tegmental area: cellular heterogeneity, connectivity and behaviour. *Nature Reviews Neuroscience, 18*(2), 73–85.

Nakanishi, H., Kita, H., & Kitai, S. (1987). Intracellular study of rat substantia nigra pars reticulata neurons in an in vitro slice preparation: Electrical membrane properties and response characteristics to subthalamic stimulation. *Brain Research, 437*(1), 45–55.

Nakano, K. (2000). Neural circuits and topographic organization of the basal ganglia and related regions. *Brain and Development, 22,* 5–16.

Nambu, A., Tokuno, H., & Takada, M. (2002). Functional significance of the cortico–subthalamo–pallidal 'hyperdirect' pathway. *Neuroscience Research, 43*(2), 111–117.

Nicola, S. M., Surmeier, D. J., & Malenka, R. C. (2000). Dopaminergic modulation of neuronal excitability in the striatum and nucleus accumbens. *Annual Review of Neuroscience, 23*(1), 185–215.

Oliva, I., & Wanat, M. J. (2016). Ventral tegmental area afferents and drug-dependent behaviors. *Frontiers in psychiatry*, 7.

Park, C., Worth, R. M., & Rubchinsky, L. L. (2010). Fine temporal structure of beta oscillations synchronization in subthalamic nucleus in Parkinson's disease. *Journal of Neurophysiology, 103*(5), 2707–2716.

Park, C., Worth, R. M., & Rubchinsky, L. L. (2011). Neural dynamics in Parkinsonian brain: The boundary between synchronized and nonsynchronized dynamics. *Physical Review E, 83*(4), 042901.

Plenz, D., & Kitai, S. T. (1998). Up and down states in striatal medium spiny neurons simultaneously recorded with spontaneous activity in fast-spiking interneurons studied in cortex–striatum–substantia nigra organotypic cultures. *The Journal of Neuroscience, 18*(1), 266–283.

Plenz, D., & Kital, S. T. (1999). A basal ganglia pacemaker formed by the subthalamic nucleus and external globus pallidus. *Nature, 400*(6745), 677–682.

Rashid, A. J., So, C. H., Kong, M. M., Furtak, T., El-Ghundi, M., Cheng, R., … George, S. R. (2007). D1–D2 dopamine receptor heterooligomers with unique pharmacology are coupled to rapid activation of Gq/11 in the striatum. *Proceedings of the National Academy of Sciences, 104*(2), 654–659.

Raz, A., Vaadia, E., & Bergman, H. (2000). Firing patterns and correlations of spontaneous discharge of pallidal neurons in the normal and the tremulous 1-methyl-4-phenyl-1, 2, 3, 6-tetrahydropyridine vervet model of Parkinsonism. *The Journal of Neuroscience, 20*(22), 8559–8571.

Reig, R., & Silberberg, G. (2014). Multisensory integration in the mouse striatum. *Neuron, 83*(5), 1200–1212.

Robledo, P., & Féger, J. (1990). Excitatory influence of rat subthalamic nucleus to substantia nigra pars reticulata and the pallidal complex: Electrophysiological data. *Brain Research, 518*(1), 47–54.

Rodriguez-Oroz, M. C., López-Azcárate, J., Garcia-Garcia, D., Alegre, M., Toledo, J., Valencia, M., … Obeso, J. A. (2010). Involvement of the subthalamic nucleus in impulse control disorders associated with Parkinson's disease. *Brain*, awq301.

Rubin, J. E., & Terman, D. (2004). High frequency stimulation of the subthalamic nucleus eliminates pathological thalamic rhythmicity in a computational model. *Journal of Computational Neuroscience, 16*(3), 211–235.

Sato, F., Lavallée, P., Lévesque, M., & Parent, A. (2000). Single-axon tracing study of neurons of the external segment of the globus pallidus in primate. *Journal of Comparative Neurology, 417*(1), 17–31.

Schrag, A., & Quinn, N. (2000). Dyskinesias and motor fluctuations in Parkinson's disease. *Brain, 123*(11), 2297–2305.

Schroll, H., Vitay, J., & Hamker, F. H. (2012). Working memory and response selection: A computational account of interactions among cortico-basalganglio-thalamic loops. *Neural Networks, 26,* 59–74.

Schultz, W. (1998). Predictive reward signal of dopamine neurons. *Journal of neurophysiology, 80*(1), 1–27.

Seeman, P. (1980). Brain dopamine receptors. *Pharmacological Reviews, 32*(3), 229–313.

Singleton, A., Farrer, M., Johnson, J., Singleton, A., Hague, S., Kachergus, J., … Nussbaum, R. (2003). α-Synuclein locus triplication causes Parkinson's disease. *Science, 302*(5646), 841–841.

Stamatakis, A. M., Jennings, J. H., Ung, R. L., Blair, G. A., Weinberg, R. J., Neve, R. L., … Deisseroth, K. (2013). A unique population of ventral tegmental area neurons inhibits the lateral habenula to promote reward. *Neuron, 80*(4), 1039–1053.

Steiner, H., & Tseng, K. Y. (2010). *Handbook of Basal Ganglia Structure and Function: A Decade of Progress* (Vol. 20), Access Online via Elsevier.

Surmeier, D. J., Ding, J., Day, M., Wang, Z., & Shen, W. (2007). D1 and D2 dopamine-receptor modulation of striatal glutamatergic signaling in striatal medium spiny neurons. *Trends in Neurosciences, 30*(5), 228–235.

Surmeier, D. J., Song, W.-J., & Yan, Z. (1996). Coordinated expression of dopamine receptors in neostriatal medium spiny neurons. *The Journal of Neuroscience, 16*(20), 6579–6591.

Swanson, L. (1982). The projections of the ventral tegmental area and adjacent regions: a combined fluorescent retrograde tracer and immunofluorescence study in the rat. *Brain research bulletin, 9*(1), 321–353.

Tachibana, Y., Iwamuro, H., Kita, H., Takada, M., & Nambu, A. (2011). Subthalamo-pallidal interactions underlying Parkinsonian neuronal oscillations in the primate basal ganglia. *European Journal of Neuroscience, 34*(9), 1470–1484.

Tepper, J., Martin, L., & Anderson, D. (1995). GABA ~ A Receptor-Mediated Inhibition of Rat Substantia Nigra Dopaminergic Neurons by Pars Reticulata Projection Neurons. *Journal of Neuroscience, 15*(4), 3092–3103.

Weinberger, M., & Dostrovsky, J. O. (2011). A basis for the pathological oscillations in basal ganglia: the crucial role of dopamine. *NeuroReport, 22*(4), 151.

Willshaw, D., & Li, Z. (2002). Subthalamic–pallidal interactions are critical in determining normal and abnormal functioning of the basal ganglia. *Proceedings of the Royal Society of London, Series B: Biological Sciences, 269*(1491), 545–551.

Wilson, C. J., & Bevan, M. D. (2011). Intrinsic dynamics and synaptic inputs control the activity patterns of subthalamic nucleus neurons in health and in Parkinson's disease. *Neuroscience, 198,* 54–68.

Wood-Kaczmar, A., Gandhi, S., & Wood, N. (2006). Understanding the molecular causes of Parkinson's disease. *Trends in Molecular Medicine, 12*(11), 521–528.

Xia, R., & Mao, Z.-H. (2012). Progression of motor symptoms in Parkinson's disease. *Neuroscience Bulletin, 28*(1), 39–48.

Yelnik, J. (2002). Functional anatomy of the basal ganglia. *Movement Disorders, 17*(S3), S15–S21.

Yamaguchi, T., Wang, H.-L., Li, X., Ng, T. H., & Morales, M. (2011). Mesocorticolimbic glutamatergic pathway. *Journal of Neuroscience, 31*(23), 8476–8490.

Yucelgen, C., Denizdurduran, B., Metin, S., Elibol, R., & Sengor, N. S. (2012). A biophysical network model displaying the role of basal ganglia pathways in action selection. In *Artificial neural networks and machine learning–ICANN 2012* (pp. 177–184), Berlin: Springer.

Chapter 3
The Motor, Cognitive, Affective, and Autonomic Functions of the Basal Ganglia

Ahmed A. Moustafa, Alekhya Mandali, Pragathi Priyadharsini Balasubramani and V. Srinivasa Chakravarthy

Abstract The basal ganglia are involved in several processes, ranging from motor to cognitive ones. This chapter briefly discusses the role of the basal ganglia in motor (including reaching, handwriting, precision grip, gait, saccade generation, and speech), cognitive (action selection, decision making, attention, working memory, sequence learning, and sleep regulation), mood/emotion (negative and positive affect), and autonomic (gastrointestinal and cardiovascular) processes. The chapter summarizes key experimental studies explaining the role of the basal ganglia in all of these motor, cognitive, and affective processes. Accordingly, this chapter provides a background on the function of the basal ganglia, which is key information that guides the reader to understand the following computational modeling efforts to understand the role of the basal ganglia in several functional processes.

3.1 Motor Processes of the Basal Ganglia

The basal ganglia influence motor control mainly via at least two pathways: the cortico-thalamic network and the basal ganglia–brainstem networks (Takakusaki, Tomita, & Yano, 2008).

The BG and Cortico-thalamic Loop: There are multiple loops in the cortico-basal ganglia network including the cognitive, motor, and limbic loops which control planned and automatic movements (Hikosaka et al., 1999; Marsden, 1982). The motor cortical areas which project to the putamen in the striatum are thought to be involved in discrete voluntary movements (Takakusaki et al., 2008). Similarly, the prefrontal areas projecting to the caudate nucleus regulate complex visually guided movements (Takakusaki et al., 2008). The parallel nature of the cortico-basal ganglia networks aids in effective integration of information from various sensory resources to plan motor actions (Hikosaka et al., 1999; Nakahara, Doya, & Hikosaka, 2001). In rats, optogenetic stimulation of the indirect pathway of the BG led to an increase in freezing and bradykinetic movements which was rescued completely by activating the direct pathway leading to increased

V. S. Chakravarthy and A. Moustafa, *Computational Neuroscience Models of the Basal Ganglia*, Cognitive Science and Technology,
https://doi.org/10.1007/978-981-10-8494-2_3

locomotion (Kravitz et al., 2010). The BG output nuclei also target the brainstem regions for automatic control of movements and along with these loops evaluate the necessary motor programs for a given context.

Before we discuss computational models of the basal ganglia, below, we briefly discuss the motor, cognitive, mood/emotion, and autonomic processes of the basal ganglia, respectively.

3.1.1 Hand—Reaching, Handwriting, Precision Grip

Hand movements involve reaching, writing, and precision grip, among others. As we discuss below, the BG plays a role in all of these processes.

Reaching movements reveal key information about how the brain plans and execute movement. Studies have found that neural activity indicates that the motor commands of reaching have a causal role on reaching characteristics such as velocity and arm position (Harris & Wolpert, 1998). The optimal control of any movement-related activity is found to be through a feedback control methodology (Schaal & Schweighofer, 2005; Shadmehr & Krakauer, 2008). Especially, the basal ganglia (BG) are involved in the learning of new actions and sequences of movements mediated by the midbrain dopaminergic signals (Hikosaka, Nakamura, Sakai, & Nakahara, 2002). This understanding applies for other sophisticated motor activity of hand such as handwriting and exerting precision grip. Handwriting activity is an interesting interplay of accurate reaches, executing a sequence of strokes which involves scaling of movements while planning for the subsequent ones (Teulings, Contreras-Vidal, Stelmach, & Adler, 1997). Precision grip, on the other hand, is the act of gripping objects between forefinger and thumb, and this also incurs high sensory-motor control, mediated by basal ganglia network (Fellows, Noth, & Schwarz, 1998; Ingvarsson, Gordon, & Forssberg, 1997; Napier, 1956).

Parkinson's disease shows impairments in the kinematics of simple reaching movement (Majsak, Kaminski, Gentile, & Flanagan, 1998) and that contributes to impaired goal-directed movement or target-tracking in pursuit-related tasks (Soliveri, Brown, Jahanshahi, Caraceni, & Marsden, 1997). PD may fall short of reaching the goal and suffer tremor or rigidity. Hypometric behavior and micrographia are common characteristics of handwriting in many PD patients (Broderick, Van Gemmert, Shill, & Stelmach, 2009; Teulings et al., 1997; Tucha et al., 2006). Patients on dopamine medication show high precision grip than controls and OFF-medication patients (Fellows et al., 1998; Ingvarsson et al., 1997; Müller & Abbs, 1990). Some inferences gained from most of the above-described symptoms suggest that the damage is beyond mere sensory deficit and is extended to basal ganglia-related network dysfunction. We see that evaluations based on coordinated arm, wrist, and finger movements on reaching, handwriting, or gripping contribute as reliable behavioral markers to quantify the damage caused by dopaminergic dysfunction in PD.

3.1.2 Gait

BG and Brainstem Interaction: The BG–brainstem interaction is thought to be responsible for the control of postural tone and the rhythmicity during gait (Takakusaki, Saitoh, Harada, & Kashiwayanagi, 2004). The influence of the BG was found to be through the GABAergic output from the substantia nigra pars reticulata (SNr), which is one of the output nuclei of the BG, and has dense projections to the mesenpontine tegmentum (Beckstead, Domesick, & Nauta, 1993; Inglis & Winn, 1995; Moriizumi, Nakamura, Tokuno, Kitao, & Kudo, 1988), where the MLR and PPN are present. The introduction of GABAA antagonists into the ventral part of MLR and PPN induced locomotion and muscle atonia (Takakusaki, Habaguchi, Ohtinata-Sugimoto, Saitoh, & Sakamoto, 2003). It was observed that a repetitive stimulation of the SNr region altered locomotion by controlling the MLR region and decreasing the step cycles and increasing the duration of the stance phase, which is equivalent to reducing the velocity of locomotion (Takakusaki et al., 2008). The SNr stimulation also decreased the PPN-mediated muscle tone inhibition. SNr activation had a considerable effect on the amplitude and duration of the rhythmic membrane oscillations of both the flexor and extensor motor neurons, suggesting the role of BG in rhythm modulation (Takakusaki, Ohta, & Harada, 2007). As a whole, the BG outputs suppress the inhibitory mechanisms of locomotion and increase the excitatory effects leading to termination of locomotion.

Many features of walking such as stride length and velocity, foot strike pattern, and the associated postural stability are highly influenced by the functioning of basal ganglia. This subcortical piece of control over the spinal cord's central pattern generators forms an essential neural circuit component for gait, in coordination with the cortical activity (Sahyoun, Floyer-Lea, Johansen-Berg, & Matthews, 2004; Takakusaki et al., 2008). Specifically, vision, space, and other context-driven cortical activity influence the gait (Lewis & Barker, 2009; Maruyama & Yanagisawa, 2006) in association with the subcortical counterpart (the basal ganglia). They are well identified in PD condition that particularly suffers an abnormality of basal ganglia control due to dopaminergic cell loss. Some symptoms include reduced stride velocity and length, flat foot strike, postural sway and shuffling steps (Hausdorff, Cudkowicz, Firtion, Wei, & Goldberger, 1998; Kimmeskamp & Hennig, 2001; Morris, Iansek, Matyas, & Summers, 1998) and the more debilitating context-driven freezing of gait (Almeida & Lebold, 2010; Cowie, Limousin, Peters, & Day, 2010).

The freezing of gait phenomenon is context dependent suggesting a definite role for the cortical and subcortical components in this impairment. They are characterized by start hesitation, destination hesitation and obstacle avoidance (Maruyama & Yanagisawa, 2006) increased cognitive load caused due to multiple simultaneous task goals, working memory has been proposed to facilitate freezing of gait; and poor availability of resource pool containing dopamine has been hypothesized to be their root cause (Lewis & Barker, 2009). Some studies also show an impaired connectivity between cortical and subcortical areas including basal ganglia, for facilitating freezing of gait (Shine et al., 2013). Medications are shown to improve the strides and reduce freezing of gait (Almeida & Lebold, 2010; Cowie et al., 2010). Patients are quite assisted by stimulating some basal ganglia nuclei such as STN in addition to administration of medications (Faist et al., 2001; Lubik et al., 2006). Freezing symptoms can manifest in upper limb in few patients.

3.1.3 Saccades

Saccades are rapid movements of both eyes, interspersed by momentary fixation of the eyes on objects of attention. Cortical substrates of saccade generation include the frontal eye fields and the lateral intraparietal sulcus, while the subcortical substrates include the Superior Colliculus (SC) and the BG. The contributions of the BG to saccade generation seem to be mediated predominantly by the SC. It was shown that neurons in one of the output ports of the BG, the substantia nigra pars reticulata (SNr), inhibit SC and maintain high firing levels during periods of fixation, and pause for some saccades (Basso & Wurtz, 2002; Hikosaka & Wurtz, 1983). Hikosaka et al. (2000) hypothesize that the BG output controls the SC in two complementary ways: by disinhibiting the SC when the direct pathway is activated, and inhibiting the saccade when the indirect pathway is activated (Hikosaka, Takikawa, & Kawagoe, 2000). The role of the SNr in modulating reward-oriented saccadic tasks was studied by using a one-direction-rewarded version of the memory-guided saccade task (Sato & Hikosaka, 2002). One study showed that certain neurons in SNr exhibited positive reward modulation, suggesting that neurons of the SNr-SC pathway promote reward modulation. Kori et al. (1995) studied the effects of the unilateral infusion of MPTP into the monkey caudate nucleus on visually guided and memory-guided saccades (Kori et al., 1995). They found that saccade latency was prolonged, while the amplitude and velocity are decreased. The role of the BG in saccade generation is further confirmed by impaired saccades in PD conditions, both in animals and humans. Studies with MPTP monkeys showed prolonged saccades, longer reaction times, and smaller peak velocities and amplitudes. The animals also showed fewer spontaneous saccades (Kato et al., 1995). PD patients exhibited a peculiar class of saccades known as 'square wave' jerks, in which a small saccade of amplitude 0.5°–3° transiently moved the eye away from the point of fixation, only to return to the original point of fixation after a couple of hundred milliseconds. Square wave jerks are also found in other syndromes like Progressive Supranuclear Palsy (PSP) and other multisystem Parkinsonian syndromes (Rascol et al., 1991).

3.1.4 Speech and Language

Many imaging and lesion studies (Cappa & Abutalebi, 1999; Svennilson, Torvik, Lowe, & Leksell, 1960; Van Buren, Li, & Ojemann, 1966) provide substantial evidences for the role of basal ganglia in language processing and production (phonology, syntax, lexical semantics, prosody, and pragmatics). Apart from the motor aspects of language involving Broca's area's control of speech production (Alm, 2004), several cognitive aspects related to the choice sequence of syntax generation, syntactic processing, and their perception (predictability) from auditory language (Kotz, Schwartze, & Schmidt-Kassow, 2009; Nenadic et al., 2003) have been related to the basal ganglia functioning. Some studies relate language processing and control to the framework of decision making that is mediated by cortico-basal ganglia circuits. Further, complex use of language involves working memory (Grossman, Carvell, Stern, Gollomp, & Hurtig, 1992), which is also actively controlled by nuclei such as basal ganglia. Selective attention contribution of the nuclei helps in syntactic processing (e.g., the BG model of (Brown & Marsden, 1988) as well. PD patients show abnormality in many aspects of language processing and production (Grossman et al., 2002; Kotz, Frisch, Von Cramon, & Friederici, 2003; Kotz et al., 2009; Schirmer, 2004). Several studies suggest that language is evolved from motor processes, which explain why the BG play a role in speech production (Lieberman, 1991).

Parkinson patients often experience difficulty in orofacial and articulatory movements that could affect their speech. Symptoms constitute bradykinetic articulatory movements as well as orofacial hypomimy (Hartelius & Svensson, 1994). The patients show less variability in fundamental frequency, face start or context-dependent speech hesitations and stuttering, very similar to freezing of gait (Canter, 1963; Cantiniaux et al., 2010; Harel, Cannizzaro, & Snyder, 2004; Kegl, Cohen, & Poizner, 1999). Speech velocity is often decreased, and the interpause speech duration is shortened, just like stride velocity and step length in PD (Cantiniaux et al., 2010). Oral festinations occur in some patients, and it can be correlated to gait festinations (Moreau et al., 2007). Studies focusing on the effects of medications in PD patients did not show much improvement in speech parameters (Wolfe, Garvin, Bacon, & Waldrop, 1975) or stuttering (Anderson, Hughes, Rothi, Crucian, & Heilman, 1999; Benke, Hohenstein, Poewe, & Butterworth, 2000). Studies on deep brain stimulation of the STN may have some negative effects on the prosody (Santens, De Letter, Van Borsel, De Reuck, & Caemaert, 2003; Wang, Metman, Bakay, Arzbaecher, & Bernard, 2003).

3.2 Cognitive Processes of the Basal Ganglia

3.2.1 Action Selection/Decision Making

Decision-making process is quite complex as the cue/stimulus should be able to predict the expected reward based on not only the action but also context/environment which is continuously changing. In other words, to make an optimal decision we need to weigh all the available options, compare them with the current goal, and choose the most rewarding one. The two main components that lead to an optimal decision making are choosing either the most rewarding choice, play it safe (i.e., 'exploit') or trying something new (i.e., 'Explore'). The basal ganglia's anatomical position and input projections make it the most suitable candidate to be involved in action selection, which is also supported by evolutionary analysis (Grillner, Robertson, & Stephenson-Jones, 2013). The striatum, the major input structure of BG, receives input from almost all the cortical areas (Packard & Knowlton, 2002) through the multiple cortico-subcortical functional loops. Moreover, dopaminergic projections, the key player to reinforce learning, modulate cortico-striatal plasticity (Kreitzer & Malenka, 2008) which also aids in the stimulus-response learning.

Generally, the final 'action selection' is assumed to be based on the combined contributions of the basal ganglia direct and indirect pathways at output nuclei (Packard & Knowlton, 2002; Smith, Beyan, Shink, & Bolam, 1998), whereas the hyperdirect mainly functions as a global stop signal. Specifically, the direct pathway of BG acts as facilitator by disinhibiting the excitatory thalamo-cortical circuitry via reduction of GPi activity giving rise to 'Go' scenario. Contrastingly, the indirect pathway is known to act as inhibitor due to odd number of inhibitory stages and by increasing the GPi activity (via the STN) leading to 'NoGo' situation. In classical accounts, the effect of dopamine (DA) on BG pathways has been described in simple Go/NoGo terms (Rogers, 2010). Under low DA conditions, IP is more active than DP leading to 'NoGo' or action inhibition (Frank, 2005), whereas in high DA conditions DP is more active than IP leading to 'Go' or facilitation of action (Chevalier & Deniau, 1990; Packard & Knowlton, 2002). Early measurements of event-related potentials in PD patients have shown that caudate nucleus (which is in the dorsal striatum) is activated in response to salient and meaningful stimuli (Kropotov & Etlinger, 1999). By using fMRI technology, experimental studies were conducted to understand the individual roles of the BG nuclei. These studies showed that dorsal striatum activity encodes reward expectation (O'Doherty, Dayan, Friston, Critchley, & Dolan, 2003; O'Doherty, 2004) and punishments (Seymour, Daw, Dayan, Singer, & Dolan, 2007). The ventral striatum also encodes risk anticipation in addition to rewards (Preuschoff, Bossaerts, & Quartz, 2006). Using fMRI studies, Tanaka et al. (2004) observed that ventro-anterior part of striatum is involved in predicting immediate rewards and dorsoposterior regions in future rewards (Tanaka et al., 2004). To understand role of other BG nuclei in action selection, Kropotov and Etlinger (1999) conducted a digit recognition task on PD patients where the authors observed that non-recognition of symbols led to inhibition of GPi activity (Kropotov & Etlinger, 1999). Apart from this, the authors also observed a selection activation of basal ganglia-thalamic circuits during set switching.

Although the neural correlates for exploitation–exploration at the cortical level have been proposed and identified, their counterparts at the subcortical level are still under debate. Using a modified version of n-arm bandit task and fMRI measurements, Daw and colleagues observed that fronto-polar cortex and intraparietal sulcus were active during exploratory choices while ventromedial prefrontal cortex (VmPFC) during exploitative (Daw, O'Doherty, Dayan, Seymour, & Dolan, 2006). They have also suggested that striatum could be the subcortical counterpart of VmPFC. Jepma and colleagues proposed the role of locus coeruleus–norepinephrine in establishing the Explore–Exploit trade off by measuring the changes in pupil diameter in human subjects who participated in a utility-based task (Jepma & Nieuwenhuis, 2011). It has also been suggested that the pallidum, in its interactions with the noradrenergic system, controls the balance between exploration–exploitation (Aston-Jones, Rajkowski, Kubiak, & Alexinsky, 1994; Doya, 2002; Russell, Allin, Lamm, & Taljaard, 1992). Monchi, Petrides, Strafella, Worsley, and Doyon (2006) observed an increase in dorsal striatum (caudate and putamen) during planning of a set shift, whereas an increase in STN activity was observed during actual shifting conditions irrespective of its planning (Monchi et al., 2006).

3.2.2 Attention

Various studies show that the basal ganglia play a key role in attentional processes (Carbon & Marie, 2003; Hayes, Davidson, Keele, & Rafal, 1998; Isoda & Hikosaka, 2008; Moustafa & Gluck, 2011; Nieoullon, 2002; Robbins, 2007; Saint-Cyr, 2003). The role of BG in attentional processes is evident by many neuropsychological studies on patients with BG dysfunction, such as PD (Allcock et al., 2009; Botha & Carr, 2012; Hall et al., 2016; Moustafa et al., 2016; Moustafa, Sherman, & Frank, 2008; Yogev et al., 2005). It is suggested that the BG's role in attentional processes includes resisting interference from distractors (Bocquillon et al., 2012). Specifically, studies found that the STN and GP play a key role in attention (Schmalbach et al., 2014; Witt, Kopper, Deuschl, & Krack, 2006). In addition, neural theories and experimental data point to a role for dopamine in attentional processes (Boulougouris & Tsaltas, 2008; Ungless, 2004). It is argued that stimulus-locked phasic responses of dopamine neurons are related to the saliency of stimuli and thus attentional allocation requirements for those stimuli. The exact role of the basal ganglia direct and indirect pathways is not known; however, it is argued that the direct pathway plays a role in paying attention to salient stimuli in the environment and the indirect pathway in learned inattention, that is, ignoring stimuli that are task irrelevant. It was also found that the hyperdirect pathway, which includes the subthalamic nucleus, plays a role in focused attention (Beck et al., 2017; Bockova et al., 2011; Schmalbach et al., 2014; Witt et al., 2006). Its exact function is not known, although it is suggested that it sends a global signal to other structures to decide whether to pay attention or not.

3.2.3 Working Memory

Working memory, a term that refers to temporary storage of information in the brain, is often used synonymously with short-term memory. The life of this form of memory is of the order of seconds. Some of the earliest findings about the neural substrates of working memory came from the lesion studies of (Jacobsen) who showed that lesions of prefrontal cortex (PFC) impaired working memory performance in monkeys. Subsequently, it was shown that neurons of PFC hold on to information in the form of sustained activity during the delay period in a delayed matching task (Fuster, 1973; Goldman-Rakic, 1991). Dopamine projections to PFC seem to play a key role in working memory functions of PFC since direct application of dopamine agonists or antagonists to PFC resulted in degradation of working memory performance (Sawaguchi & Goldman-Rakic, 1994; Zahrt, Taylor, Mathew, & Arnsten, 1997). The role of basal ganglia in working memory has been demonstrated by functional neuroimaging studies (Tomasi, Chang, Caparelli, & Ernst, 2007). Prefrontal cortex and basal ganglia control access to working memory (McNab et al., 2008) and animal electrophysiology (Lewis, Dove, Robbins, Barker, & Owen, 2004; Menon, Anagnoson, Glover, & Pfefferbaum, 2000; Postle & D'Esposito, 1999). The role of the basal ganglia in working memory is further confirmed by the fact that working memory performance is impaired in Parkinson's patients (Beato et al., 2008; Fallon, Mattiesing, Muhammed, Manohar, & Husain, 2017; Fournet, Moreaud, Roulin, Naegele, & Pellat, 2000; Hodgson, Dittrich, Henderson, & Kennard, 1999; Lewis, Slabosz, Robbins, Barker, & Owen, 2005; Moustafa, Bell, Eissa, & Hewedi, 2013; Moustafa, Herzallah, & Gluck, 2013; Moustafa, Sherman, & Frank, 2008; Owen, Doyon, Dagher, Sadikot, & Evans, 1998). Several studies have shown that working memory is a key function underlying several other cognitive processes, such as sequential movement (for discussion, see Moustafa et al., 2016).

3.2.4 Sequence Learning

In sequential arm movements, it was shown that pallidal neurons are selectively activated for certain components of the movement (Mushiake & Strick, 1995). By studying hand–eye coordination in trained monkeys, Kermadi and Joseph (1995) have shown that neurons in the caudate nucleus are tuned to specific sequences of eye–hand movements (Kermadi & Joseph, 1995). The involvement of both caudate and putamen in sequence learning was confirmed in humans using functional Magnetic Resonance Imaging (Rauch et al., 1997).

Benecke, Rothwell, Dick, Day, and Marsden (1987) have noted impairment in execution of movement sequences in patients with Parkinson's disease (Benecke et al., 1987). By analyzing the interonset and movement durations, the authors conclude that the peculiar difficulty faced by PD patients can be traced to a deficit in the capacity to switch from one motor program to another in a motor sequence. Similar studies were performed by Harrington and Haaland (1991) with PD patients performing reaction time experiments involving sequential hand movements. PD patients exhibited more errors and performed slower than controls in executing long and complex sequences, suggestive of deficit in switching hand postures (Harrington & Haaland, 1991). It is argued that the basal ganglia learn sequential movements step by step via dopaminergic projections from the substantia nigra pars compacta.

3.2.5 Sleep Regulation

Considering the well-known effect of psychostimulants on sleep and wakefulness, and the link between such drugs and the brain's reward system, a role for the basal ganglia in sleep regulation is anticipated. Another reason that points to a link between BG and sleep regulation is the fact that Parkinson's disease patients are typically plagued by sleep disturbances. Studies on rats involving chemical lesioning of the striatum revealed that the dorsal striatum plays a role in wakefulness, whereas the nAcc plays a role in sleep (Lazarus, Chen, Urade, & Huang, 2013). Striatal lesioning also affected the dynamics of sleep leading to fragmentation, which is the amount of movement or restlessness in a sleep period. Particularly, lesioning of nAcc led to shortened duration of bouts of Non-Rapid Eye Movement (NREM) sleep. The Nucleus Accumbens Core/Shell plays a role in Sleep-Wake Regulation and Modafinil-Induced Arousal. Lesioning of GPe resulted in a radical increase in wakefulness manifesting as insomnia. Dopamine levels in various brain regions are also known to fluctuate with sleep stages. For example, extracellular dopamine levels are lower in medial prefrontal (mPFC) and nAcc during NREM sleep, but higher during waking and Rapid Eye Movement (REM) sleep (Lena et al., 2005). Consistent with the previous finding, it was shown that modafinil, a wakefulness-promoting drug, increases extracellular levels of dopamine in mPFC and nAcc (Murillo-Rodríguez, Haro, Palomero-Rivero, Millán-Aldaco, & Drucker-Colín, 2007). Similar PD drugs like piribedil and pramipexole, which are D2R agonists, cause sudden sleep attacks in humans (Lipford & Silber, 2012; Tan, 2003).

3.3 Mood and Emotional Processes of the Basal Ganglia

3.3.1 Negative and Positive Affect

In addition to motor and cognitive processes, various studies show that the basal ganglia also play a key role in emotional processes, including positive and negative ones (Dannlowski et al., 2013; Eitan et al., 2013; Espinosa-Parrilla, Baunez, & Apicella, 2013; Levy & Dubois, 2006; Subramanian, Hindle, Jackson, & Linden, 2010). Specifically, it has been shown that the ventral striatum is involved in reward and positive affective processes (Dannlowski et al., 2013; Moretti & Signori, 2016; Steele, Kumar, & Ebmeier, 2007) as well negative affective processes (Correia, McGrath, Lee, Graybiel, & Goosens, 2016; Graham et al., 2016; Reznikov, Binko, Nobrega, & Hamani, 2016). Similarly, the subthalamic nucleus is also involved in positive and negative emotional processes (Altug, Acar, Acar, & Cavlak, 2011; Eitan et al., 2013; Karachi et al., 2005; Schneider et al., 2003) as well as mood processes (Czernecki et al., 2008; Pinsker, Amtage, Berger, Nikkhah, & van Elst, 2013).

Several studies have also shown that depressive episodes are related to basal ganglia dysfunction (Laasonen-Balk et al., 1999; Pan et al., 2017). Depression is not uncommon in PD patients (Herzallah et al., 2010). It has been suggested that reduced activity in the ventral striatum is related to the occurrence of depression in PD patients (Remy, Doder, Lees, Turjanski, & Brooks, 2005).

3.4 Autonomic Processes of the Basal Ganglia

The basal ganglia have also been found to play an essential role in autonomous activity. The classic features of PD symptoms shed light on the autonomic control of dopaminergic projections and their target nuclei such as the basal ganglia. These symptoms include autonomic gastrointestinal dysfunction, abnormal salivation, dysphagia, constipation (Edwards, Quigley, Hofman, & Pfeiffer, 1993), bladder dysfunction (Murnaghan, 1961; Porter & Bors, 1971), and thermoregulation (Appenzeller & Goss, 1971; Gubbay & Barwick, 1966). Several cardiovascular and renal functions are also disturbed in PD, leading to disturbances to heart rate variability (Goldstein, Holmes, Dendi, Bruce, & Li, 2002; Kallio et al., 2000; Senard, Brefel-Courbon, Rascol, & Montastruc, 2001) and abnormal blood pressure regulation. Studies manipulating chemical signals through microinjections at the striatum have provided direct link between the functioning of basal ganglia to blood pressure regulation (Pazo & Medina, 1983). The Nucleus Tractus Solitarii projects, via a set of cortical intermediary stages, to the nAcc, an important nucleus in the brain's reward system. Neurons in the Nucleus Tractus Solitarii also receive sensory signals from the circulatory system via vagus and glossopharyngeal nerves. Such anatomical data are related to the neural hierarchy that controls systemic circulation; it is compelling to speculate that the brain's reward network is involved in regulation of systemic circulation (Cechetto & Shoemaker, 2009; Neafsey, 1991; Resstel & Correa, 2006; Verberne & Owens, 1998). Further, research using direct electrical stimulation in the ganglia has provided evidences for the basal ganglia control of bladder functions (Pazo, 1976).

References

Allcock, L. M., Rowan, E. N., Steen, I. N., Wesnes, K., Kenny, R. A., & Burn, D. J. (2009). Impaired attention predicts falling in Parkinson's disease. *Parkinsonism & Related Disorder, 15*(2), 110–115. https://doi.org/10.1016/j.parkreldis.2008.03.010 S1353-8020(08)00111-9 [pii].

Alm, P. A. (2004). Stuttering and the basal ganglia circuits: A critical review of possible relations. *Journal of Communication Disorders, 37*(4), 325–369.

Almeida, Q. J., & Lebold, C. A. (2010). Freezing of gait in Parkinson's disease: A perceptual cause for a motor impairment? *Journal of Neurology, Neurosurgery and Psychiatry, 81*(5), 513–518.

Altug, F., Acar, F., Acar, G., & Cavlak, U. (2011). The influence of subthalamic nucleus deep brain stimulation on physical, emotional, cognitive functions and daily living activities in patients with Parkinson's disease. *Turkish Neurosurgery, 21*(2), 140–146. https://doi.org/10.5137/1019-5149.JTN.3956-10.0.

Anderson, J. M., Hughes, J. D., Rothi, L. J. G., Crucian, G. P., & Heilman, K. (1999). Developmental stuttering and Parkinson's disease: The effects of levodopa treatment. *Journal of Neurology, Neurosurgery and Psychiatry, 66*(6), 776–778.

Appenzeller, O., & Goss, J. E. (1971). Autonomic deficits in Parkinson's syndrome. *Archives of Neurology, 24*(1), 50–57.

Aston-Jones, G., Rajkowski, J., Kubiak, P., & Alexinsky, T. (1994). Locus coeruleus neurons in monkey are selectively activated by attended cues in a vigilance task. *The Journal of Neuroscience, 14*(7), 4467–4480.

Basso, M. A., & Wurtz, R. H. (2002). Neuronal activity in substantia nigra pars reticulata during target selection. *Journal of Neuroscience, 22*(5), 1883–1894.

Beato, R., Levy, R., Pillon, B., Vidal, C., du Montcel, S. T., Deweer, B., … Cardoso, F. (2008). Working memory in Parkinson's disease patients: Clinical features and response to levodopa. *Arquivos de Neuro-Psiquiatria, 66*(2A), 147–151.

Beck, A. K., Lutjens, G., Schwabe, K., Dengler, R., Krauss, J. K., & Sandmann, P. (2017). Thalamic and basal ganglia regions are involved in attentional processing of behaviorally significant events: Evidence from simultaneous depth and scalp EEG. *Brain Structure and Function*. https://doi.org/10.1007/s00429-017-1506-z.

Beckstead, R. M., Domesick, V. B., & Nauta, W. J. (1993). Efferent connections of the substantia nigra and ventral tegmental area in the rat. *Neuroanatomy* (pp. 449–475). Berlin: Springer.

Benecke, R., Rothwell, J., Dick, J., Day, B., & Marsden, C. (1987). Disturbance of sequential movements in patients with Parkinson's disease. *Brain, 110*(2), 361–379.

Benke, T., Hohenstein, C., Poewe, W., & Butterworth, B. (2000). Repetitive speech phenomena in Parkinson's disease. *Journal of Neurology, Neurosurgery and Psychiatry, 69*(3), 319–324.

Bockova, M., Chladek, J., Jurak, P., Halamek, J., Balaz, M., & Rektor, I. (2011). Involvement of the subthalamic nucleus and globus pallidus internus in attention. *Journal of Neural Transmission (Vienna), 118*(8), 1235–1245. https://doi.org/10.1007/s00702-010-0575-4.

Bocquillon, P., Bourriez, J. L., Palmero-Soler, E., Destee, A., Defebvre, L., Derambure, P., et al. (2012). Role of basal ganglia circuits in resisting interference by distracters: A swLORETA study. *PLoS ONE, 7*(3), e34239. https://doi.org/10.1371/journal.pone.0034239.

Botha, H., & Carr, J. (2012). Attention and visual dysfunction in Parkinson's disease. *Parkinsonism & Related Disorder* https://doi.org/10.1016/j.parkreldis.2012.03.004 S1353-8020(12)00080-6 [pii].
Boulougouris, V., & Tsaltas, E. (2008). Serotonergic and dopaminergic modulation of attentional processes. *Progress in Brain Research, 172,* 517–542.
Broderick, M. P., Van Gemmert, A. W., Shill, H. A., & Stelmach, G. E. (2009). Hypometria and bradykinesia during drawing movements in individuals with Parkinson's disease. *Experimental Brain Research, 197*(3), 223–233.
Brown, R., & Marsden, C. (1988). 'Subcorttcal dementia': The neuropsychological evidence. *Neuroscience, 25*(2), 363–387.
Canter, G. J. (1963). Speech characteristics of patients with Parkinson's disease: I. Intensity, pitch, and duration. *Journal of Speech & Hearing Disorders.*
Cantiniaux, S., Vaugoyeau, M., Robert, D., Horrelou-Pitek, C., Mancini, J., Witjas, T., et al. (2010). Comparative analysis of gait and speech in Parkinson's disease: Hypokinetic or dysrhythmic disorders? *Journal of Neurology, Neurosurgery and Psychiatry, 81*(2), 177–184.
Cappa, S., & Abutalebi, J. (1999). Subcortical aphasia. *The Concise Encyclopedia of Language Pathology*, 319–327.
Carbon, M., & Marie, R. M. (2003). Functional imaging of cognition in Parkinson's disease. *Current Opinion in Neurology, 16*(4), 475–480.
Cechetto, D. F., & Shoemaker, J. K. (2009). Functional neuroanatomy of autonomic regulation. *Neuroimage, 47*(3), 795–803.
Chevalier, G., & Deniau, J. (1990). Disinhibition as a basic process in the expression of striatal functions. *Trends in Neurosciences, 13*(7), 277–280.
Correia, S. S., McGrath, A. G., Lee, A., Graybiel, A. M., & Goosens, K. A. (2016). Amygdala-ventral striatum circuit activation decreases long-term fear. *Elife, 5.* https://doi.org/10.7554/elife.12669.
Cowie, D., Limousin, P., Peters, A., & Day, B. L. (2010). Insights into the neural control of locomotion from walking through doorways in Parkinson's disease. *Neuropsychologia, 48*(9), 2750–2757.
Czernecki, V., Schupbach, M., Yaici, S., Levy, R., Bardinet, E., Yelnik, J., … Agid, Y. (2008). Apathy following subthalamic stimulation in Parkinson disease: A dopamine responsive symptom. *Movement Disorder, 23*(7), 964–969.
Dannlowski, U., Domschke, K., Birosova, E., Lawford, B., Young, R., Voisey, J., … Zwanzger, P. (2013). Dopamine D(3) receptor gene variation: Impact on electroconvulsive therapy response and ventral striatum responsiveness in depression. *International Journal of Neuropsychopharmacology, 16*(7), 1443–1459. https://doi.org/10.1017/s1461145711001659 S1461145711001659 [pii].
Daw, N. D., O'Doherty, J. P., Dayan, P., Seymour, B., & Dolan, R. J. (2006). Cortical substrates for exploratory decisions in humans. *Nature, 441*(7095), 876–879.
Doya, K. (2002). Metalearning and neuromodulation. *Neural Networks, 15*(4), 495–506.
Edwards, L., Quigley, E., Hofman, R., & Pfeiffer, R. (1993). Gastrointestinal symptoms in parkinson disease: 18-month follow-up study. *Movement Disorders, 8*(1), 83–86.
Eitan, R., Shamir, R. R., Linetsky, E., Rosenbluh, O., Moshel, S., Ben-Hur, T., … Israel, Z. (2013). Asymmetric right/left encoding of emotions in the human subthalamic nucleus. *Frontiers in Systems Neuroscience, 7*, 69. https://doi.org/10.3389/fnsys.2013.00069.
Espinosa-Parrilla, J. F., Baunez, C., & Apicella, P. (2013). Linking reward processing to behavioral output: Motor and motivational integration in the primate subthalamic nucleus. *Frontiers in Computational Neuroscience, 7,* 175. https://doi.org/10.3389/fncom.2013.00175.
Faist, M., Xie, J., Kurz, D., Berger, W., Maurer, C., Pollak, P., et al. (2001). Effect of bilateral subthalamic nucleus stimulation on gait in Parkinson's disease. *Brain, 124*(8), 1590–1600.
Fallon, S. J., Mattiesing, R. M., Muhammed, K., Manohar, S., & Husain, M. (2017). Fractionating the neurocognitive mechanisms underlying working memory: Independent effects of dopamine and Parkinson's disease. *Cerebral Cortex*, 1–12. https://doi.org/10.1093/cercor/bhx242.

Fellows, S. J., Noth, J., & Schwarz, M. (1998). Precision grip and Parkinson's disease. *Brain: A Journal of Neurology, 121*(9), 1771–1784.
Fournet, N., Moreaud, O., Roulin, J. L., Naegele, B., & Pellat, J. (2000). Working memory functioning in medicated Parkinson's disease patients and the effect of withdrawal of dopaminergic medication. *Neuropsychology, 14*(2), 247–253.
Frank, M. J. (2005). Dynamic dopamine modulation in the basal ganglia: A neurocomputational account of cognitive deficits in medicated and nonmedicated Parkinsonism. *Journal of Cognitive Neuroscience, 17*(1), 51–72.
Fuster, J. M. (1973). Unit activity in prefrontal cortex during delayed-response performance: Neuronal correlates of transient memory. *Journal of Neurophysiology*.
Goldman-Rakic, P. S. (1991). Cellular and circuit basis of working memory in prefrontal cortex of nonhuman primates. *Progress in Brain Research, 85*, 325–336.
Goldstein, D., Holmes, C., Dendi, R., Bruce, S., & Li, S.-T. (2002). Orthostatic hypotension from sympathetic denervation in Parkinson's disease. *Neurology, 58*(8), 1247–1255.
Graham, A. M., Buss, C., Rasmussen, J. M., Rudolph, M. D., Demeter, D. V., Gilmore, J. H., … Fair, D. A. (2016). Implications of newborn amygdala connectivity for fear and cognitive development at 6-months-of-age. *Developmental Cognitive Neuroscience, 18*, 12–25. https://doi.org/10.1016/j.dcn.2015.09.006.
Grillner, S., Robertson, B., & Stephenson-Jones, M. (2013). The evolutionary origin of the vertebrate basal ganglia and its role in action selection. *The Journal of Physiology, 591*(22), 5425–5431.
Grossman, M., Carvell, S., Stern, M. B., Gollomp, S., & Hurtig, H. I. (1992). Sentence comprehension in Parkinson's disease: The role of attention and memory. *Brain and Language, 42*(4), 347–384.
Grossman, M., Zurif, E., Lee, C., Prather, P., Kalmanson, J., Stern, M. B., et al. (2002). Information processing speed and sentence comprehension in Parkinson's disease. *Neuropsychology, 16*(2), 174.
Gubbay, S., & Barwick, D. (1966). Two cases of accidental hypothermia in Parkinson's disease with unusual EEG findings. *Journal of Neurology, Neurosurgery and Psychiatry, 29*(5), 459.
Hall, J. M., O'Callaghan, C., Shine, J. M., Muller, A. J., Phillips, J. R., Walton, C. C., … Moustafa, A. A. (2016). Dysfunction in attentional processing in patients with Parkinson's disease and visual hallucinations. *Journal of Neural Transmission (Vienna), 123*(5), 503–507. https://doi.org/10.1007/s00702-016-1528-3.
Harel, B., Cannizzaro, M., & Snyder, P. J. (2004). Variability in fundamental frequency during speech in prodromal and incipient Parkinson's disease: A longitudinal case study. *Brain and Cognition, 56*(1), 24–29.
Harrington, D. L., & Haaland, K. Y. (1991). Sequencing in Parkinson's disease: Abnormalities in programming and controlling movement. *Brain, 114*(1), 99–115.
Harris, C. M., & Wolpert, D. M. (1998). Signal-dependent noise determines motor planning. *Nature, 394*(6695), 780–784.
Hartelius, L., & Svensson, P. (1994). Speech and swallowing symptoms associated with Parkinson's disease and multiple sclerosis: a survey. *Folia Phoniatrica et Logopaedica, 46*(1), 9–17.
Hausdorff, J. M., Cudkowicz, M. E., Firtion, R., Wei, J. Y., & Goldberger, A. L. (1998). Gait variability and basal ganglia disorders: Stride-to-stride variations of gait cycle timing in Parkinson's disease and Huntington's disease. *Movement Disorders, 13*(3), 428–437.
Hayes, A. E., Davidson, M. C., Keele, S. W., & Rafal, R. D. (1998). Toward a functional analysis of the basal ganglia. *Journal of Cognitive Neuroscience, 10*(2), 178–198.
Herzallah, M. M., Moustafa, A. A., Misk, A. J., Al-Dweib, L. H., Abdelrazeq, S. A., Myers, C. E., et al. (2010). Depression impairs learning whereas anticholinergics impair transfer generalization in Parkinson patients tested on dopaminergic medications. *Cognitive and Behavioral Neurology, 23*(2), 98–105. https://doi.org/10.1097/WNN.0b013e3181df3048.

Hikosaka, O., Nakahara, H., Rand, M. K., Sakai, K., Lu, X., Nakamura, K., … Doya, K. (1999). Parallel neural networks for learning sequential procedures. *Trends in Neurosciences, 22*(10), 464–471.

Hikosaka, O., Nakamura, K., Sakai, K., & Nakahara, H. (2002). Central mechanisms of motor skill learning. *Current Opinion in Neurobiology, 12*(2), 217–222.

Hikosaka, O., Takikawa, Y., & Kawagoe, R. (2000). Role of the basal ganglia in the control of purposive saccadic eye movements. *Physiological Reviews, 80*(3), 953–978.

Hikosaka, O., & Wurtz, R. H. (1983). Effects on eye movements of a GABA agonist and antagonist injected into monkey superior colliculus. *Brain Research, 272*(2), 368–372.

Hodgson, T. L., Dittrich, W. H., Henderson, L., & Kennard, C. (1999). Eye movements and spatial working memory in Parkinson's disease. *Neuropsychologia, 37*(8), 927–938.

Inglis, W. L., & Winn, P. (1995). The pedunculopontine tegmental nucleus: Where the striatum meets the reticular formation. *Progress in Neurobiology, 47*(1), 1–29.

Ingvarsson, P. E., Gordon, A. M., & Forssberg, H. (1997). Coordination of manipulative forces in Parkinson's disease. *Experimental Neurology, 145*(2), 489–501.

Isoda, M., & Hikosaka, O. (2008). Role for subthalamic nucleus neurons in switching from automatic to controlled eye movement. *Journal of Neuroscience, 28*(28), 7209–7218. https://doi.org/10.1523/jneurosci.0487-08.2008 28/28/7209 [pii].

Jepma, M., & Nieuwenhuis, S. (2011). Pupil diameter predicts changes in the exploration–exploitation trade-off: Evidence for the adaptive gain theory. *Journal of Cognitive Neuroscience, 23*(7), 1587–1596.

Kallio, M., Haapaniemi, T., Turkka, J., Suominen, K., Tolonen, U., Sotaniemi, K., … Myllylä, V. (2000). Heart rate variability in patients with untreated Parkinson's disease. *European Journal of Neurology, 7*(6), 667–672.

Karachi, C., Yelnik, J., Tande, D., Tremblay, L., Hirsch, E. C., & Francois, C. (2005). The pallidosubthalamic projection: An anatomical substrate for nonmotor functions of the subthalamic nucleus in primates. *Movement Disorders, 20*(2), 172–180.

Kato, M., Miyashita, N., Hikosaka, O., Matsumura, M., Usui, S., & Kori, A. (1995). Eye movements in monkeys with local dopamine depletion in the caudate nucleus. I. Deficits in spontaneous saccades. *Journal of Neuroscience, 15*(1), 912–927.

Kegl, J., Cohen, H., & Poizner, H. (1999). Articulatory consequences of Parkinson's disease: Perspectives from two modalities. *Brain and Cognition, 40*(2), 355–386.

Kermadi, I., & Joseph, J. (1995). Activity in the caudate nucleus of monkey during spatial sequencing. *Journal of Neurophysiology, 74*(3), 911–933.

Kimmeskamp, S., & Hennig, E. M. (2001). Heel to toe motion characteristics in Parkinson patients during free walking. *Clinical Biomechanics, 16*(9), 806–812.

Kori, A., Miyashita, N., Kato, M., Hikosaka, O., Usui, S., & Matsumura, M. (1995). Eye movements in monkeys with local dopamine depletion in the caudate nucleus. II. Deficits in voluntary saccades. *Journal of Neuroscience, 15*(1), 928–941.

Kotz, S. A., Frisch, S., Von Cramon, D. Y., & Friederici, A. D. (2003). Syntactic language processing: ERP lesion data on the role of the basal ganglia. *Journal of the International Neuropsychological Society, 9*(7), 1053–1060.

Kotz, S. A., Schwartze, M., & Schmidt-Kassow, M. (2009). Non-motor basal ganglia functions: A review and proposal for a model of sensory predictability in auditory language perception. *Cortex, 45*(8), 982–990.

Kravitz, A. V., Freeze, B. S., Parker, P. R., Kay, K., Thwin, M. T., Deisseroth, K., et al. (2010). Regulation of parkinsonian motor behaviors by optogenetic control of basal ganglia circuitry. *Nature, 466*(7306), 622.

Kreitzer, A. C., & Malenka, R. C. (2008). Striatal plasticity and basal ganglia circuit function. *Neuron, 60*(4), 543–554.

Kropotov, J. D., & Etlinger, S. C. (1999). Selection of actions in the basal ganglia–thalamocortical circuits: Review and model. *International Journal of Psychophysiology, 31*(3), 197–217.

Laasonen-Balk, T., Kuikka, J., Viinamaki, H., Husso-Saastamoinen, M., Lehtonen, J., & Tiihonen, J. (1999). Striatal dopamine transporter density in major depression. *Psychopharmacology (Berl), 144*(3), 282–285.

Lazarus, M., Chen, J. F., Urade, Y., & Huang, Z. L. (2013). Role of the basal ganglia in the control of sleep and wakefulness. *Current Opinion in Neurobiology, 23*(5), 780–785. https://doi.org/10.1016/j.conb.2013.02.001.

Lena, I., Parrot, S., Deschaux, O., Muffat-Joly, S., Sauvinet, V., Renaud, B., ... Gottesmann, C. (2005). Variations in extracellular levels of dopamine, noradrenaline, glutamate, and aspartate across the sleep–wake cycle in the medial prefrontal cortex and nucleus accumbens of freely moving rats. *Journal of Neuroscience Research, 81*(6), 891–899.

Levy, R., & Dubois, B. (2006). Apathy and the functional anatomy of the prefrontal cortex-basal ganglia circuits. *Cerebral Cortex, 16*(7), 916–928.

Lewis, S. J., & Barker, R. A. (2009). A pathophysiological model of freezing of gait in Parkinson's disease. *Parkinsonism & Related Disorders, 15*(5), 333–338.

Lewis, S. J., Dove, A., Robbins, T. W., Barker, R. A., & Owen, A. M. (2004). Striatal contributions to working memory: A functional magnetic resonance imaging study in humans. *European Journal of Neuroscience, 19*(3), 755–760.

Lewis, S. J., Slabosz, A., Robbins, T. W., Barker, R. A., & Owen, A. M. (2005). Dopaminergic basis for deficits in working memory but not attentional set-shifting in Parkinson's disease. *Neuropsychologia, 43*(6), 823–832.

Lieberman, P. (1991). *Uniquely human: The evolution of speech, thought, and selfless behavior.* Cambridge, MA: Harvard University Press.

Lipford, M. C., & Silber, M. H. (2012). Long-term use of pramipexole in the management of restless legs syndrome. *Sleep Medicine, 13*(10), 1280–1285.

Lubik, S., Fogel, W., Tronnier, V., Krause, M., König, J., & Jost, W. (2006). Gait analysis in patients with advanced Parkinson disease: Different or additive effects on gait induced by levodopa and chronic STN stimulation. *Journal of Neural Transmission (Vienna), 113*(2), 163–173.

Majsak, M. J., Kaminski, T., Gentile, A. M., & Flanagan, J. R. (1998). The reaching movements of patients with Parkinson's disease under self-determined maximal speed and visually cued conditions. *Brain: A Journal of Neurology, 121*(4), 755–766.

Marsden, C. (1982). The mysterious motor function of the basal ganglia: The Robert Wartenberg Lecture. *Neurology.*

Maruyama, T., & Yanagisawa, N. (2006). Cognitive impact on freezing of gait in Parkinson's disease. *Parkinsonism & Related Disorders, 12,* S77–S82.

McNab, F., Leroux, G., Strand, F., Thorell, L., Bergman, S., & Klingberg, T. (2008). Common and unique components of inhibition and working memory: An fMRI, within-subjects investigation. *Neuropsychologia, 46*(11), 2668–2682.

Menon, V., Anagnoson, R. T., Glover, G. H., & Pfefferbaum, A. (2000). Basal ganglia involvement in memory-guided movement sequencing. *NeuroReport, 11*(16), 3641–3645.

Monchi, O., Petrides, M., Strafella, A. P., Worsley, K. J., & Doyon, J. (2006). Functional role of the basal ganglia in the planning and execution of actions. *Annals of Neurology, 59*(2), 257–264.

Moreau, C., Ozsancak, C., Blatt, J. L., Derambure, P., Destee, A., & Defebvre, L. (2007). Oral festination in Parkinson's disease: Biomechanical analysis and correlation with festination and freezing of gait. *Movement Disorders, 22*(10), 1503–1506.

Moretti, R., & Signori, R. (2016). Neural correlates for apathy: Frontal-prefrontal and parietal cortical-subcortical circuits. *Frontiers in Aging Neuroscience, 8,* 289. https://doi.org/10.3389/fnagi.2016.00289.

Moriizumi, T., Nakamura, Y., Tokuno, H., Kitao, Y., & Kudo, M. (1988). Topographic projections from the basal ganglia to the nucleus tegmenti pedunculopontinus pars compacta of the cat with special reference to pallidal projections. *Experimental Brain Research, 71*(2), 298–306.

Morris, M., Iansek, R., Matyas, T., & Summers, J. (1998). Abnormalities in the stride length-cadence relation in parkinsonian gait. *Movement Disorders, 13*(1), 61–69.

Moustafa, & Gluck, M. A. (2011). A neurocomputational model of dopamine and prefrontal-striatal interactions during multicue category learning by Parkinson patients. *Journal of Cognitive Neuroscience, 23*(1), 151–167. https://doi.org/10.1162/jocn.2010.21420.

Moustafa, A. A., Bell, P., Eissa, A. M., & Hewedi, D. H. (2013a). The effects of clinical motor variables and medication dosage on working memory in Parkinson's disease. *Brain and Cognition, 82*(2), 137–145. https://doi.org/10.1016/j.bandc.2013.04.001.

Moustafa, A. A., Chakravarthy, S., Phillips, J. R., Crouse, J. J., Gupta, A., Frank, M. J., ... Jahanshahi, M. (2016). Interrelations between cognitive dysfunction and motor symptoms of Parkinson's disease: Behavioral and neural studies. *Reviews in the Neurosciences*. https://doi.org/10.1515/revneuro-2015-0070.

Moustafa, A. A., Herzallah, M. M., & Gluck, M. A. (2013b). Dissociating the cognitive effects of levodopa versus dopamine agonists in a neurocomputational model of learning in Parkinson's disease. *Neurodegenerative Diseases, 11*(2), 102–111. https://doi.org/10.1159/000341999.

Moustafa, A. A., Sherman, S. J., & Frank, M. J. (2008). A dopaminergic basis for working memory, learning and attentional shifting in Parkinsonism. *Neuropsychologia, 46*(13), 3144–3156. https://doi.org/10.1016/j.neuropsychologia.2008.07.011 S0028-3932(08)00297-2 [pii].

Müller, F., & Abbs, J. H. (1990). Precision grip in parkinsonian patients. *Advances in Neurology, 53,* 191.

Murillo-Rodríguez, E., Haro, R., Palomero-Rivero, M., Millán-Aldaco, D., & Drucker-Colín, R. (2007). Modafinil enhances extracellular levels of dopamine in the nucleus accumbens and increases wakefulness in rats. *Behavioural Brain Research, 176*(2), 353–357.

Murnaghan, G. (1961). Neurogenic disorders of the bladder in Parkinsonism. *BJU International, 33*(4), 403–409.

Mushiake, H., & Strick, P. L. (1995). Pallidal neuron activity during sequential arm movements. *Journal of Neurophysiology, 74*(6), 2754–2758.

Nakahara, H., Doya, K., & Hikosaka, O. (2001). Parallel cortico-basal ganglia mechanisms for acquisition and execution of visuomotor sequences—A computational approach. *Journal of Cognitive Neuroscience, 13*(5), 626–647.

Napier, J. R. (1956). The prehensile movements of the human hand. *Bone & Joint Journal, 38*(4), 902–913.

Neafsey, E. J. (1991). Prefrontal cortical control of the autonomic nervous system: Anatomical and physiological observations. *Progress in Brain Research, 85,* 147–166.

Nenadic, I., Gaser, C., Volz, H.-P., Rammsayer, T., Häger, F., & Sauer, H. (2003). Processing of temporal information and the basal ganglia: New evidence from fMRI. *Experimental Brain Research, 148*(2), 238–246.

Nieoullon, A. (2002). Dopamine and the regulation of cognition and attention. *Progress in Neurobiology, 67*(1), 53–83.

O'Doherty, J. P., Dayan, P., Friston, K., Critchley, H., & Dolan, R. J. (2003). Temporal difference models and reward-related learning in the human brain. *Neuron, 38*(2), 329–337.

O'Doherty, J. P. (2004). Reward representations and reward-related learning in the human brain: Insights from neuroimaging. *Current Opinion in Neurobiology, 14*(6), 769–776.

Owen, A. M., Doyon, J., Dagher, A., Sadikot, A., & Evans, A. C. (1998). Abnormal basal ganglia outflow in Parkinson's disease identified with PET. *Brain: A Journal of Neurology, 121*(5), 949–965.

Packard, M. G., & Knowlton, B. J. (2002). Learning and memory functions of the basal ganglia. *Annual Review of Neuroscience, 25*(1), 563–593.

Pan, P. M., Sato, J. R., Salum, G. A., Rohde, L. A., Gadelha, A., Zugman, A., ... Stringaris, A. (2017). Ventral striatum functional connectivity as a predictor of adolescent depressive disorder in a longitudinal community-based sample. *American Journal of Psychiatry, 174*(11), 1112–1119. https://doi.org/10.1176/appi.ajp.2017.17040430.

Pazo, J., & Medina, J. (1983). Cholinergic mechanisms within the caudate nucleus mediate changes in blood pressure. *Neuropharmacology, 22*(6), 717–720.

Pazo, J. H. (1976). Caudate-putamen and globus pallidus influences on a visceral reflex. *Acta physiologica latino americana, 26*(4), 260–266.
Pinsker, M., Amtage, F., Berger, M., Nikkhah, G., & van Elst, L. T. (2013). Psychiatric side-effects of bilateral deep brain stimulation for movement disorders. *Acta Neurochirurgica Supplementum, 117,* 47–51. https://doi.org/10.1007/978-3-7091-1482-7_8.
Porter, R. W., & Bors, E. (1971). Neurogenic bladder in Parkinsonism: Effect of thalamotomy. *Journal of Neurosurgery, 34*(1), 27–32.
Postle, B. R., & D'Esposito, M. (1999). Dissociation of human caudate nucleus activity in spatial and nonspatial working memory: An event-related fMRI study. *Cognitive Brain Research, 8* (2), 107–115.
Preuschoff, K., Bossaerts, P., & Quartz, S. R. (2006). Neural differentiation of expected reward and risk in human subcortical structures. *Neuron, 51*(3), 381–390.
Rascol, O., Sabatini, U., Simonetta-Moreau, M., Montastruc, J., Rascol, A., & Clanet, M. (1991). Square wave jerks in parkinsonian syndromes. *Journal of Neurology, Neurosurgery and Psychiatry, 54*(7), 599–602.
Rauch, S. L., Whalen, P. J., Savage, C. R., Curran, T., Kendrick, A., Brown, H. D., ... Rosen, B. R. (1997). Striatal recruitment during an implicit sequence learning task as measured by functional magnetic resonance imaging. *Human Brain Mapping, 5*(2), 124-132.
Remy, P., Doder, M., Lees, A., Turjanski, N., & Brooks, D. (2005). Depression in Parkinson's disease: Loss of dopamine and noradrenaline innervation in the limbic system. *Brain, 128*(Pt 6), 1314–1322. https://doi.org/10.1093/brain/awh445.
Resstel, L., & Correa, F. (2006). Involvement of the medial prefrontal cortex in central cardiovascular modulation in the rat. *Autonomic Neuroscience, 126,* 130–138.
Reznikov, R., Binko, M., Nobrega, J. N., & Hamani, C. (2016). Deep brain stimulation in animal models of fear, anxiety, and posttraumatic stress disorder. *Neuropsychopharmacology, 41*(12), 2810–2817. https://doi.org/10.1038/npp.2016.34.
Robbins, T. W. (2007). Shifting and stopping: Fronto-striatal substrates, neurochemical modulation and clinical implications. *Philosophical Transactions of the Royal Society of London: Series B, Biological Sciences, 362*(1481), 917–932.
Rogers, R. D. (2010). The roles of dopamine and serotonin in decision making: Evidence from pharmacological experiments in humans. *Neuropsychopharmacology, 36*(1), 114–132.
Russell, V., Allin, R., Lamm, M., & Taljaard, J. (1992). Regional distribution of monoamines and dopamine D1-and D2-receptors in the striatum of the rat. *Neurochemical Research, 17*(4), 387–395.
Sahyoun, C., Floyer-Lea, A., Johansen-Berg, H., & Matthews, P. (2004). Towards an understanding of gait control: Brain activation during the anticipation, preparation and execution of foot movements. *Neuroimage, 21*(2), 568–575.
Saint-Cyr, J. A. (2003). Frontal-striatal circuit functions: Context, sequence, and consequence. *Journal of the International Neuropsychological Society, 9*(1), 103–127.
Santens, P., De Letter, M., Van Borsel, J., De Reuck, J., & Caemaert, J. (2003). Lateralized effects of subthalamic nucleus stimulation on different aspects of speech in Parkinson's disease. *Brain and Language, 87*(2), 253–258.
Sato, M., & Hikosaka, O. (2002). Role of primate substantia nigra pars reticulata in reward-oriented saccadic eye movement. *Journal of Neuroscience, 22*(6), 2363–2373.
Sawaguchi, T., & Goldman-Rakic, P. S. (1994). The role of D1-dopamine receptor in working memory: Local injections of dopamine antagonists into the prefrontal cortex of rhesus monkeys performing an oculomotor delayed-response task. *Journal of Neurophysiology, 71*(2), 515–528.
Schaal, S., & Schweighofer, N. (2005). Computational motor control in humans and robots. *Current Opinion in Neurobiology, 15*(6), 675–682.
Schirmer, A. (2004). Timing speech: A review of lesion and neuroimaging findings. *Cognitive Brain Research, 21*(2), 269–287.

Schmalbach, B., Gunther, V., Raethjen, J., Wailke, S., Falk, D., Deuschl, G., et al. (2014). The subthalamic nucleus influences visuospatial attention in humans. *Journal of Cognitive Neuroscience, 26*(3), 543–550. https://doi.org/10.1162/jocn_a_00502.

Schneider, F., Habel, U., Volkmann, J., Regel, S., Kornischka, J., Sturm, V., & Freund, H. J. (2003). Deep brain stimulation of the subthalamic nucleus enhances emotional processing in Parkinson disease. *Archives of General Psychiatry, 60*(3), 296–302. yoa10144 [pii].

Senard, J.-M., Brefel-Courbon, C., Rascol, O., & Montastruc, J.-L. (2001). Orthostatic hypotension in patients with Parkinson's disease. *Drugs and Aging, 18*(7), 495–505.

Seymour, B., Daw, N., Dayan, P., Singer, T., & Dolan, R. (2007). Differential encoding of losses and gains in the human striatum. *Journal of Neuroscience, 27*(18), 4826–4831.

Shadmehr, R., & Krakauer, J. W. (2008). A computational neuroanatomy for motor control. *Experimental Brain Research, 185*(3), 359–381.

Shine, J. M., Matar, E., Ward, P. B., Bolitho, S. J., Pearson, M., Naismith, S. L., & Lewis, S. J. (2013). Differential neural activation patterns in patients with Parkinson's disease and freezing of gait in response to concurrent cognitive and motor load. *PLoS One, 8*(1), e52602.

Smith, Y., Beyan, M. D., Shink, E., & Bolam, J. P. (1998). Microcircuitry of the direct and indirect pathways of the basal ganglia. *Neuroscience (Oxford), 86,* 353–388.

Soliveri, P., Brown, R., Jahanshahi, M., Caraceni, T., & Marsden, C. (1997). Learning manual pursuit tracking skills in patients with Parkinson's disease. *Brain: A Journal of Neurology, 120* (8), 1325–1337.

Steele, J. D., Kumar, P., & Ebmeier, K. P. (2007). Blunted response to feedback information in depressive illness. *Brain, 130*(Pt 9), 2367–2374. https://doi.org/10.1093/brain/awm150.

Subramanian, L., Hindle, J. V., Jackson, M. C., & Linden, D. E. (2010). Dopamine boosts memory for angry faces in Parkinson's disease. *Movement Disorders, 25*(16), 2792–2799.

Svennilson, E., Torvik, A., Lowe, R., & Leksell, L. (1960). Treatment of parkinsonism by stereotactic thermolesions in the pallidal region. A clinical evaluation of 81 cases. *Acta Psychiatrica Scandinavica, 35*(3), 358–377.

Takakusaki, K., Habaguchi, T., Ohtinata-Sugimoto, J., Saitoh, K., & Sakamoto, T. (2003). Basal ganglia efferents to the brainstem centers controlling postural muscle tone and locomotion: A new concept for understanding motor disorders in basal ganglia dysfunction. *Neuroscience, 119*(1), 293–308.

Takakusaki, K., Ohta, R., & Harada, H. (2007). *Modulation of the excitability of hindlimb motor neurons during fictive locomotion by the basal ganglia efferents to the brainstem in decerebrate cats.* Paper Presented at the Social Neuroscience Abstract.

Takakusaki, K., Saitoh, K., Harada, H., & Kashiwayanagi, M. (2004). Role of basal ganglia–brainstem pathways in the control of motor behaviors. *Neuroscience Research, 50*(2), 137–151.

Takakusaki, K., Tomita, N., & Yano, M. (2008). Substrates for normal gait and pathophysiology of gait disturbances with respect to the basal ganglia dysfunction. *Journal of Neurology, 255,* 19–29.

Tan, E. (2003). Piribedil-induced sleep attacks in Parkinson's disease. *Fundamental & Clinical Pharmacology, 17*(1), 117–119.

Tanaka, S. C., Doya, K., Okada, G., Ueda, K., Okamoto, Y., & Yamawaki, S. (2004). Prediction of immediate and future rewards differentially recruits cortico-basal ganglia loops. *Nature Neuroscience, 7*(8), 887–893.

Teulings, H.-L., Contreras-Vidal, J. L., Stelmach, G. E., & Adler, C. H. (1997). Parkinsonism reduces coordination of fingers, wrist, and arm in fine motor control. *Experimental Neurology, 146*(1), 159–170.

Tomasi, D., Chang, L., Caparelli, E., & Ernst, T. (2007). Different activation patterns for working memory load and visual attention load. *Brain Research, 1132,* 158–165.

Tucha, O., Mecklinger, L., Thome, J., Reiter, A., Alders, G., Sartor, H., … Lange, K. (2006). Kinematic analysis of dopaminergic effects on skilled handwriting movements in Parkinson's disease. *Journal of Neural Transmission (Vienna), 113*(5), 609–623.

Ungless, M. A. (2004). Dopamine: The salient issue. *Trends in Neurosciences, 27*(12), 702–706.

Van Buren, J., Li, C., & Ojemann, G. (1966). The fronto-striatal arrest response in man. *Electroencephalography and Clinical Neurophysiology, 21*(2), 114–130.
Verberne, A. J., & Owens, N. C. (1998). Cortical modulation of thecardiovascular system. *Progress in Neurobiology, 54*(2), 149–168.
Wang, E., Metman, L. V., Bakay, R., Arzbaecher, J., & Bernard, B. (2003). The effect of unilateral electrostimulation of the subthalamic nucleus on respiratory/phonatory subsystems of speech production in Parkinson's disease—A preliminary report. *Clinical Linguistics & Phonetics, 17* (4–5), 283–289.
Witt, K., Kopper, F., Deuschl, G., & Krack, P. (2006). Subthalamic nucleus influences spatial orientation in extra-personal space. *Movement Disorders, 21*(3), 354–361. https://doi.org/10.1002/mds.20728.
Wolfe, V., Garvin, J., Bacon, M., & Waldrop, W. (1975). Speech changes in Parkinson's disease during treatment with L-dopa. *Journal of Communication Disorders, 8*(3), 271–279.
Yogev, G., Giladi, N., Peretz, C., Springer, S., Simon, E. S., & Hausdorff, J. M. (2005). Dual tasking, gait rhythmicity, and Parkinson's disease: Which aspects of gait are attention demanding? *European Journal of Neuroscience, 22*(5), 1248–1256. https://doi.org/10.1111/j.1460-9568.2005.04298.x EJN4298 [pii].
Zahrt, J., Taylor, J. R., Mathew, R. G., & Arnsten, A. F. (1997). Supranormal stimulation of D1 dopamine receptors in the rodent prefrontal cortex impairs spatial working memory performance. *Journal of Neuroscience, 17*(21), 8528–8535.

Chapter 4
Classical Computational Approaches to Modeling the Basal Ganglia

Ahmed A. Moustafa and V. Srinivasa Chakravarthy

Abstract There have been several modeling approaches to simulate BG structure and function. In this chapter, we discuss major modeling frameworks that have been proposed to simulate many functions of the BG. Many of such modeling studies are classical approaches in the field of BG modeling, which have been repeatedly to simulate many BG functions. In short, here we discuss the following model approaches: dimensionality reduction models, action section selection models, Go/NoGo models, reinforcement learning (RL) models of the basal ganglia, and Actor–Critic models. Importantly, this chapter mainly provides an overview of main architectures used to simulate the BG structure and function. In addition, we discuss many other models, such as those of gait, reaching, and other in the following chapters.

Below we review the following BG models: dimensionality reduction, action section selection models, Go/NoGo, reinforcement learning (RL), and Actor–Critic models.

4.1 Dimensionality Reduction Models

Most of the research on BG's functionality and anatomy was oriented toward understanding its role in action selection. This methodology started as the box diagrams explaining various anatomical areas, pathways, and connections (Albin, Young, & Penney, 1989; Gurney, Prescott, & Redgrave, 2001a). Bergman and associates differed from this main track and studied the BG's role in dimensionality reduction of the information from the cortex using computational methods which were further validated using the neural activity of the primate BG while performing behavioral tasks (Bar-Gad, Havazelet-Heimer, Goldberg, Ruppin, & Bergman, 2000; Bar-Gad, Morris, & Bergman, 2003). The concept of dimensionality reduction came from the observation that a very large number of cortical neurons project to input port, i.e., striatum which is consistent across the species. It has been reported that around 17×10^6 cortical neurons project to 1.7×10^6 bringing the convergence ratio to 10, and this ratio is even higher in primates

V. S. Chakravarthy and A. Moustafa, *Computational Neuroscience Models of the Basal Ganglia*, Cognitive Science and Technology,
https://doi.org/10.1007/978-981-10-8494-2_4

(571) and humans (347) (Bar-Gad et al., 2003). This trend is further observed between striatum and pallidal neurons. The GPi and thalamus then project back to cortex which is similar to divergence. It is highly important for the information to be preserved, updated via learning and resent back to the cortex without any losses.

To study this aspect, Bergman and associates proposed the reinforcement-driven dimensionality reduction (RDDR) model which uses the dopaminergic reinforcement signal to modulate the learning between the layers (Bar-Gad et al., 2000). The basic RDDR model consists of multilayer feedforward representing the cortex, striatum, and the GPi. The feedforward weights are updated using Hebbian learning and the lateral connections within the layers use anti-Hebbian learning methodology. The input layer has more number of neurons (higher dimension) than the output layer (lower dimension) to mimic the cortico-BG anatomy. The output neuron activity is a function of both inputs received from the first layer and the lateral input within its own layer. The reinforcement signal as a reward modulates the learning of feedforward weights (Bar-Gad et al., 2000).

Using this network architecture, they studied the activity of each layer during encoding of the information and finally calculated the reconstruction error which is the mean-squared difference between the original and reconstructed elements (from the output layer) overall input patterns (Bar-Gad et al., 2000). The model predicts no/low correlation in the activity of the output neurons when compared to the input layer correlation. This model prediction was in agreement with the neural recordings from primate's pallidum which show irregular activity despite a higher correlated cortical input. The cause for the irregular and low correlated output layer activity is attributed to be by the dynamics of lateral connection within the layer. Since the reinforcing dopaminergic signal modulates Hebbian learning in feedforward layers in the model, the authors suggest that this dynamics leads to a network that not only encodes the maximal variability of its input space but also encodes the variability of the reward-distorted space.

Finally, they propose an advanced RDDR model which overcomes the drawbacks of the basic model (Bar-Gad et al., 2003). The first and foremost constraint is restricting the weights to be either positive or negative which was absent in the basic model. Second, the individual neurons in the basic model were linear but since most of BG neurons show a nonlinearity in their firing rate, the individual units were proposed to be changed to sigmoidal nonlinear units. Furthermore, they plan on including and expanding the model into multiple loop, sparsely connected system which is more biologically realistic.

4.2 Action Selection Models

Organisms respond to different actions to different sensory stimuli received from the world. However, when two sensory stimuli are received simultaneously, it is often impossible to express the two corresponding actions, since the two actions may be incompatible, for example, flight or fight when facing a predator. Therefore, while sensory streams from the external world may be simultaneous, the organism's actions in response to those stimuli must first go through a certain arbitration process that selects the most optimal actions in a given context. By analyzing the anatomical position of the BG with respect to the sensory-motor cortical pathways, Gurney and colleagues (Bogacz & Gurney, 2007; Gurney et al., 2001a; Gurney, Prescott, & Redgrave, 2001b; Humphries, Stewart, & Gurney, 2006) had proposed that the BG perform some sort of action selection. Along with this line of thinking, the BG system has been proposed to be the vertebrate solution to the action selection problem (Redgrave, Prescott, & Gurney, 1999). Several studies suggest that the basal ganglia play a key role in action selection (Seo, Lee, & Averbeck, 2012).

In an attempt to explain how the BG structures implement the putative action selection function, Redgrave et al. (1999) proposed that the striatum computes some sort of salience of multiple action alternatives that influence the motor cortex at a given instant. Actions with the highest salience win the competition and have the highest chance to be expressed. Local inhibition among the GABAergic medium spiny neurons (MSNs) is thought to provide the cellular-level mechanism for competition among the representations of multiple actions at the level of striatum. Furthermore, the focused inhibitory projections from the striatum to the GPi combined with the diffuse excitatory projections from the STN to the GPi are thought to create a feedforward off-center/on-surround input/output response at the level of the output nucleus of BG, viz. GPi. Similar proposals have been made by others as well (Mink, 1996). These ideas have been further substantiated since they been realized in a robotic system driven by a BG-like control architecture (Prescott, 2002).

The idea that the lateral inhibitory network of MSNs within the striatum can implement competition among actions has been explored by other researchers also. For instance, Wickens (1997) investigated the exact topological and other connectivity parameters that enable the striatal network to perform the desired function of action competition. Another attractive feature of this network model is the observed change in the dynamics under the condition of Parkinsonian-like reduction in dopamine. When the dopamine levels are reduced below 'normal,' the network dynamics switched from competition to co-activation, whereby multiple incompatible actions are selected, a situation that can naturally explain Parkinsonian motor impairments like rigidity.

Based on the action selection theory of the BG developed by Gurney and others since the 90s, Humphries et al. (2006) developed a detailed spiking neuron model of the BG system, incorporating all the major nuclei. The model exhibits action selection dynamics as has been intuitively proposed and demonstrated in simpler rate-coded models earlier. Another prominent feature of the model is the synchronized oscillations of STN–GPe system, which can be modulated by dopamine. In this model, under normal or high dopamine conditions, STN and GPe are decoupled and exhibit desynchronized oscillations; under reduced dopamine conditions, the two nuclei are dynamically coupled, displaying pathologically synchronized oscillations as under Parkinson's conditions.

In one interesting study, Amos (2000) proposes a model that simulates performance in the Wisconsin Card Sorting Test (WCST). The model incorporates interaction between the prefrontal cortex (PFC) and the basal ganglia. The model provides a computational validity of how patients with PFC disorders show perseverative errors and patients with basal ganglia disorders show random errors in performing the WCST, in which the agent learns to categorize cards based on a certain rule. Further, after making some number of correct trials, the agent must relearn to sort the cards based on a different rule. The model incorporates closed and open cortex-basal-ganglia loops. PFC subserves active maintenance of the sorting rule, while the basal ganglia subserve selecting motor responses. The striatum subserves integration of information encoded in the cortex and maps cortical activity into motor responses. The basal ganglia receive input from PFC and sensory association cortex (which encodes represents target and input cards). The function of basal ganglia input to PFC is to provide feedback that the response made is correct (not modeled). If the response is not correct, PFC changes its sorting rule based on feedback from the basal ganglia. Note that this assumption is different from the assumption that basal ganglia input to PFC subserves gating of information into WM in that in the former, maintaining of information in WM is not dependent on the integrity of the basal ganglia. In the Amos model, DA projected to striatum or PFC is assumed to increase the signal-to-noise ratio (Cohen & Servan-Schreiber, 1992). That is, DA projected to a striatal or prefrontal neuron decreases the effect of noise and thus increases neural responses to stimuli. DA depletion in a brain area is modeled by decreasing the gain of sigmoidal units representing that area (also see Cohen & Servan-Schreiber, 1992). In this model, lesioning of brain area is simulated by decreasing the output of neurons representing that area.

The simulation results show that DA reduction in PFC is associated with the occurrence of perseverative responses, while DA reduction in the striatum is associated with the occurrence of random errors. Frontal dysfunction is associated with the occurrence of perseverative responses because it is only sensory association areas that project information to the striatum. Because the representation of the sorting rule is not maintained in frontal cortex, it is only motor response that is associated with the striatal unit that has the highest activation is selected in the entire experiment (I am assuming that noise is fixed throughout the whole experiment; note that there is no learning in the model). The model has some limitations. It is not trained to perform the task. Also, the basal ganglia indirect pathway is not incorporated in the model. Also, the model does not account for the finding that damage to the basal ganglia leads to the occurrence of perseverative responses in WM tasks, including the WCST.

Based on prior work, some recent models also show that the dopamine projection to the striatum plays a key role in action section while dopamine projection to the prefrontal cortex is key for attentional learning, that is, learning to pay attention to key information in the environment based on corrective feedback (Moustafa & Gluck, 2011a, 2011b; Moustafa, Herzallah, & Gluck, 2014).

There have been several recent models building on the action selection hypothesis of BG function. For example, a recent extension by Gurney and colleagues also showed that the neuropeptide Substance P (SP) and encephalin in the striatum play a role on action selection and sequential processing (Buxton, Bracci, Overton, & Gurney, 2017). In another study, the same group has simulated the relationship between action selection and gamma and beta oscillations (Blenkinsop, Anderson, & Gurney, 2017).

4.3 Go/NoGo Models

Frank provided a neurocomputational model of how the BG, thalamus, and cortex, along with DA and other neurotransmitters, interact with reward-based, motor, and cognitive learning tasks. Frank simulated performance in different motor and cognitive tasks subserved by the BG, including WM and decision making. Frank's framework suggests that the BG modulate both motor and cognitive actions, which are encoded in the motor and prefrontal cortices (Frank, Loughry, & O'Reilly; Houk, 2005; O'Reilly & Frank, 2006). In the motor domain, these models assume that BG output to the premotor cortex is responsible for action selection. Similarly, in the cognitive domain, the BG modulates representations encoded in the prefrontal cortex (Frank et al., 2001; Middleton & Strick, 2000, 2002; O'Reilly & Frank, 2006).

Most importantly, Frank suggests that DA subserves motor, cognitive, and reward-based learning and performance in the BG, which is supported by prior research (Delgado, Miller, Inati, & Phelps, 2005; Schultz, 1998; Schultz, Dayan, & Montague, 1997; Shohamy, Myers, Geghman, Sage, & Gluck, 2006). DA bursts and dips facilitate learning to select the most adaptive response and avoid the least adaptive ones via changes in synaptic plasticity in the BG direct (Go) and indirect (NoGo) pathways. Subsequent experiments supported a core prediction of the models, where PD patients off medication were impaired at learning from positive relative to negative reinforcement, whereas the same patients on medication showed the opposite pattern of learning bias (Frank, Seeberger, & O'Reilly, 2004). In Frank's models, the simulated non-medicated PD condition show enhanced negative, but not positive, reinforcement learning, and simulated DA medications reverse this bias (Frank et al., 2004). Frank also confirmed his models' predictions in subsequent empirical studies in ADHD patients (Frank, Santamaria, O'Reilly, & Willcutt), and in normal healthy subjects who were administered dopamine agonists and antagonists (Frank & O'Reilly, 2006).

More recent extensions of these models explore the specific role of the STN in decision making. In particular, Frank's model suggests that the STN provides a dynamic global NoGo signal on motor and cognitive actions. Empirical evidence for this hypothesis is reported in an fMRI study with human subjects (Aron & Poldrack, 2006). Simulation results also suggest that the STN plays an important role in high-conflict decisions, that is, decisions in which both alternatives are equally good or equally bad (Hershey et al., 2004). In Frank's model, the representation of multiple competing responses in premotor and cingulate regions (Braver et al., 2001; Frank, 2005), by activating the STN, is critical for slowing responding and preventing premature responding during difficult decisions. In these models, simulated STN lesions lead to premature responding in choice paradigms, consistent with evidence for this behavior in STN-lesioned rats (Baunez & Robbins, 1997). Frank's model suggests that by tonically stimulating the STN at unnaturally high frequencies, DBS effectively eliminates dynamic functions of the STN (similar to a lesion) and therefore removes the global NoGo signal (Benazzouz & Hallett, 2000; Limousin et al., 1997; Meissner et al., 2005). Specifically, STN is more active when response conflict is high (i.e., both responses have similar reinforcement values). This increase in activity slows responding (as evidenced by gradual thalamic activity increases) and prevents the model from making hasty decisions in high-conflict conditions. Both intact networks and those with STN–DBS successfully learn to select the appropriate response in the training phase. However, in the test phase, STN–DBS impaired selection in the high-conflict condition, in which the stimuli have comparable reinforcement values (80% vs. 70%).

Further, the sometimes reported DBS side effects of emotional hyperactivity or uncontrolled laughter (Czernecki et al., 2002; Funkiewiez et al., 2003; Krack et al., 2001) or distractibility (Saint-Cyr, Trepanier, Kumar, Lozano, & Lang, 2000) could result from reduction in NoGo activity possibly due to current spread to the limbic or associative STN (see Karachi et al., 2005; Krack et al., 2001). Frank's framework suggests that distractibility is a higher level cognitive analogue of premature responding and could occur due to excessive gating of information into the prefrontal cortex (Frank et al., 2001). This hypothesis was tested using a cognitive task known as the AX-CPT. The AX-CPT is a working memory task in which subjects are presented with sequential letter stimuli (A, B, X, Y; printed in red) and are instructed to press one of two keys to each letter presentation (Cohen, Barch, Carter, & Servan-Schreiber, 1999; Servan-Schreiber, Cohen, & Steingard, 1996). Subjects are to instructed to press key on the right side of the keyboard ('m') when A is followed by X (AX 'target' trials) and to press a left key ('z') otherwise (AY, BX, and BY trials). In short, Frank's modeling framework suggests that the prefrontal cortex plays a key role in the active maintenance of information in WM, whereas the BG is key for modulating when and when not to update information into WM, a function that becomes further.

Beiser and Houk (1998) proposed a gating model which is conceptually similar to the Frank et al. (2001) model. In both models, input stimuli are transiently represented in prefrontal cortex and gating of information into WM is subserved by PFC-basal-ganglia-segregated loops. The Beiser and Houk model simulates performance in a delayed sequence learning task. In this task, the model is presented with a sequence of key illuminations in some order. After a delay, the simulated subject is supposed to press the keys in the same order they were presented. The Beiser and Houk model assumes that the sequences of key illuminations (which are presented in a temporal order) are spatially represented within PFC. The model simulates gating of stimuli into WM (termed the encoding problem). The model shows that different key illumination sequences have different spatial representations (i.e., patterns of activity) in PFC. this model hypothesizes that input stimuli transiently activate PFC neurons, which in turn activate the caudate nucleus. Then, disinhibition of the thalamus leads to maintenance of the stimulus in frontocorticothalamic loops. The model has some limitations. The Beiser and Houk model does not simulate how the model presses the keys in the same order they are presented (termed the decoding problem). Another limitation of this model is that it is not trained to perform the task. Also, the indirect pathway of the basal ganglia is not incorporated in the model.

Building on Go/NoGo models of the basal ganglia, one recent model simulated the role basal ganglia interaction with the spinal cord in action selection (Kim et al., 2017). Unlike prior models, this model simulates arm reaching in dynamical environments that involve choosing between multiple actions. One other model also built on previous Go/NoGo models but further incorporated a two-term Hebb rule to train synapses in the striatum (Baston & Ursino, 2015).

4.4 RL Models of Basal Ganglia

The most commonly used models to simulate BG function are RL models.

Reinforcement learning (RL) is an unsupervised machine learning approach which has a large similarity with the mechanism of brain's functioning. In RL, an agent (e.g., the BG) in a state (s_t) at time 't' makes an action (a_t) and receives a reward (r_t) from the environment. The aim of the agent is to maximize the reward by choosing an optimal policy. With its roots in the theory of instrumental conditioning in psychology, RL describes the manner in which an agent learns stimulus–response (S-R) relations based on action outcomes: S–R pairs associated with rewarding outcomes are reinforced while those that result in punishment are attenuated. Typically, since there is a delay between an action and its outcomes, RL theory proposes a surrogate to reward, known as value, which is used by the actor in selecting a potentially rewarding action (Sutton & Barto, 1998). Schultz's seminal experiment where dopaminergic activity was recorded from macaques has shown that DA codes for reward prediction error. This error term was similar to the temporal difference error ('δ') in RL. Experimental data show that BG receives reward-related information in the form of dopaminergic input to striatum (Chakravarthy, Joseph, & Bapi, 2010; Niv, 2009). Dopamine has also been known to induce cortico-striatal plasticity changes which (Reynolds & Wickens, 2002) modulate the Hebb-like plasticity of cortico-striatal synapses (Surmeier, Ding, Day, Wang, & Shen, 2007). Based on these observations, a considerable body of the modeling literature has grown around the notion that BG uses this reward-related information from DA neurons to perform various cognitive functions such as decision making and sequence generation (Chakravarthy et al., 2010; Niv, 2009).

Many models dealt before just focus on the action selection strategies in the basal ganglia (Bar-Gad et al., 2003; Gurney et al., 2001a; Humphries, Khamassi, & Gurney, 2012). The essential reinforcing properties of dopamine were not fully utilized in such models. Plenty of evidences relate dopamine activity to long-term potentiation and depression of synapses (Schultz, 1998; Wickens, Horvitz, Costa, & Killcross, 2007; Wise, 2004; Wise & Rompre, 1989). Perhaps, the three-factor rule for synapse learning has dopamine as one important factor along with pre- and post-synaptic information to control the synaptic strength and their learning (Wickens & Kötter, 1995). Furthermore, the classic experiments by Schultz and colleagues, and models by Houk and colleagues (Houk, Adams, & Barto, 1995; Schultz, 1998), show that dopamine does not just have rewarding aspects, but also reward prediction properties. Detailed experiments prove that a mathematical quantity called reward prediction error (Sutton & Barto, 1998) well matches the signaling of dopaminergic neurons.

The experimental evidences begin with the classic one by Schultz and colleagues (Schultz, 1998) that show dopaminergic neurons increase their firing at the time of reward, 'r'. If we denote their activity as variable, 'δ', then

$$\delta \propto r$$

Bringing in the concept of rewards for learning and decision making begs the necessity to impart the ideas of reinforcement learning (Sutton & Barto, 1998). Reinforcement learning is a branch of machine learning, where the agent updates knowledge about the environment and makes efficient decisions, by sampling rewards for his/her actions in a state. The goal is to maximize the rewards obtained in a state.

More specifically, dopamine was found not just to respond to rewards alone, but to reward predictions. Reward predictions are tracked by a function called 'value' in reinforcement learning. It is defined as

$$Q(t) = \sum_{i=t+1}^{\infty} r_i$$

The above formula can be improvised by accounting for the discounting of future rewards using a factor, γ:

$$Q(t) = r_{t+1} + \gamma r_{t+2} + \gamma^2 r_{t+3} + \cdots + \gamma^{n-1} r_{t+n}$$

Serotonergic neurons are suggested to correlate with the discount factor, γ (Tanaka et al., 2007). Updating the value function happens by:

$$Q(t+1) = Q(t) + \eta_Q \delta_t$$

The algorithm can be defined to explain state-action-reward-next state-next action (SARSA) as a sequence, and Q function can be written as $Q(s_t, a_t)$, where 's_t'

is the state at time 't', 'a_t' is the action performed at time 't', and 'η_Q' is the learning rate of the action value function ($0 < \eta_Q < 1$). The temporal difference (TD) error measure of DA is defined by δ_t in the following equation for the case of immediate reward problems ($\gamma = 0$).

$$\delta_t = r_t - Q(s_t, a_t)$$

This is called reward prediction error, and the quantity closely matches with the dopaminergic firing than just the reward or reward prediction value. It also represents Rescorla Wagner's (RW) rule that brings about an association of unconditioned stimulus (reward, US) and the conditioned stimulus (states, CS). In case of a nonzero discount factor, the temporal prediction (TD) is defined as:

$$\delta_t = r_t + \gamma Q(s_{t+1}, a_{t+1}) - Q(s_t, a_t)$$

The striatal neurons are suggested to compute and keep track of value functions and reward prediction-related quantities (Balleine, Delgado, & Hikosaka, 2007; Delgado, 2007; O'Doherty et al., 2004; Samejima, Ueda, Doya, & Kimura, 2005). In RL language, this approximates to the function of *critic* module (Joel, Niv, & Ruppin, 2002).

A companion to critic is called *actor* module. This module uses the evaluations of states and actions as computed by the critic, for executing a choice selection. Action selection could be exploratory or exploitatory (Sutton & Barto, 1998). The function that decides the amount of randomness in making a choice is called policy, π. Some famous policies in RL include epsilon greedy—where at any time for a state, random action is executed with probability, ϵ; soft-max, where there is a temperature parameter, β, controlling exploration. Lower values of β less differentiate value function, Q and result in more exploration than higher values of β.

$$\pi_{a1} = \frac{e^{-\beta Q_{a1}}}{e^{-\beta Q_{a1}} + e^{-\beta Q_{a2}}}$$

Here, π_{a1} denotes the probability of choosing action, a_1, from a state, s, at time, t. The policy equivalent in the basal ganglia composes the dynamics executed by the direct and indirect pathways.

Actor–Critic Models

Actor–Critic models, on the other hand, focus on simulating (perceptually- and memory-guided) motor actions. These BG models simulate behavior in motor learning tasks, such as instrumental conditioning (e.g., Houk, 1995a), S-R (e.g., Berns & Sejnowski, 1996; Khamassi, Girard, Berthoz, & Guillot, 2004; Suri, Bargas, & Arbib, 2001), sequence learning (Suri & Schultz, 1998), and delayed-response tasks (Suri & Schultz, 1999).

Actor–Critic models dissociate motor learning from reward prediction learning in motor tasks (Houk, 1995b; O'Doherty et al., 2004). In an instrumental

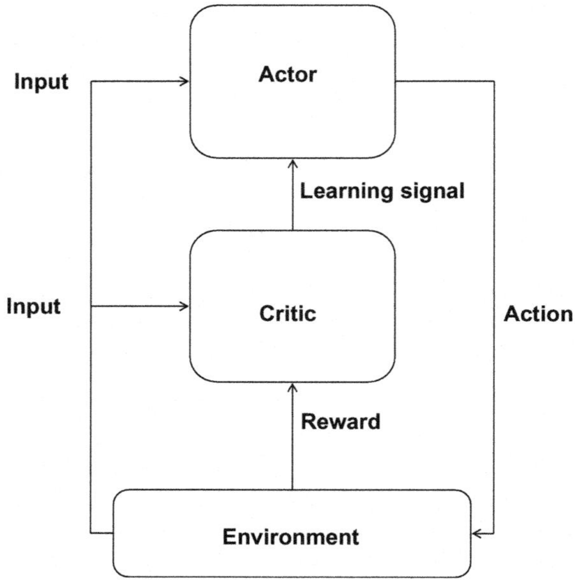

Fig. 4.1 Actor–Critic architecture. Both the Actor and Critic receive inputs from the environment (which includes the experimenter). The Actor is responsible for making motor responses. If it makes the correct motor response, the model is rewarded. The Critic is responsible for reward prediction learning. The Critic receives a reward from the environment and sends a learning signal to the Actor, informing it whether the action made has rewarding consequences or not (Barto, 2003)

conditioning task, for example, the animal is triggered (e.g., by illumination of a light or presentation of an auditory signal) to make a certain motor response in order to receive a reward. Actor–Critic models assume that in these tasks, the animal learns (a) that the trigger stimulus predicts the occurrence of a reward (reward prediction) and (b) how to make the motor response that is followed by the reward.

In the Actor–Critic architecture (Fig. 4.1), the Critic is responsible for learning how to predict a reward, while the Actor is responsible for adapting motor actions based on instructions from the Critic (Barto, 1995, 2003). The Critic sends a learning signal (also known as reinforcement or prediction error signal) to the Actor informing it whether the motor response it has made had rewarding consequences. A positive signal informs the Actor to increase the likelihood of making (i.e., reinforce) the action it has just made, whereas a negative one informs the Actor not to make the motor response it has just made. The Critic, on the other hand, does not receive a signal from the Actor (Fig. 4.1). However, it is informed whether the motor response that the Actor made had rewarding consequences.

The Houk and colleagues models

Houk and colleagues (Houk, 1995a; Houk et al., 1995) provided perhaps the first model to suggest that the basal ganglia are structurally and functionally similar to an Actor–Critic architecture (Joel et al., 2002). They suggested that the basic structure of the Actor–Critic architecture can be mapped onto the striatum. They suggested that the striosomes (and their efferent targets) are functionally equivalent to the Critic. This is based on the fact that (a) the striosomes are reciprocally connected to the SNc (a brain area subserving reward-based learning) and

(b) synaptic modification in the cortex-striosome pathway is DA-dependent, presumably subserving reward-based learning. Houk et al. also suggested that the matrisomes (and their efferent targets) are functionally equivalent to the Actor. This is based on the fact that matrisomes (via the GPi and thalamus) send projections to motor cortex.

Houk et al. (1995) proposed a conceptual model (i.e., without simulation studies) that simulates behavior in an instrumental conditioning task. In this task, the monkey learns to press a lever in order to receive a reward. A stimulus (CS) triggers the monkey to make a motor response. The model assumes that learning to predict the US is subserved by the cortex, striosomes, and SNc. The model also assumes that learning to make the motor response (pressing the lever) is subserved by the matrisomes, cortex, GPi, and thalamus. Most importantly, Houk et al. (1995) argue that the striosomes–SNc pathways compute the TD error. The inhibitory pathway, the excitatory pathway, and lateral hypothalamus input to SNc, respectively, compute $P(t-1), P(t), R(t)$.

The Houk et al. (1995) model has some limitations. One limitation is that the model does not explain the role of corticocortical connectivity in learning instrumental conditioning tasks (though it is depicted). Similarly, the model does not explain the role of the basal ganglia indirect pathway in learning to perform motor tasks. Also, the model is a conceptual, rather than a simulation, model. Houk et al.'s assumptions that the striosomes are functionally equivalent to the Critic, while the matrisomes are functionally equivalent to the Actor, are incorporated in models simulating functions subserved by the basal ganglia, such as S-R learning (Khamassi et al., 2004; Suri et al., 2001), a spatial delayed-response task (Suri & Schultz, 1999), and sequence learning tasks (Suri & Schultz, 1998; Tian, Arnold, Sejnowski, & Jabri, 2003).

The Suri and Colleagues Models

Suri and Schultz (1998) propose an Actor–Critic model that simulates behavior in a sequence learning task. In this task, the model is trained to sequentially associate the presence of different stimuli (A, B, C, D, E, F, and G) with making different motor responses (Q, R, S, T, U, V, and W). In other words, on each trial, the model learns to associate the presentation of A with making response Q, the presentation of B with making response R, and so on. Similar to the Houk et al. model, this model assumes that the striosomes subserve reward prediction learning, while the matrisomes subserve action learning. Each compartment is simulated using a one-layer network. The cortex connectivity to the matrisomes and striosomes is fully connected. The model assumes that action selection (Selection in this study is construed as choosing an action from several, potential actions) is achieved in the matrisomes. A WTA network, presumably simulating lateral inhibition between matrisomal units, selects the unit with the highest activity. Each matrisomal unit corresponds to an action.

The Suri and Schultz (1998) model is trained using the TD algorithm. To my knowledge, there are no recording studies from DA neurons while the animal is performing a sequence learning task. The model, however, assumes that DA phasic signal shifts to the time of CS's, until it reaches the earliest CS. This assumption is based on the fact that DA phasic responses move back in time in DRTs (Schultz, Apicella, & Ljungberg, 1993) and instrumental conditioning tasks (Schultz et al., 1997).

They also studied the effect of training the model using an unconditional reinforcement signal, which means that the DA signal does not shift to the time of conditioned stimuli and always associated with the reward whether predicted or not —hence its name. Simulation results suggest that non-shifting of the DA signal is associated with impaired learning, providing evidence that shifting of the DA phasic signal is associated with enhanced learning to perform the task. Tian et al. (2003) propose an Actor–Critic model, which is conceptually similar to the Suri and Schultz (1998) model. The Tian et al. model simulates a sequence learning task and is applied to the field of robotics. One limitation of the model is that it assumes the model will make actions only when a stimulus is presented. This is not plausible because animal studies show that during learning, animals sometimes make motor responses before or much after the trigger presentation.

Suri and Schultz (1999) propose another Actor–Critic model that simulates behavior in a spatial DRT (Schultz et al., 1993). This model's assumptions are similar to those of the Suri and Schultz (1998) model. This model simulates memory-guided motor responses. The model assumes that memory-guided motor responses are subserved by the PFC-basal ganglia pathway. It also assumes that learning to make memory-guided motor responses is subserved by DA projections to the striatum. Suri and Schultz (1999) also simulated the occurrence of preservative responses in PD patients. They assumed that striatal DA reduction in PD leads to non-shifting of the DA phasic signal to the time of conditioned stimuli. They trained their model using an unconditional reinforcement signal. Their model suggested that inappropriate time shifting of the DA phasic signal is responsible for the occurrence of perseverative responses in PD. Simulation results also show that not presenting the predicted reward after learning leads to behavioral extinction. The Suri and Schultz (1999) model has some limitations. This model does not simulate how input information is gated into WM. Further, this model does not deal with the presentation of distractors.

Suri et al. (2001) propose another Actor–Critic model that simulates behavior in a S-R task, namely a T-maze task. In this task, the rat learns to associate turning right or left, with facing the green or red stimuli, respectively. The rat is rewarded if it turns right and reaches the green stimulus. The rat is not rewarded if it turns left. The assumptions of this model are similar to those of Suri and Schultz (1998, 1999), though it is more physiologically detailed.

As conceptually similar to the Suri and Schultz (1998) model, Baldassarre and colleagues proposed Actor–Critic models that simulate performance in a foraging task (Baldassarre, 2002; Baldassarre & Parisi, 2000). In this task, an agent searches for many food pellets put in a two-dimensional board (The task modeled in the Baldassarre and Parisi (2000) study is slightly different from that used in the Baldassarre (2000) study in that in the former, ten reward objects are used while only three reward objects are used in the latter). The model assumes that the basal ganglia subserve S-R learning. It is trained using the TD algorithm. A WTA network selects the action that has the highest activation. The Actor is modeled using sigmoidal units with noisy threshold. The model uses a unit, called a matcher, which computes the motivation of the model, such as move when hungry and food is present. The underlying biological mechanism of the matcher is not specified

The simulation studies show that the Actor–Critic architecture is powerful enough to simulate a relatively complex task. In this task, the model learns to make many correct motor responses before receiving a reward (i.e., to move in the direction of food). The results show that the model learns to find food in about 30 steps (randomly it takes about 100 steps). One limitation of these models is that one is trained for 200,000 trials and the other for 10,000 trials to perform the task.

The Berns and Sejnowski (1996) Model

Berns and Sejnowski (1996) proposed an Actor–Critic model that assumes that the basal ganglia subserve action selection. Similar to the Houk et al. (1995) model, in this model the striosomes subserve the reward-based learning aspect of the task, while the matrisomes subserve making motor processes. Neurons are simulated using sigmoidal units. The model is trained using the TD algorithm. The model assumes that both the VTA and SNc projection to the striatum subserve TD learning; this, however, did not add any computational power to the model than that of Suri and Schultz's models which only uses the SNc.

The neural substrate of action selection in this model is different from that of Suri and colleagues (see above). Suri and colleagues assume that lateral connectivity of matrisomal neurons (simulated by a WTA network) subserve action selection. In this model, however, action selection is achieved in the GPi. The Berns and Sejnowski model assumes that both the striatal and STN input to the GPi subserve action selection. The Berns and Sejnowski model uses the term winner-lose-all instead of winner-take-all because the winning GPi unit is inhibited, not stimulated. In this model, the striatum is sparsely connected to the GPi, while the STN sends excitatory, diffuse projections to the GPi. The model assumes that each GPi unit subserves making a different motor response. In this model, the STN prevents all but the winning GPi unit from being inhibited. The simulation results found—and actually predicted—that damage to the STN leads to inability to stop selected actions. They modeled perseveration as related to DA reduction. However, it is not known how DA reduction is simulated. Or what kind of perseveration they modeled.

One limitation of this model is that it does not incorporate in the model. Another limitation of the model is that it assumes that the same striatal neurons send projections to both the GPe (the indirect pathway) and GPi/SNr. Wilson (2004) noted that this is not biologically plausible.

O'Reilly (2003) proposed a model very similar to that of Berns and Sejnowski (1996). The model simulates performance in the 1-2-AX task. This task is an extension of the AX–CPT task (described above), in which the agent (human subjects or computer model) learns to maintain two items in working memory. The model incorporates interactions between the basal ganglia and PFC. The model is trained using a combination of temporal difference and supervised learning (Leabra) algorithms. The model assumes that the basal ganglia subserve action selection. This model did not simulate gating of information into WM, though earlier work by the same author assumes that the basal ganglia subserve gating of information into WM. Following this work, one model has incorporated the role of the basal ganglia in both action section and gating of information into WM (Moustafa & Maida, 2007).

Several recent models have incorporated reinforcement learning methods to simulate various functions of the basal ganglia. For example, Shivkumar, Muralidharan, and Chakravarthy (2017) have proposed a model to solve context-based RL processes. The model assumes that the striosomes control the function of matrisome by selecting most optimal action. The model made several predictions that can be tested in future experimental studies. Stocco (2017) has also extended RL and action selection function of the basal ganglia to simulate more complex decision-making tasks that cannot be simulated using simpler RL models. Importantly, most RL models of the BG focus on action learning but often ignore the representation of time. Accordingly, a recent study has incorporated a time-sensitive action selection mechanism. This model accounts for interval timing processes, which is the perception of time duration (Gershman, Moustafa, & Ludvig, 2014). For similar work on representation of time in RL models, see Moustafa, Cohen, Sherman, and Frank (2008).

4.5 Conclusions

The modeling approaches described here show some promises in simulating some of BG functions. However, they have many limitations as they are there many functions supported by the BG that cannot be simulated by these models. In the following chapter, we discuss a somewhat novel modeling approach that can be used to simulate most, if not all, BG functions. Further, in the next few chapters, we discuss several other BG models that build on the neural architectures presented here.

References

Albin, R. L., Young, A. B., & Penney, J. B. (1989). The functional anatomy of basal ganglia disorders. *Trends in Neurosciences, 12*(10), 366–375.

Amos, A. (2000). A computational model of information processing in the frontal cortex and basal ganglia. *Journal of Cognitive Neuroscience, 12*(3), 505–519.

Aron, A. R., & Poldrack, R. A. (2006). Cortical and subcortical contributions to Stop signal response inhibition: Role of the subthalamic nucleus. *Journal of Neuroscience, 26*(9), 2424–2433.

Baldassarre, G. (2002). A modular neural-network model of the basal ganglia's role in learning and selecting motor behaviours. *Journal of Cognitive Systems Research, 3,* 5–13.

Baldassarre, G., & Parisi, D. (2000). Classical and instrumental conditioning: From laboratory phenomena to integrated mechanisms for adaptation.

Balleine, B. W., Delgado, M. R., & Hikosaka, O. (2007). The role of the dorsal striatum in reward and decision-making. *Journal of Neuroscience, 27*(31), 8161–8165. https://doi.org/10.1523/JNEUROSCI.1554-07.2007.

Bar-Gad, I., Havazelet-Heimer, G., Goldberg, J. A., Ruppin, E., & Bergman, H. (2000). Reinforcement-driven dimensionality reduction-a model for information processing in the basal ganglia. *Journal of Basic and Clinical Physiology and Pharmacology, 11*(4), 305–320.

Bar-Gad, I., Morris, G., & Bergman, H. (2003). Information processing, dimensionality reduction and reinforcement learning in the basal ganglia. *Progress in Neurobiology, 71*(6), 439–473. https://doi.org/10.1016/j.pneurobio.2003.12.001.

Barto, A. G. (1995). Adaptive critics and the basal ganglia. In J. C. Houk, J. L. Davis & D. G. Beiser (Eds.), *Models of information processing in the basal ganglia* (pp. xii, 382p). Cambridge, MA: MIT Press.

Barto, A. G. (2003). Reinforcement learning. In M. A. Arbib (Ed.), *The handbook of brain theory and neural networks* (pp. 963–968). Cambridge, MA: MIT Press.

Baston, C., & Ursino, M. (2015). A biologically inspired computational model of basal ganglia in action selection. *Computational Intelligence and Neuroscience, 2015,* 187417. https://doi.org/10.1155/2015/187417.

Baunez, C., & Robbins, T. W. (1997). Bilateral lesions of the subthalamic nucleus induce multiple deficits in an attentional task in rats. *European Journal of Neuroscience, 9*(10), 2086–2099.

Beiser, D. G., & Houk, J. C. (1998). Model of cortical-basal ganglionic processing: Encoding the serial order of sensory events. *Journal of Neurophysiology, 79*(6), 3168–3188. https://doi.org/10.1152/jn.1998.79.6.3168.

Benazzouz, A., & Hallett, M. (2000). Mechanism of action of deep brain stimulation. *Neurology, 55*(12 Suppl 6), S13–S16.

Berns, G. S., & Sejnowski, T. J. (1996). How the basal ganglia make decisions. In A. Damasio, H. Damasio, & Y. Christen (Eds.), *The neurobiology of decision making*. Berlin: Springer.

Blenkinsop, A., Anderson, S., & Gurney, K. (2017). Frequency and function in the basal ganglia: The origins of beta and gamma band activity. *Journal of Physiology, 595*(13), 4525–4548. https://doi.org/10.1113/JP273760.

Bogacz, R., & Gurney, K. (2007). The basal ganglia and cortex implement optimal decision making between alternative actions. *Neural Computation, 19*(2), 442–477.

Braver, T. S., Barch, D. M., Keys, B. A., Carter, C. S., Cohen, J. D., Kaye, J. A., … Reed, B. R. (2001). Context processing in older adults: Evidence for a theory relating cognitive control to neurobiology in healthy aging. *Journal of Experimental Psychology: General, 130*(4), 746–763.

Buxton, D., Bracci, E., Overton, P. G., & Gurney, K. (2017). Striatal neuropeptides enhance selection and rejection of sequential actions. *Frontiers in Computational Neuroscience, 11,* 62. https://doi.org/10.3389/fncom.2017.00062.

Chakravarthy, V., Joseph, D., & Bapi, R. S. (2010). What do the basal ganglia do? A modeling perspective. *Biological Cybernetics, 103*(3), 237–253.

Cohen, J. D., Barch, D. M., Carter, C., & Servan-Schreiber, D. (1999). Context-processing deficits in schizophrenia: Converging evidence from three theoretically motivated cognitive tasks. *Journal of Abnormal Psychology, 108*(1), 120–133.

Cohen, J. D., & Servan-Schreiber, D. (1992). Context, cortex, and dopamine: A connectionist approach to behavior and biology in schizophrenia. *Psychological Review, 99*(1), 45–77.

Czernecki, V., Pillon, B., Houeto, J. L., Pochon, J. B., Levy, R., & Dubois, B. (2002). Motivation, reward, and Parkinson's disease: Influence of dopatherapy. *Neuropsychologia, 40*(13), 2257–2267.

Delgado, M. R. (2007). Reward-related responses in the human striatum. *Annals of the New York Academy of Sciences, 1104*(1), 70–88.

Delgado, M. R., Miller, M. M., Inati, S., & Phelps, E. A. (2005). An fMRI study of reward-related probability learning. *Neuroimage, 24*(3), 862–873.

Frank, M. J. (2005). Dynamic dopamine modulation in the basal ganglia: A neurocomputational account of cognitive deficits in medicated and nonmedicated Parkinsonism. *Journal of Cognitive Neuroscience, 17*(1), 51–72.

Frank, M. J., Loughry, B., & O'Reilly, R. C. (2001). Interactions between frontal cortex and basal ganglia in working memory: A computational model. *Cognitive, Affective, & Behavioral Neuroscience, 1*(2), 137–160.

Frank, M. J., & O'Reilly, R. C. (2006). A mechanistic account of striatal dopamine function in human cognition: Psychopharmacological studies with cabergoline and haloperidol. *Behavioral Neuroscience, 120*(3), 497–517.

Frank, M. J., Seeberger, L. C., & O'Reilly, R. C. (2004). By carrot or by stick: Cognitive reinforcement learning in parkinsonism. *Science, 306*(5703), 1940–1943.

Funkiewiez, A., Ardouin, C., Krack, P., Fraix, V., Van Blercom, N., Xie, J., … Pollak, P. (2003). Acute psychotropic effects of bilateral subthalamic nucleus stimulation and levodopa in Parkinson's disease. *Movement Disorders, 18*(5), 524–530.

Gershman, S. J., Moustafa, A. A., & Ludvig, E. A. (2014). Time representation in reinforcement learning models of the basal ganglia. *Frontiers in Computational Neuroscience, 7,* 194. https://doi.org/10.3389/fncom.2013.00194.

Gurney, K., Prescott, T. J., & Redgrave, P. (2001a). A computational model of action selection in the basal ganglia. I. A new functional anatomy. *Biological Cybernetics, 84*(6), 401–410.

Gurney, K., Prescott, T. J., & Redgrave, P. (2001b). A computational model of action selection in the basal ganglia. II. Analysis and simulation of behaviour. *Biological Cybernetics, 84*(6), 411–423.

Hershey, T., Revilla, F. J., Wernle, A., Gibson, P. S., Dowling, J. L., & Perlmutter, J. S. (2004). Stimulation of STN impairs aspects of cognitive control in PD. *Neurology, 62*(7), 1110–1114.

Houk, J. C. (1995a). Information processing in modular circuits linking basal ganglia and cerebral Cortex. In J. C. Houk, J. L. Davis & D. G. Beiser (Eds.), *Models of information processing in the basal ganglia* (pp. xii, 382p). Cambridge, MA: MIT Press.

Houk, J. C. (1995b). A model of how the basal ganglia generate and use neural signals that predict reinforcement. In J. C. Houk, J. L. Davis & D. G. Beiser (Eds.), *Models of information processing in the basal ganglia* (pp. xii, 382p). Cambridge, MA: MIT Press.

Houk, J. C. (2005). Agents of the mind. *Biological Cybernetics, 92*(6), 427–437.

Houk, J. C., Adams, J. L., & Barto, A. G. (1995). A model of how the basal ganglia generate and use neural signals that predict reinforcement. *Models of Information Processing in the Basal Ganglia*, 249–270.

Humphries, M. D., Khamassi, M., & Gurney, K. (2012). Dopaminergic control of the exploration-exploitation trade-off via the basal ganglia. *Frontiers in Neuroscience, 6.*

Humphries, M. D., Stewart, R. D., & Gurney, K. N. (2006). A physiologically plausible model of action selection and oscillatory activity in the basal ganglia. *The Journal of Neuroscience, 26* (50), 12921–12942.

Joel, D., Niv, Y., & Ruppin, E. (2002). Actor–critic models of the basal ganglia: New anatomical and computational perspectives. *Neural Networks, 15*(4), 535–547.

Karachi, C., Yelnik, J., Tande, D., Tremblay, L., Hirsch, E. C., & Francois, C. (2005). The pallidosubthalamic projection: An anatomical substrate for nonmotor functions of the subthalamic nucleus in primates. *Movement Disorders, 20*(2), 172–180.

Khamassi, M., Girard, B., Berthoz, A., & Guillot, A. (2004). *Comparing three Critic models of reinforcement learning in the basal ganglia connected to a detailed actor part in a S-R task.* Paper presented at the Proceedings of the Eighth International Conference on Intelligent Autonomous Systems IAS-8, Amsterdam, The Netherlands.

Kim, T., Hamade, K. C., Todorov, D., Barnett, W. H., Capps, R. A., Latash, E. M., … Molkov, Y. I. (2017). Reward based motor adaptation mediated by basal ganglia. *Frontiers in Computational Neuroscience, 11*, 19. https://doi.org/10.3389/fncom.2017.00019.

Krack, P., Kumar, R., Ardouin, C., Dowsey, P. L., McVicker, J. M., Benabid, A. L., & Pollak, P. (2001). Mirthful laughter induced by subthalamic nucleus stimulation. *Movement Disorders, 16*(5), 867–875.

Limousin, P., Greene, J., Pollak, P., Rothwell, J., Benabid, A. L., & Frackowiak, R. (1997). Changes in cerebral activity pattern due to subthalamic nucleus or internal pallidum stimulation in Parkinson's disease. *Annals of Neurology, 42*(3), 283–291.

Meissner, W., Leblois, A., Hansel, D., Bioulac, B., Gross, C. E., Benazzouz, A., & Boraud, T. (2005). Subthalamic high frequency stimulation resets subthalamic firing and reduces abnormal oscillations. *Brain, 128*(10), 2372–2382.

Middleton, F. A., & Strick, P. L. (2000). Basal ganglia output and cognition: Evidence from anatomical, behavioral, and clinical studies. *Brain and Cognition, 42*(2), 183–200.

Middleton, F. A., & Strick, P. L. (2002). Basal-ganglia 'projections' to the prefrontal cortex of the primate. *Cerebral Cortex, 12*(9), 926–935.

Mink, J. W. (1996). The basal ganglia: Focused selection and inhibition of competing motor programs. *Progress in Neurobiology, 50*(4), 381.

Moustafa, A. A., & Gluck, M. A. (2011a). A neurocomputational model of dopamine and prefrontal-striatal interactions during multicue category learning by Parkinson patients. *Journal of Cognitive Neuroscience, 23*(1), 151–167. https://doi.org/10.1162/jocn.2010.21420.

Moustafa, A. A., & Gluck, M. A. (2011b). Computational cognitive models of prefrontal-striatal-hippocampal interactions in Parkinson's disease and schizophrenia. *Neural Netw, 24*(6), 575–591. https://doi.org/10.1016/j.neunet.2011.02.006.

Moustafa, A. A., & Maida, A. S. (2007). Using TD learning to simulate working memory performance in a model of the prefrontal cortex and basal ganglia. *Cognitive Systems Research, 8,* 262–281.

Moustafa, A. A., Cohen, M. X., Sherman, S. J., & Frank, M. J. (2008). A role for dopamine in temporal decision making and reward maximization in parkinsonism. *Journal of Neuroscience, 28*(47), 12294–12304. https://doi.org/10.1523/JNEUROSCI.3116-08.2008.

Moustafa, A. A., Herzallah, M. M., & Gluck, M. A. (2014). A model of reversal learning and working memory in medicated and unmedicated patients with Parkinson's disease. *Journal of Mathematical Psychology, 59,* 120–131.

Niv, Y. (2009). Reinforcement learning in the brain. *Journal of Mathematical Psychology, 53*(3), 139–154.

O'Doherty, J., Dayan, P., Schultz, J., Deichmann, R., Friston, K., & Dolan, R. J. (2004). Dissociable roles of ventral and dorsal striatum in instrumental conditioning. *Science, 304* (5669), 452–454.

O'Reilly, R. C., & Frank, M. J. (2006). Making working memory work: A computational model of learning in the prefrontal cortex and basal ganglia. *Neural Computation, 18*(2), 283–328.

O'Reilly, R. C. (2003). Making working memory work: A computational model of learning in the prefrontal cortex and basal ganglia. ICS Technical Report, (pp. 1–23).

Prescott, T. J. (2002). Basal ganglia. In M. A. Arbib (Ed.), *The handbook of brain theory and neural networks* (pp. xvii, 1290p). Cambridge, MA: MIT Press.

Redgrave, P., Prescott, T. J., & Gurney, K. (1999). The basal ganglia: A vertebrate solution to the selection problem? *Neuroscience, 89*(4), 1009–1023.

Reynolds, J. N. J., & Wickens, J. R. (2002). Dopamine-dependent plasticity of corticostriatal synapses. *Neural Networks, 15*(4), 507–521.

Saint-Cyr, J. A., Trepanier, L. L., Kumar, R., Lozano, A. M., & Lang, A. E. (2000). Neuropsychological consequences of chronic bilateral stimulation of the subthalamic nucleus in Parkinson's disease. *Brain, 123*(10), 2091–2108.

Samejima, K., Ueda, Y., Doya, K., & Kimura, M. (2005). Representation of action-specific reward values in the striatum. *Science, 310*(5752), 1337–1340.

Schultz, W. (1998). Predictive reward signal of dopamine neurons. *Journal of Neurophysiology, 80*(1), 1–27.

Schultz, W., Apicella, P., & Ljungberg, T. (1993). Responses of monkey dopamine neurons to reward and conditioned stimuli during successive steps of learning a delayed response task. *Journal of Neuroscience, 13*(3), 900–913.

Schultz, W., Dayan, P., & Montague, P. R. (1997). A neural substrate of prediction and reward. *Science, 275*(5306), 1593–1599.

Seo, M., Lee, E., & Averbeck, B. B. (2012). Action selection and action value in frontal-striatal circuits. *Neuron, 74*(5), 947–960. https://doi.org/10.1016/j.neuron.2012.03.037.

Servan-Schreiber, D., Cohen, J. D., & Steingard, S. (1996). Schizophrenic deficits in the processing of context. A test of a theoretical model. *Archives of General Psychiatry, 53*(12), 1105–1112.

Shivkumar, S., Muralidharan, V., & Chakravarthy, V. S. (2017). A biologically plausible architecture of the striatum to solve context-dependent reinforcement learning tasks. *Frontiers in Neural Circuits, 11*(45). https://doi.org/10.3389/fncir.2017.00045.

Shohamy, D., Myers, C. E., Geghman, K. D., Sage, J., & Gluck, M. A. (2006). L-dopa impairs learning, but spares generalization, Parkinson's disease. *Neuropsychologia, 44*(5), 774–784.

Stocco, A. (2017). A biologically plausible action selection system for cognitive architectures: Implications of basal ganglia anatomy for learning and decision-making models. *Cognitive Science* https://doi.org/10.1111/cogs.12506.

Suri, R. E., Bargas, J., & Arbib, M. A. (2001). Modeling functions of striatal dopamine modulation in learning and planning. *Neuroscience, 103*(1), 65–85.

Suri, R. E., & Schultz, W. (1998). Learning of sequential movements by neural network model with dopamine-like reinforcement signal. *Experimental Brain Research, 121*(3), 350–354.

Suri, R. E., & Schultz, W. (1999). A neural network model with dopamine-like reinforcement signal that learns a spatial delayed response task. *Neuroscience, 91*(3), 871–890.

Surmeier, D. J., Ding, J., Day, M., Wang, Z., & Shen, W. (2007). D1 and D2 dopamine-receptor modulation of striatal glutamatergic signaling in striatal medium spiny neurons. *Trends in Neurosciences, 30*(5), 228–235.

Sutton, R. S., & Barto, A. G. (1998). *Reinforcement learning: An introduction* (Vol. 1). Cambridge: Cambridge University Press.

Tanaka, S. C., Schweighofer, N., Asahi, S., Shishida, K., Okamoto, Y., Yamawaki, S., & Doya, K. (2007). Serotonin differentially regulates short- and long-term prediction of rewards in the ventral and dorsal striatum. *PLoS One, 2*(12), e1333. https://doi.org/10.1371/journal.pone.0001333.

Tian, L., Arnold, M., Sejnowski, T., & Jabri, M. (2003). *A biologically inspired computational model of the block copying task.* Paper presented at the Proceedings of the third international workshop on Epigenetic robotics, Lund University Cognitive Studies.

Wickens, J., & Kötter, R. (1995). Cellular models of reinforcement.

Wickens, J. R. (1997). Basal Ganglia: Structure and computations [Invited Review]. *Network: Computation in Neural Systems, 8,* R77–R109.

Wickens, J. R., Horvitz, J. C., Costa, R. M., & Killcross, S. (2007). Dopaminergic mechanisms in actions and habits. *Journal of Neuroscience, 27*(31), 8181–8183.

Wilson, C. J. (2004). Basal ganglia. In G. M. Shepherd (Ed.), *The synaptic organization of the 136 brain* (pp. 361–413). New York: Oxford University Press.

Wise, R. A. (2004). Dopamine, learning and motivation. *Nature Reviews Neuroscience, 5*(6), 483–494.

Wise, R. A., & Rompre, P.-P. (1989). Brain dopamine and reward. *Annual Review of Psychology, 40*(1), 191–225.

Chapter 5
The Basal Ganglia System as an Engine for Exploration

V. Srinivasa Chakravarthy
and Pragathi Priyadharsini Balasubramani

Abstract One of the earliest attempts at building a theory of the basal ganglia (BG) is based on the clinical findings that lesions to the direct and indirect pathways of the BG produce quite opposite motor manifestations (Albin et al., in Trends Neurosci 12(10):366–375, 1989). While lesions of the direct pathway (DP), affecting particularly the projections from the striatum to GPi, are associated with hypokinetic disorders (distinguished by a paucity of movement), lesions of the indirect pathway (IP) produce hyperkinetic disorders, such as chorea and tremor. In this chapter, we argue that describing the two BG pathways as having mutually opponent actions has limitations. We argue that the BG indirect pathway also plays a role in exploration. We should evidence from various motor learning and decision-making tasks that exploration is a necessary process in various behavioral processes. Importantly, we use the exploration mechanism explained here to simulate various processes of the basal ganglia which we discuss in the following chapters.

5.1 Introduction

Subsequent investigations into the function of the direct and indirect pathways seemed to confirm this dual effect of the two pathways on the output nuclei of the BG. Neurons of the output nuclei of the BG (GPi and SNr) exhibit tonically high firing patterns. Under normal, resting conditions, the output nuclei of the BG tend to inhibit the thalamus and further on the motor cortex, thereby inhibiting movement. However, when inhibitory inputs (GABA) from the striatum, acting via the direct pathway, suppress the activity of the BG output nuclei, the baseline inhibition of the motor cortex by the BG output nuclei is released, and movement is enabled. On the contrary, if the indirect pathway, particularly the STN, is activated, it excites the output nuclei further, thereby inhibiting the movement. These findings let to the thinking that activation of the direct pathway facilitates movement, earning it the name of Go pathway. Contrarily, the indirect pathway was dubbed the NoGo pathway since its activation typically inhibits movement.

V. S. Chakravarthy and A. Moustafa, *Computational Neuroscience Models of the Basal Ganglia*, Cognitive Science and Technology,
https://doi.org/10.1007/978-981-10-8494-2_5

The idea of a simple complementarity between the direct and indirect pathways is further supported by the manner in which dopamine affects striatal neurons. The projection neurons of the striatum, viz. the medium spiny neurons (MSNs) express D1- and D2-type dopamine receptors. To quote from Chap. 43 in the classic neuroscience book by Kandel, Schwartz, and Jessell (2000) 'striatal neurons that project directly to the two output nuclei have D1 dopamine receptors, while those that project in the indirect pathway have D2 receptors that reduce transmission.' Thus under low dopamine conditions in the striatum, the cortico-striatal inputs are allowed to pass directly via the direct pathway, while at lower dopamine levels the cortical inputs are routed via the indirect pathway.

Such descriptions that reinforce the simple complementarity between the two functional BG pathways are now commonly found in clinical, neurobiological, and computational accounts of BG (Houk, Davis, & Beiser, 1995; Kandel et al., 2000).

However, the attempt to describe the two BG pathways as having simple mutually opponent actions on the BG targets has its limitations. There is a tendency in classical neuroscience to describe interactions among brain areas in simple binary terms: Area A excites area B, or area A inhibits area B. But the developments in computational neuroscience that occurred over the last three decades offer a rich repertoire of dynamical concepts, at single neuronal and network level—attractors, chaos, synchronization etc.,—that can be used to describe brain function. By applying these notions to attempt to understand the functional architecture of BG, we notice a certain weakness in some of the popular accounts of function of the indirect pathway. Let us consider, for example, an explanation of why the activation of the indirect pathway inhibits movement, presented in chapter on BG (Chap. 43, in Kandel, Schwartz, & Jessel, 2000). '...phasic activation of the indirect pathway transiently increases inhibition of the thalamus, as *can be determined by considering the polarity of connections* between the striatum and the external pallidal segment, between the external segment and the subthalamic nucleus, and between the subthalamic nucleus and the internal pallidal segment.' The essence of the argument is as follows: Since the projections from the striatum to GPe are inhibitory (GABA), those from GPe to STN are inhibitory (GABA), STN to GPi are excitatory (glutamate) and the output projections of GPi are also inhibitory (GABA), an odd number of inhibitory projections have an inhibitory net effect, and activation of the indirect pathway inhibits movement. However, the argument omits an important anatomical fact related to the indirect pathway: the feedback projection from STN to GPe. The STN–GPe system becomes an excitatory–inhibitory loop that is capable of exhibiting a rich variety of neural dynamics, with important functional implications. These ideas will be discussed in greater detail further on in this chapter (Fig. 5.1).

Another component of the BG functional anatomy that does not easily lend itself to simplified 'binary' descriptions is the firing activity of the mesencephalic dopamine neurons. Since dopamine is not a neurotransmitter, but a neuromodulator, capable of modulating synaptic strength, its action on target structures cannot be easily reduced to excitation or inhibition. An elegant line of experiments performed by Schultz and colleagues (Schultz, Dayan, & Montague, 1997) on the firing

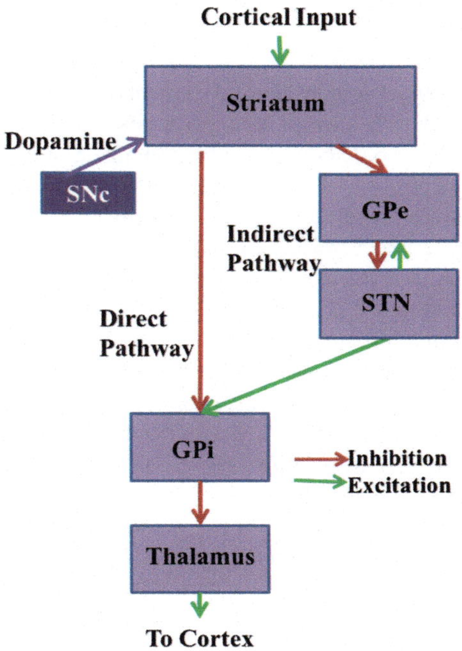

Fig. 5.1 Schematic showing the direct and indirect pathways of basal ganglia

properties of mesencephalic dopamine cells revealed a novel informational significance of dopamine firing. Although activities of dopaminergic cells have been linked to reward sensing for a long time, experiments by Schultz et al. (1997) specifically showed that dopamine neurons of Ventral Tegmental Area (VTA) respond to unexpected rewards (food or juice). Furthermore, when a sensory stimulus (like a sound or a light flash) consistently precedes the appearance of reward, such that the stimulus is predictive of the reward, then dopamine cells fire significantly in response to the stimulus and not so much in response to the reward. Furthermore, dopamine firing rate actually dropped when the reward was omitted at the time when the animal was expecting the reward. Such findings led to the insight that dopamine cell activity is analogous to a quantity known as temporal difference error (TD error) (Montague, Dayan, & Sejnowski, 1996) which appears in reinforcement learning (RL) theory, a branch of machine learning (Sutton, & Barto, 1998). The recognition of the analogy between mesencephalic dopamine signals and TD error signal of RL had inspired a much larger effort to draw parallels between other elements of RL theory and anatomical components of BG. Although the effort to explain various functions of BG using RL concepts is a story in the making, it is believed that RL holds the promise to create a comprehensive theory of BG in long term (Chakravarthy, Joseph, & Bapi, 2010).

5.1.1 The Indirect Pathway and Exploration

Now let us consider the elements of RL in simple terms, as a pedagogic exercise, and make an attempt to superimpose those elements onto the anatomy of BG circuit. Figure 5.2 shows a simple schematic of learning by reinforcement. The aim of RL is to perform actions that maximize rewards from the environment. In Fig. 5.2a, the system, representing an organism, receives a stimulus S_+ and responds with an action A_+, which results in a (positive) reward feedback. Therefore, RL leads to further reinforcement of the connection between S_+ and A_+. In Fig. 5.2b, the system receives a stimulus S_-, responds with A_-, and receives a punishment (or negative reward). This time the connection between S_- and A_- is attenuated.

Thus, the system that must learn by RL needs to have access to three types of information: stimulus, action, and reward/punishment. Now consider the simplified schematic of the BG circuit with respect to the cortex and the dopaminergic projections from SNc (Fig. 5.3). Note that the input port of the BG, the striatum, conveniently receives inputs related to the sensory-motor state from the corresponding cortical areas via the cortico-striatal projections, and the reward-related information via the nigrostriatal projections from the SNc. The putative role of BG in RL can be given a preliminary justification by invoking its anatomy (Fig. 5.3).

Let us proceed further and present arguments in favor of the role of the BG circuit as an elaborate RL engine. To this end, consider an animal faced with the challenge of pressing two buttons: a white one and a black one. The white button when pressed leads to reward in the form of delivery of a drop of juice, whereas the

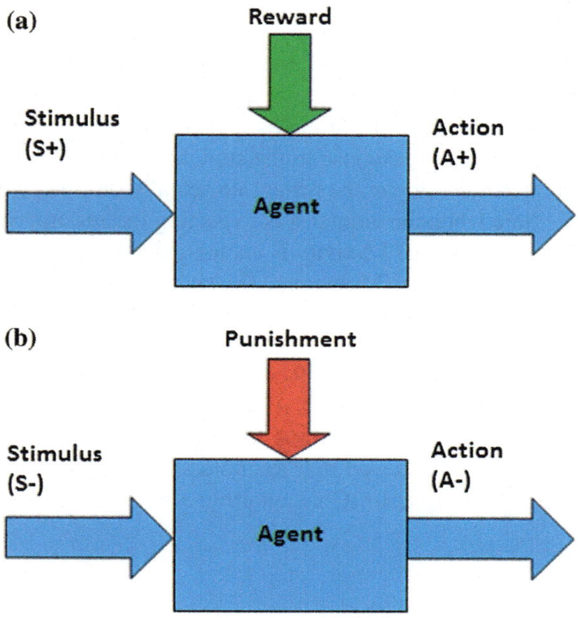

Fig. 5.2 A schematic diagram that shows how stimulus–action relationships can be learnt by reinforcement learning. **a** If the action (A_+) performed in response to a stimulus (S_+) leads to reward, the S_+–A_+ relationship is reinforced. **b** If the action (A_-) performed in response to a stimulus (S_-) leads to punishment, the S_-–A_- relationship is attenuated

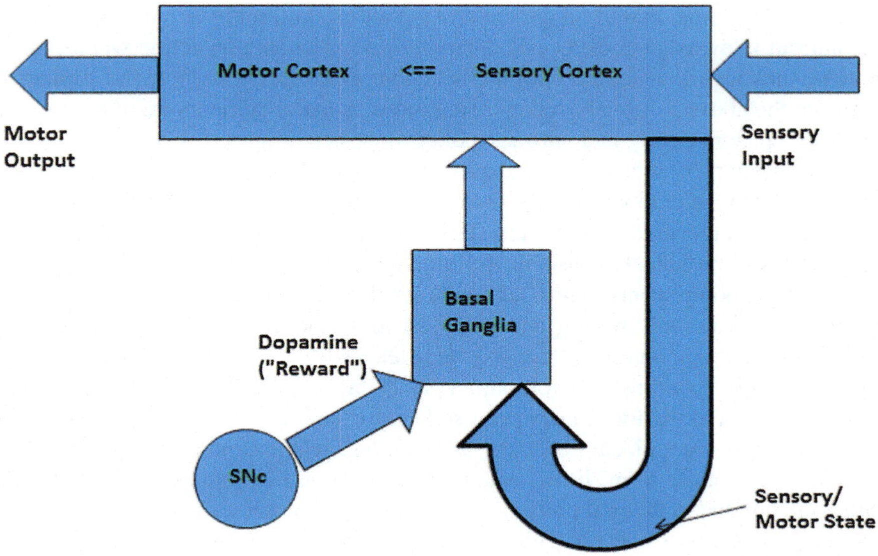

Fig. 5.3 A schematic of the anatomical location of the BG with respect to the cortex. See text for explanation

black button gives a punishment of a brief electric shock. Initially, the animal is in a 'naïve' state about the two choices that it is facing; therefore, its 'weightages' (w_{white} and w_{black}) about the reward giving potential of the two actions are the same (say, $w_{white} = 0.5 = w_{black}$). Imagine the animal presses white first and duly gets its juice reward. Its 'weightage' for 'white' now increases to, say, 0.7. It then presses the black button and is shocked by the punitive feedback. Its weightage for the gray drops to, say, 0.3. As it presses the two buttons a few more times, its attraction for the white button is reinforced, leading to further increase in w_{white}, whereas its aversion to the black is also strengthened, leading to further reduction in w_{black}. Thus, the 'weightages' (w_{white} and w_{black}) that the animal had learnt from its experience, will aid the animal in making an informed choice in the future: prefer the button with higher weightage, and avoid the one with a smaller weightage. The weightages, thus, act as some sort of surrogates to reward or punishment. Since rewards or punishment arrive *after* the actions are performed, and the animal has an obvious need to know, a priori, what is going to happen after it presses a certain button, it consults the weightages to make a decision. Notice that these 'weightages' are simply the values that were more formally defined earlier in Sect. 5.4 of Chap. 4.

Imagine that a third button, a gray one, is introduced into the experiment at this point. The animal can now adopt one of two strategies. Keep pressing the white, avoiding the black forever; or try out the new gray. The curious animal chooses to press gray and discovers, to its delight, that the choice fetches it a large piece of apple. In RL terms, making a choice corresponding to the highest value, as far as the current knowledge goes, is called *exploitation*; choosing an action whose value is either unknown, or not the optimal, is known as *exploration*. Note the similarity between the above informal account and a more formal treatment of exploration (in terms of epsilon-greedy and softmax policies) earlier in Sect. 5.4 of Chap. 4.

The question of substrates for exploration in BG does not seem to have been given adequate attention. In Actor–Critic models of BG, the Actor and Critic components in the striatum are often identified and discussed (Joel, Niv, & Ruppin, 2002); dopamine signal is also in the limelight thanks to its role in training the Actor–Critic components. Thus, although exploitation and exploration are complementary processes, the yin and yang, so to speak, of reinforcement learning, substrates for exploitation in the BG were eagerly sought but not so much the substrates corresponding to exploration. Such omission is perhaps not surprising since, in the Actor–Critic framework itself (see Chap. 4 for discussion), the Actor and Critic are recognized explicitly as *modules*, while exploration is merely a *mechanism*, a 'noise term' in RL equations. Since noise is ubiquitous in the brain, either arising due to thermal noise or chaotic neural dynamics, perhaps no need was felt to search for a specific substrate for exploration in BG. Or alternatively, exploration is modeled by invoking the stochastic softmax policy, but it remains to be explained how exactly the neural dynamics of the brain implements the stochastic dynamics of the softmax policy.

Interestingly, even the experimental literature reflects the same partial view of the neural substrates of RL components in the BG. Cortical substrates for both exploitation and exploration have indeed been discovered: value computations in orbitofrontal cortex (Knutson, Adams, Fong, & Hommer, 2001); exploration in anterior frontopolar cortex and intraparietal sulcus (Daw, O'Doherty, Dayan, Seymour, & Dolan, 2006). Functional imaging studies in humans suggest that the anterior cingulate cortex (ACC) could be involved in balancing between exploitation and exploration (Rushworth & Behrens, 2008). Similar fMRI studies by Yoshida and Ishii (2006) found activation in prefrontal cortex and ACC when subjects are exploring a maze (Yoshida & Ishii, 2006). When it comes to subcortical substrates, imaging studies by O'Doherty and colleagues suggest that the ventral and dorsal striata correspond to the Critic and Actor, respectively (O'Doherty, Dayan, Schultz, Deichmann, Friston, & Dolan, 2004). Thus, though both cortical and subcortical substrates of exploitation have been discovered, no corresponding subcortical substrates for exploration have been found.

However, is there a compelling reason for subcortical substrates of exploration to exist? It was shown that decorticated kittens can exhibit exploratory and goal-oriented behavior (Stein, Grillner, Selverston, & Stuart, 1997). STN-lesioned rats were shown to exhibit perseverative behavior, or reduced exploration of new options (Baunez et al., 2001). When bicuculine, an antagonist of GABA, was injected into anterior GPe, it elicited stereotypic movements, while injections into dorsolateral GPe produced hyperactivity which includes exploratory or foraging movements (Grabli et al., 2004).

Thus, it appears compelling that the STN–GPe system, which constitutes the IP of BG, might be the subcortical substrate for exploratory behavior. The STN–GPe system and its intriguing oscillatory activity do not seem to occupy a prominent place in Actor–Critic class of the modeling literature. The STN–GPe oscillations have assumed a special significance since they have been linked to Parkinsonian tremor (Hurtado, Gray, Tamas, & Sigvardt, 1999). There is a line of modeling work that presents the STN–GPe system as a pacemaker in the brain, in reference to the strong oscillatory activity of this system (Terman, Rubin, Yew, & Wilson, 2002). Using a simplified model of STN–GPe system, Gillies et al. (Willshaw & Li, 2002) showed that removal of the cortical input to STN results in increased firing rates. They also studied the importance of the balance between the cortical input to STN and striatal input to GPe in determining the STN oscillatory activity. A higher inhibition from the striatal input onto GPe is predicted to produce oscillations in STN and GPe even with a weaker excitatory drive from the cortex. Similarly, a model of STN–GPe using Izhikevich spiking neurons accounted for spiking dynamics seen in the system by using variables that code for the membrane voltage, activation of K^+ ionic currents, and inactivation of Na^+ ionic currents (Michmizos & Nikita, 2011). But the aforementioned STN–GPe models are not based on RL and therefore do not address RL-related features like value computation in striatum and dopamine signaling. However, these models have been able to explain behavioral effects of pathological oscillations of STN–GPe, in connection with Parkinsonian tremor (Hurtado, Gray, Tamas, & Sigvardt, 1999; Terman et al., 2002).

Under dopamine-deficient or Parkinsonian conditions, the firing patterns of neurons of both STN and GPe were found to show dramatically increased correlation, though unaccompanied by significant decrease in firing rate (Bergman, Wichmann, Karmon, & DeLong, 1994; Brown et al., 2001). Since exploration is driven by noise in RL models, a brain region that drives exploration is expected to be a source of noise, generated perhaps by the complex neural dynamics of that region. Thus, considering the neural activity with low correlation found in STN–GPe under normal conditions, and an increase in correlation, or loss of complexity, found in pathological conditions, it is plausible to assume that the STN–GPe is a subcortical substrate for exploration.

In the subsequent chapters of this book, we will be presenting a line of BG models that embody the idea that STN–GPe system is a subcortical substrate for exploration. We have outlined how by extended application of RL concepts it is possible to build comprehensive models of BG in which the value computations of striatum, and the exploitative dynamics subserved by the DP, can be combined with the oscillations of STN–GPe that drive exploratory behavior (Chakravarthy et al., 2010). Thus emerges a view that while DP supports exploitation, IP subserves exploration. Such a view seems to be at variance to the classical view that describes the DP and IP as Go and NoGo pathways, respectively. We will show in this chapter that the exploitation (DP) versus exploration (IP) view can be reconciled with Go (DP) versus NoGo (IP) view, by inserting a third regime dubbed the Explore regime, corresponding to exploration, between the older Go and NoGo regimes (Kalva, Rengaswamy, Chakravarthy, & Gupte, 2012). A series of BG models based on this view have been developed to account for a wide variety of BG-related motor and cognitive behaviors [such as spatial navigation, saccades, reaching, reward–punishment learning (Gangadhar, Joseph, & Chakravarthy, 2008; Krishnan, Ratnadurai, Subramanian, Chakravarthy, & Rengaswamy, 2011; Magdoom et al., 2011; Pragathi Priyadharsini, Ravindran, & Srinivasa Chakravarthy, 2012; Sridharan, Prashanth, & Chakravarthy, 2006; Sukumar, Rengaswamy, & Chakravarthy, 2012)].

The outline of the chapter is as follows: Sect. 5.2 presents a simple network model of BG. In this model, the STN and GPe modules are modeled as an excitatory–inhibitory loop, capable of producing oscillations. Dynamics of STN–GPe system is characterized in terms of correlations of neural activity, as parameters that control strengths of the connections within the loop are varied. In Sect. 5.3, the model of Sect. 5.2 is applied to a series of action selection problems, starting with discrete action selection, moving on to continuous action spaces. A discussion of the entire work is presented in the final section.

5.2 The Basic Model

The intuitive ideas outlined in the previous section are now embodied in a simple mathematical model of BG. This essential model captures structural aspects of BG —it explicitly represents striatum, STN, GPe and GPi, the DP, IP. Consistent with Actor–Critic models of BG, it identifies the nigrostriatal dopamine signal with TD error and uses this error to train cortico-striatal connections by RL. It models, in an elementary form, the action of dopamine in switching between DP and IP, via the differential action of dopamine on the D1 and D2 receptors of striatal medial spiny neurons. It exhibits oscillations in the STN–GPe system and also captures some of the known conditions in which these oscillations are produced. It shows how value computation occurs in the striatum. It displays the classical Go and NoGo behaviors with the explore behavior in addition.

Salience-based action selection is considered to be one of the primary functions of BG (Gurney, Prescott, & Redgrave, 2001; Redgrave, Prescott, & Gurney, 1999). We now present equations of the above-mentioned BG model and apply the model first to the simple problem of binary action selection.

5.2.1 Striatum

The binary action selection problem presently considered consists of choosing between two inputs based on their magnitude which represents their 'salience.' Thus, the input, which represents the cortico-striatal afferents, consists of a two-dimensional vector, I^{ext}. This input is presented to the striatum which consists of two 1D layers of medium spiny neurons (MSNs) of striatum (Fig. 5.4). The first layer represents neurons that express D1-type dopamine receptors, whereas the

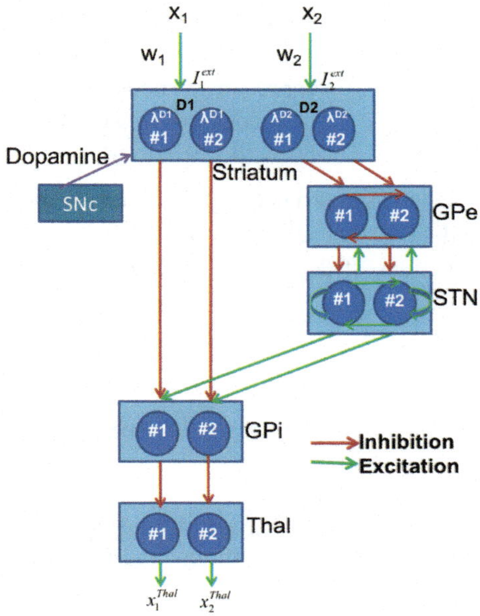

Fig. 5.4 Schematic flow of the signal in the network model

second layer represents D2-expressing neurons. The D1- and D2-expressing neuronal layers project to GPi and GPe, respectively; they also receive dopaminergic projections from SNc. Each component of I^{ext} is uniquely connected to one neuron each in D1 and D2 layers. Each neuron in D1 and D2 layers combines the dopamine signal (δ) with their respective inputs. Responses of D1 (D2) neurons, which increase (decrease) with increasing dopamine levels, are defined as follows:

$$V_i^{\text{D1\_Str}} = I_i^{\text{ext}} \lambda^{\text{D1}} \tag{5.2.1.1}$$

$$V_i^{\text{D2\_Str}} = I_i^{\text{ext}} \lambda^{\text{D2}} \tag{5.2.1.2}$$

where

$V_i^{\text{D1\_Str}}$ outputs of D1 neurons of striatum;
$V_i^{\text{D2\_Str}}$ outputs of D2 neurons of striatum; δ

λ^{D1} and λ^{D2} they are weighting factors that represent the effect of dopamine, δ, on the responses of D1 and D2 neurons, respectively. A similar approach to modeling dopamine modulation of MSN firing output was adopted in Humphries and Gurney (2002), Humphries and Prescott (2010).

Increased striatal dopamine levels increase the activation of D1 neurons while reducing the activation of D2 neurons. Therefore, we model λ^{D1} and λ^{D2} as functions of dopamine (δ) as follows:

$$\lambda^{\mathrm{D1}} = \left(\frac{1}{1 + \mathrm{e}^{a_1(\delta - \theta_{\mathrm{D1}})}}\right) \tag{5.2.1.3}$$

$$\lambda^{\mathrm{D2}} = \left(\frac{1}{1 + \mathrm{e}^{a_2(\delta - \theta_{\mathrm{D2}})}}\right) \tag{5.2.1.4}$$

Since λ^{D1} and λ^{D2} are increasing and decreasing functions of δ, it is easy to see that $a_1 < 0$ and $a_2 > 0$; θ_{D1} and θ_{D2} are bias terms.

5.2.2 *Modeling the STN–GPe System*

5.2.2.1 Modeling STN–GPe Neuron Pair

STN and GPe form a loop with excitatory projections from STN to GPe and inhibitory projections in the reverse direction. Dynamics of a single STN–GPe neuron pair is given as,

$$\tau_s \frac{\mathrm{d}x^{\mathrm{STN}}}{\mathrm{d}t} = -x^{\mathrm{STN}} + U^{\mathrm{STN}} - w_{gs}x^{\mathrm{GPe}} + I^{\mathrm{HDP}} + K^{\mathrm{STN}} \tag{5.2.2.1}$$

$$U^{\mathrm{STN}} = \tan h\left(\lambda^{\mathrm{STN}} x^{\mathrm{STN}}\right) \tag{5.2.2.2}$$

$$\tau_g \frac{\mathrm{d}x^{\mathrm{GPe}}}{\mathrm{d}t} = -x^{\mathrm{GPe}} + w_{sg} U^{\mathrm{STN}} - I^{\mathrm{D2\_Str}} \tag{5.2.2.3}$$

$$I^{\mathrm{D2\_Str}} = W_{\mathrm{Str}}^{\mathrm{GPe}} V^{\mathrm{D2\_Str}} \tag{5.2.2.4}$$

$V^{\mathrm{D2\_Str}}$ output of D2 striatum
$W_{\mathrm{Str}}^{\mathrm{GPe}}$ weight from Str to GPe
I^{HDP} cortical input to STN arriving via the hyperdirect pathway
x^{STN} internal state of STN neuron
K^{STN} a constant bias current given to STN neuron

U^{STN}	output of the STN neuron
x^{GPe}	internal state (and the output) of the GPe neuron
w_{sg}	strength of the connection between STN to GPe
w_{gs}	strength of the connection between GPe to STN.

The above system (Eqs. 5.2.2.1–5.2.2.4) represents a Lienard system and is capable of producing limit cycle oscillations for appropriate parameter values (see Appendix). Similar models of STN and GPe neural dynamics have been proposed (Gillies, Willshaw, Atherton, & Arbuthnott, 2002).

Note that the output of D2 striatum is introduced with a negative sign ($-I^{D2}$) in Eq. (5.2.2.3) in order to represent the inhibitory (GABAergic) projection from D2 striatum to GPe. Likewise, the cortical input via HDP to STN is introduced in Eq. (5.2.2.1) with a positive sign ($+I^{HDP}$) to represent the excitatory (glutamatergic) HDP projections.

Consider the effect of I^{D2} and I^{HDP} on STN–GPe activity. Let $I^{HDP} = 0$ and $I^{D2} = 0.5$. Other parameters are: $K^{STN} = -1$, $w_{sg} = w_{gs} = 1$. The STN–GPe neuron pair shows fixed point behavior (Fig. 5.5a), a property that is reflected in the corresponding phase plot (Fig. 5.5b). But oscillations are seen when I^{D2} is raised to 0.9, keeping $I^{HDP} = 0$ (Fig. 5.6a). In this case, the two nullclines intersect in the middle branch of the STN nullcline (Fig. 5.6b). Thus in the above simple STN–GPe cell pair model, oscillations are elicited by increased striatal input to GPe, a property that is corroborated by electrophysiological data (Flores-Hernandez et al., 2002). Kravitz et al. (2010) observed that increased firing of D2 MSNs in the striatum induces a state similar to Parkinson's disease, with motor symptoms like freezing, bradykinesia, and difficulty in movement initiation (Kravitz et al., 2010).

A similar pattern of dynamics is seen when I^{D2} is fixed and I^{HDP} is varied. Figures 5.7b and 5.8b show the nullclines for $I^{HDP} = 0.5$ and 0.9, respectively (with $I^{D2} = 0$), and Fig. 5.7a, b shows the corresponding neural dynamics. Note that increasing cortical input to STN triggers STN–GPe oscillations. It may be seen from Eqs. (5.2.2.1, 5.2.2.3) that there is a simple complementarity between I^{D2} and I^{HDP}. Decreasing I^{HDP} in Eq. (5.2.2.1) lowers the x^{STN}-nullcline, just as increasing

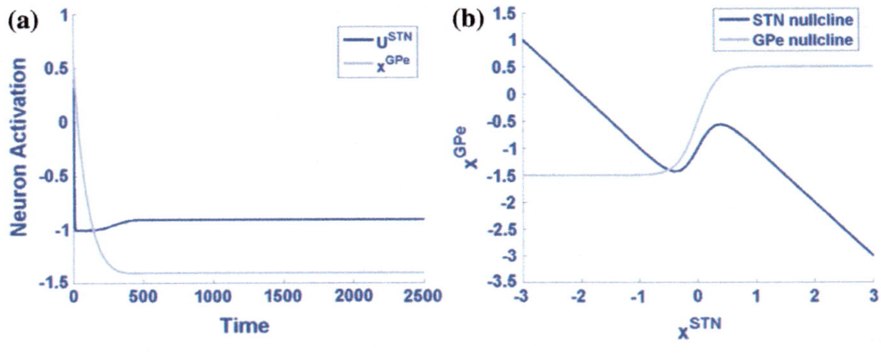

Fig. 5.5 **a** Neural activation and **b** phase plot for $I^{HDP} = 0$ and $I^{D2} = 0.5$, $w_{sg} = w_{gs} = 1$

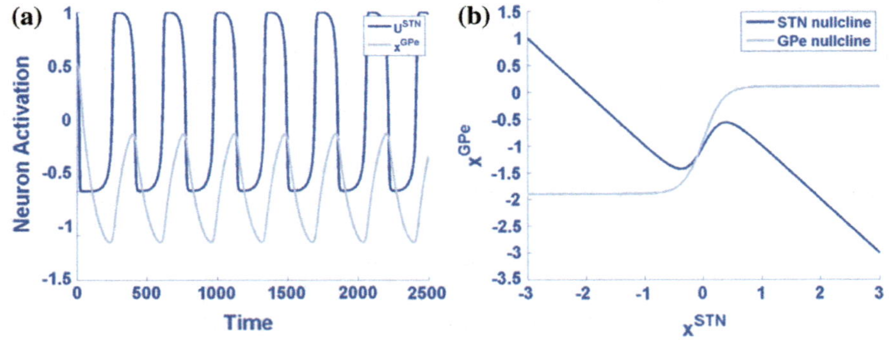

Fig. 5.6 **a** Neural activation and **b** phase plot for $I^{\mathrm{HDP}} = 0$ and $I^{\mathrm{D2}} = 0.9$, $w_{sg} = w_{gs} = 1$

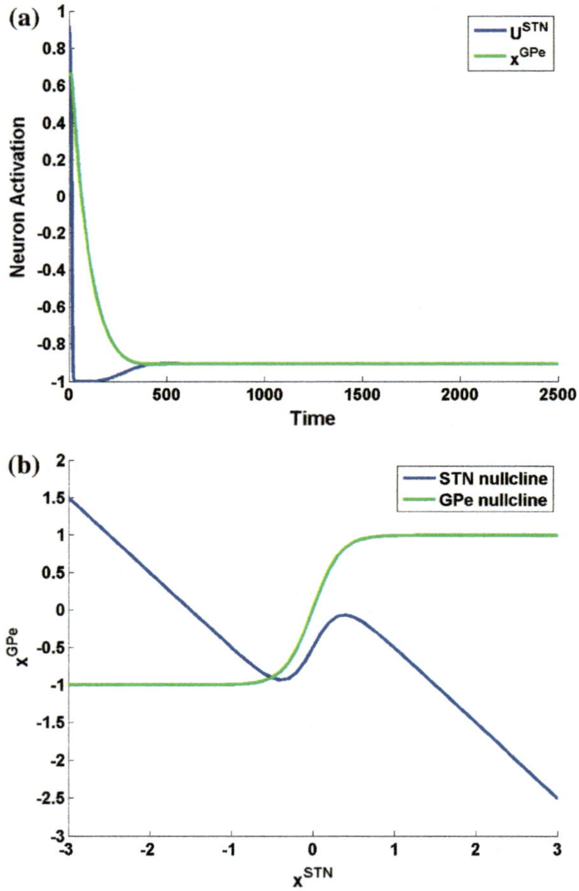

Fig. 5.7 **a** Neural activation and **b** phase plot for $I^{\mathrm{HDP}} = 0.5$ and $I^{\mathrm{D2}} = 0$, $w_{sg} = w_{gs} = 1$

the inhibition to GPe (I^{D2}) lowers the x^{GPe}-nullcline. Therefore, as inhibitory input to GPe is increased, the amount of excitation to be given to STN in order to produce oscillations is also reduced. A similar complementarity between cortical input to STN and striatal input to GPe was exhibited by a lumped model of STN–GPe interactions proposed by Gillies et al. (2002). The effect of cortical input to STN on STN–GPe oscillations has been borne out by electrophysiological studies that show that ablation of cortical areas that project to STN largely abolished the low-frequency oscillations in STN and GPe (Magill, Bolam, & Bevan, 2001).

The third factor that controls oscillations in STN–GPe system is dopamine. There is experimental evidence that the STN–GPe system exhibits low-frequency oscillations under dopamine-deficient conditions as in those of Parkinson's disease (Plenz & Kital, 1999). It appears that reduction of dopamine strengthens the coupling between STN and GPe triggering oscillations. The effect of dopamine on STN–GPe synaptic strengths can be accounted for in terms of the presence of D2-type dopamine receptors on the two nuclei (Steiner & Tseng, 2010). D2 receptors are present on the axon terminals of glutamatergic cortical projections to STN and GABAergic projections from GPe to STN. Therefore, the effect of increased dopamine on these connections is to effectively reduce their strength. Let us now consider the effect of reducing the coupling strengths between STN and GPe (w_{gs} and w_{sg}) in Eqs. (5.2.2.1, 5.2.2.3). When $w_{sg} = w_{gs} = 1$, $I^{D2} = 0.9$, and $I^{HDP} = 0$, we have seen that oscillations are produced (Figs. 5.6 and 5.8), but the oscillations disappear when $w_{sg} = w_{gs} = 0.5$, with the current inputs remaining the same (Fig. 5.9a, b).

Thus, we have seen, using the simple cell pair model of Eqs. (5.2.2.1–5.2.2.4), that STN–GPe oscillations can be triggered by (1) inhibitory striatal input to GPe, (2) excitatory cortical input to STN, and (3) strengthening of STN–GPe interactions by dopamine reduction. A complete phase-plane analysis of Eqs. (5.2.2.1–5.2.2.4) would reveal monostability and bistability in addition to limit cycle behavior that corresponds to STN–GPe oscillations. But since the above-mentioned behaviors are particularly relevant to our subsequent presentation, we chose to highlight them, while passing over a more exhaustive analysis in this context. For more exclusive analysis of similar two-variable models of STN–GPe, the reader may consult (Gillies et al., 2002).

5.2.2.2 Network Model of STN–GPe System

We now present a network model of the STN–GPe system. In the model, STN and GPe layers have equal number of neurons, with each neuron in STN uniquely connected bidirectionally to a neuron in GPe. Both STN and GPe layers are further assumed to have weak lateral connections within the layer. Dynamics of STN and GPe interactions is described as:

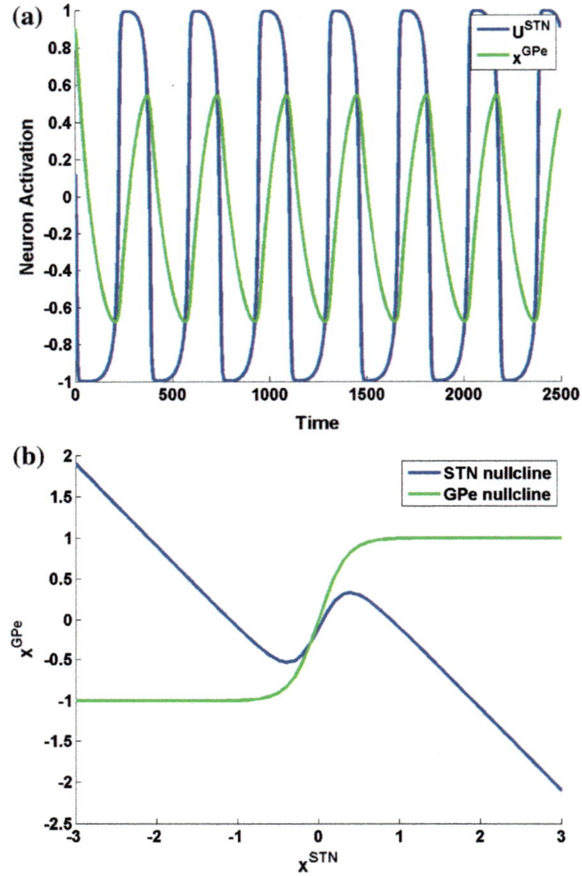

Fig. 5.8 **a** Neural activation and **b** phase plot for $I^{\mathrm{HDP}} = 0.5$ and $I^{\mathrm{D2}} = 0$, $w_{sg} = w_{gs} = 1$

$$\tau_s \frac{\mathrm{d}x_i^{\mathrm{STN}}}{\mathrm{d}t} = -x_i^{\mathrm{STN}} + \sum_j^n W_{ij}^{\mathrm{STNLat}} U_j^{\mathrm{STN}} - w_{gs} x_i^{\mathrm{GPe}} \tag{5.2.2.5}$$

$$U_i^{\mathrm{STN}} = \tan h\left(\lambda^{\mathrm{STN}} x_i^{\mathrm{STN}}\right) \tag{5.2.2.6}$$

$$\tau_g \frac{\mathrm{d}x_i^{\mathrm{GPe}}}{\mathrm{d}t} = -x_i^{\mathrm{GPe}} + \sum_{j=1}^n W_{ij}^{\mathrm{GPeLat}} x_j^{\mathrm{GPe}} + w_{sg} U_i^{\mathrm{STN}} - V_i^{\mathrm{D2}} \tag{5.2.2.7}$$

'n'	size of STN or GPe layer
x_i^{GPe}	internal state (same as the output) of the *i*th neuron in GPe
x_i^{STN}	state of *i*th neuron in STN
W_{ij}^{GPeLat}	lateral connections within GPe are defined as

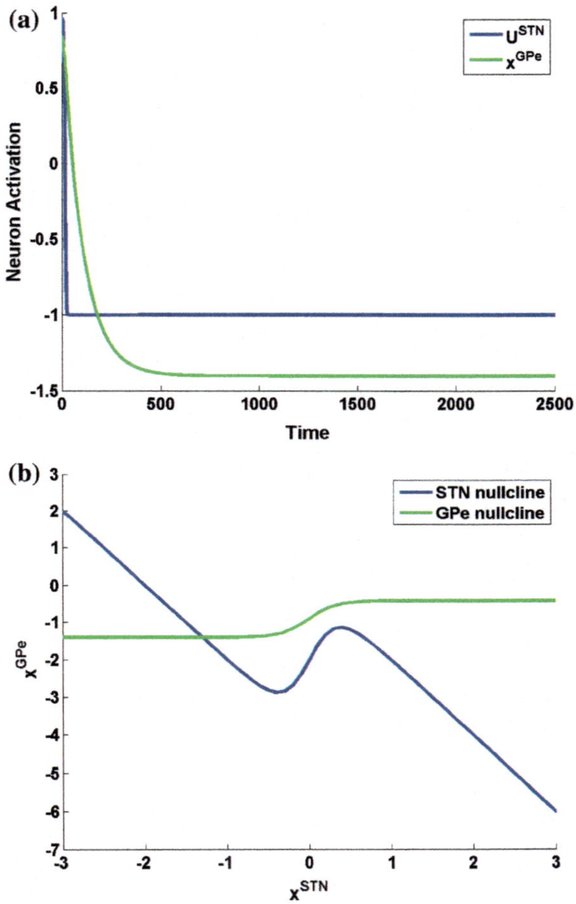

Fig. 5.9 **a** Neural activation and **b** phase plot for $I^{\mathrm{HDP}} = 0$ and $I^{\mathrm{D2}} = 0.9$, $w_{sg} = w_{gs} = 0.5$

$W_{ij}^{\mathrm{GPeLat}} = -\varepsilon_g$, for all i and j ($i \sim= j$), where ε_g is a small positive number, and the connections within STN are defined as:
$W_{ij}^{\mathrm{STNLat}} = \varepsilon_s$, for all i and j ($i \sim= j$), where ε_s is a small positive number.

We assume that both STN and GPe have complete internal connectivity, where every neuron in the nucleus is connected to every other neuron in the nucleus, with the same connection strength. That common lateral connection strength for STN is ε_s, and for GPe it is ε_g. Likewise, STN and GPe neurons are connected in a one-to-one fashion—ith neuron in STN is connected to ith neuron in GPe and vice versa. The common connection strength for STN $\rightarrow$ GPe connections is w_{sg}, whereas the GPe $\rightarrow$ STN connection strength is w_{gs}.

A key dynamic property of the STN–GPe system, which is often observed in pathological conditions, is synchronized oscillation between STN and GPe. We now examine synchronization in STN–GPe network model as various connectivity parameters are varied. We use correlation as a measure of synchronization. Particularly, we vary the within ($\varepsilon = \varepsilon_s = \varepsilon_g$) and across ($w = w_{sg} = -w_{gs}$) nuclei connection parameters and study their effect on correlation between neural activity within individual nuclei and across the two nuclei.

We simulate the above system (Eqs. 5.2.2.5–5.2.2.7) where the number of neurons in STN (GPe) is $n = 20$. The slope parameter λ^{STN} in Eq. (5.2.2.6) is 3. There are no external currents presented to STN and GPe. Note that the bias term K^{STN} in Eq. (5.2.2.1) is not present in the corresponding network model (Eq. 5.2.2.5). We calculate correlations—within STN (C^{STN}), within GPE (C^{GPe}), and between STN and GPe (C^{STN_GPe}). Correlations are calculated only when at least one neuron in the network is in the oscillatory mode, since correlation between neural activities in fixed point mode is not particularly informative. Presence of oscillations was determined as follows: Every STN and GPe neuron activity were analyzed for 150 time steps. If the standard deviation of the activity during the interval [75:150] steps exceeds 0.2, then it is labeled to show oscillatory activity.

Figures 5.10 and 5.11 show the range of parameters (ϵ_s, w_{sg}) for which the STN–GPe network oscillates—the 'oscillatory region'—which happens to be a nearly triangular patch in the (ϵ_s, w_{sg}) space, where ϵ_s varies over the range [0, 0.05] and w varies over the range [0, 1]. It appears that there is some sort of competition between w_{sg} and ϵ_s in producing oscillations. Larger values of w_{sg} tend to produce oscillations, while larger values of ϵ_s inhibit them. Therefore, within the domain of interest ($\epsilon_s \in [0, 0.05]$ and $w_{sg} \in [0, 1]$), oscillations occur for small ϵ_s and large w_{sg}, mostly in the triangular region shown in Figs. 5.10 and 5.11. However, oscillations do occur at sparsely distributed points outside the triangular region of Figs. 5.10 and 5.11. C^{STN} is found by calculating mean of the correlation coefficients for every pair of the n neurons within the STN. The self-connections/correlations are not

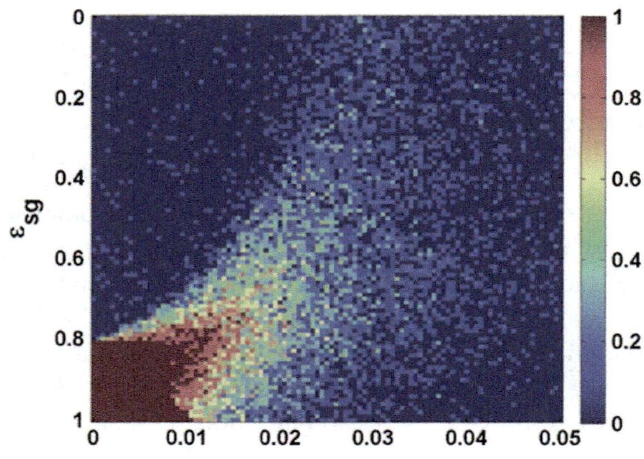

Fig. 5.10 Regions of parameter space (ϵ_s, w_{sg}) over which STN exhibits oscillations. The above figure projects the probability of oscillation as a function of (ϵ_s, w_{sg}), when run for 50 instances

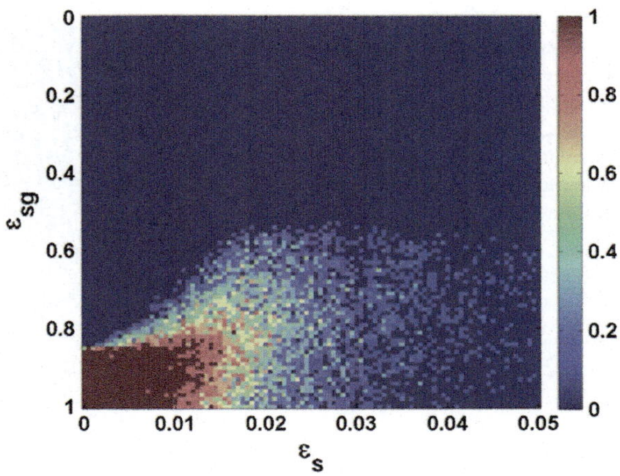

Fig. 5.11 Regions of parameter space (ϵ_s, w_{sg}) over which GPe exhibits oscillations. The above figure projects the probability of oscillation as a function of (ϵ_s, w_{sg}), when run for 50 instances

taken into account. A similar approach is taken for computing, C^{GPe}, which denotes correlations within GPe. For internucleus correlational analysis between STN and GPe, every neuron in STN is analyzed for correlation to every neuron of GPe. Finally, the mean of the coefficients is taken as the $C^{\text{STN-GPe}}$ value. The test used for correlations is Pearson's linear correlation analysis, and the coefficients are reported with p value < 0.05 for within STN (C^{STN}), within GPe (C^{GPe}), and between STN and GPe ($C^{\text{STN-GPe}}$). Figures 5.12, 5.13, and 5.14 report the test results as a function of ϵ_s and w_{sg} that broadly show a rising trend of correlation with increasing ϵ_s and w_{sg}.

If the STN–GPe system is to serve as a source of exploration, one would expect the activity of STN, which projects to GPi, to possess high spatiotemporal complexity. We presently quantify that spatiotemporal complexity using pair-wise correlations among neural activity. Kalva et al. (2012) calculated Lyapunov

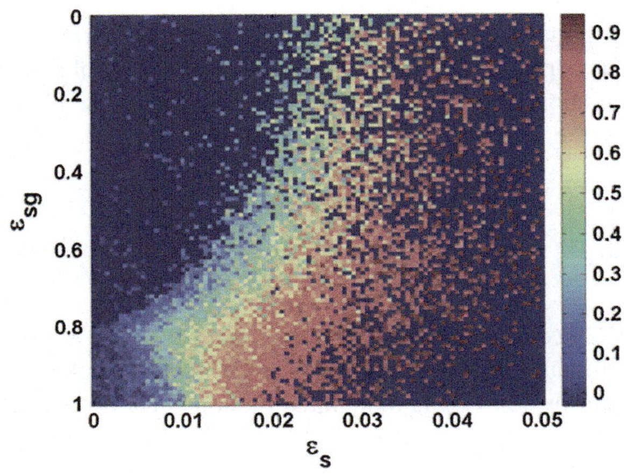

Fig. 5.12 C^{STN} over the parameter space (ϵ_s, w_{sg}) averaged for 50 instances. For given values of parameter pair (ϵ_s, w_{sg}), correlation is calculated only when there is at least one oscillating neuron in STN

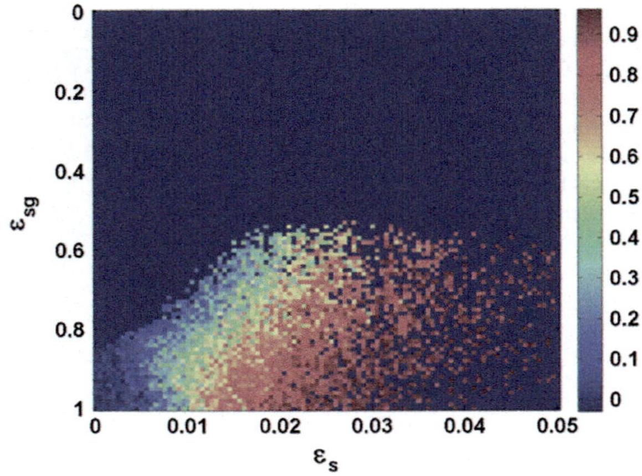

Fig. 5.13 C^{GPe} over the parameter space (ϵ_s, w_{sg}) averaged for 50 instances. For given values of parameter pair (ϵ_s, w_{sg}), correlation is calculated only when there is at least one oscillating neuron in GPe

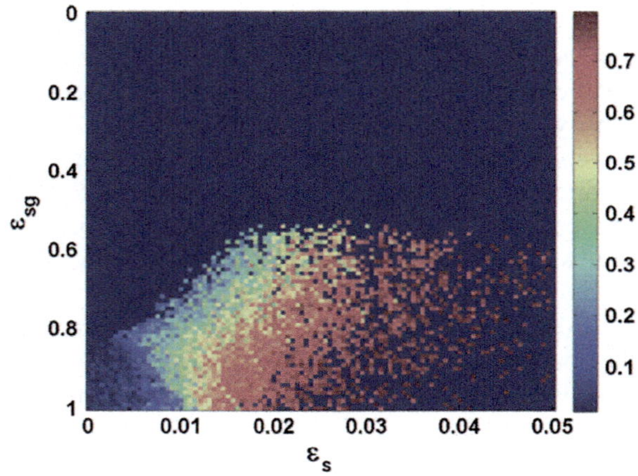

Fig. 5.14 $C^{\mathrm{STN\_GPe}}$ over the parameter space (ϵ_s, w_{sg}) averaged for 50 instances. For given values of parameter pair (ϵ_s, w_{sg}), correlation is calculated only when there is at least one oscillating neuron in GPe and STN

exponents to characterize STN dynamics in terms of chaos and showed how chaos can drive exploration. In this chapter, we exploit the ability of ϵ_s to control correlation within STN and show that ϵ_s can be used to control the level of exploration in the proposed BG model.

5.2.3 GPi

GPi combines the GABAergic striatal outflow via DP with glutamatergic STN output from IP. There is evidence to believe that this combination of DP and IP outflows in GPi is modulated by dopamine projections to GPi. Kliem et al. (2007) show that when D1 receptors in GPi, which are primarily located on the axons of

GABAergic striato-pallidal projections, are activated, firing levels of GPi neurons are reduced (Kliem et al., 2007). Since D1 receptors are activated at increased dopamine levels, this implies that at higher dopamine levels, DP outflow is facilitated over IP, consistent with the nature of switching facilitated by dopamine in the striatum. Assuming a complementary action of dopamine on the glutamatergic STN–GPi projections (the weights denoted by $w^{\text{STN\_GPe}}$), we compute the GPi output, U_i^{GPi}, as follows:

$$\dot{U}_i^{\text{GPi}} = -\lambda^{\text{DP}} V_i^{\text{D1\_Str}} + w^{\text{STN\_GPi}} \lambda^{\text{IP}} U_i^{\text{STN}} + C_i^{\text{GPi}} \tag{5.2.3.1}$$

We model λ_i^{D1} and λ_i^{D2} as functions of dopamine (δ) as follows:

$$\lambda^{\text{DP}} = \left(\frac{1}{1 + e^{a_3(\delta - \theta_{\text{DP}})}}\right) \tag{5.2.3.2}$$

$$\lambda^{\text{IP}} = \left(\frac{1}{1 + e^{a_4(\delta - \theta_{\text{IP}})}}\right) \tag{5.2.3.3}$$

C_i^{GPi} is a constant bias term.

Since λ^{DP} and λ^{IP} are increasing and decreasing functions of δ, respectively, it is easy to see that $a_3 < 0$ and $a_4 > 0$.

5.2.4 Action Selection in Thalamus

There is a body of the BG modeling literature that posits that the reward processing machinery of the nigrostriatal system computes the salience associated with competing actions (Chakravarthy et al., 2010; Joel et al., 2002; Redgrave et al., 1999). Adding to this idea, in the present model we propose that the STN–GPe system provides the exploratory drive. These two elements are combined downstream either in GPi, or further along in the thalamic nuclei which receive afferents from GPi. The competitive neural dynamics of the reticular complex of the thalamus is thought to subserve the attentional spotlight (Crick, 1984). This competitive dynamics of neurons of thalamic reticular complex has been used by Humphries and Gurney (2002) to model action selection (Humphries & Gurney, 2002). We implement an elementary form of action selection in the thalamic part of the model.

GPi neurons project to thalamus over inhibitory connections. Hence, the thalamic afferents, I_i^{Thal}, can be expressed simply as,

$$I_i^{\text{Thal}} = -U_i^{\text{GPi}} \tag{5.2.4.1}$$

These afferents activate thalamic neurons as follows,

$$\frac{dx_i^{\mathrm{Thal}}}{dt} = -x_i^{\mathrm{Thal}} + I_i^{\mathrm{Thal}} \tag{5.2.4.2}$$

where x_i^{Thal} is the state of the thalamic neuron. Action selection is done as follows:

If $x_i^{\mathrm{Thal}}(t) > x_{\mathrm{th}}$ for some i, at some time t, then 'i'th action is selected and the states of all the thalamic neurons are immediately reset, i.e., $x_i^{\mathrm{Thal}}(t) = 0$; if all $x_i^{\mathrm{Thal}}(t)$ fail to reach x_{th}, then no action is selected, a case that is considered to be 'NoGo.'

Note that there are parallel channels linking the cortico-striatal afferents to the thalamic region in the model. Each neuron in the striatum has one-to-one connections to thalamic neurons via the DP, and via the IP. In the present formulation, the only interactions occur in the STN and GPe, in the form of weak lateral connections (see Eqs. 5.2.2.5 and 5.2.2.7).

5.3 Simulation Experiments

5.3.1 Binary Action Selection

The model described in the previous section is now used to simulate a simple binary action selection task. The cortico-striatal input, I_i^{ext}, $i = 1, 2$, represents two possible actions, and the magnitudes of I_i^{ext} represent the saliencies associated with the actions. The selected action is denoted by the winning neuron in the thalamus. Due to the complex dynamics of the STN–GPe system, it is not necessary that the winning action is always the one with greater saliency. There can also be no winner at all. We thus have three types of final outcomes, classified as 'Go,' 'Explore,' and 'NoGo' as follows:

'Go'—when the winning neuron has greater salience.
'Explore'—when the winning neuron has lesser salience.
'NoGo'—when there is no winner and therefore no action selection.

We now consider the effect of δ, or dopamine, in determining the type of action selection. From classical depictions of BG function, we know that striatal dopamine level switches the transmission between DP and IP, with high dopamine activating DP and therefore selecting Go, and low dopamine activating IP resulting in a NoGo (Frank, 2005). But since the new Explore regime is the focus of the present study, we wish to see how and if dopamine, δ, has any effect on the Explore regime. Note that the model of action selection we are considering at the moment is not based on RL and δ is not TD error. In fact, there is no output error, or learning in the model described at the moment. The immediate objective is to see the effect of dopamine, δ, as defined in Eqs. (5.2.1.3, 5.2.1.4), on action selection.

To this end, we simulate a network with the following parameters:

The number of neurons in STN (GPe) is $n = 20$. The values of parameters are $a_1 = -1$, $\theta_{D1} = 0.1$ in Eq. (5.2.1.3); and $a_2 = 1$, $\theta_{D2} = 0$ in Eq. (5.2.1.4); and $w_{sg} = w_{gs} = 0.91$ in Eqs. (5.2.2.7) and (5.2.2.5), respectively. w_{STN_GPi} is taken as 0.3 (Eq. 5.2.3.1), and the threshold used in action selection at thalamus is $x_{th} = 0.715$. Taking $dt = 0.01$, τ_s is set to be 0.1 (Eq. 5.2.2.5), and τ_g is 0.3 (Eq. 5.2.2.7).

Figure 5.15 shows the probability of selection of Go/Explore/NoGo regimes as a function of dopamine (δ). Consistent with the classical picture, the Go regime is selected with high probability for large δ, and the NoGo regime for small δ. But the Explore regime is also selected, in addition to NoGo, for smaller values of δ, with the regime attaining a peak for moderate values of δ. There is no sharp boundary between regimes, but different regimes dominate for different ranges of δ. Since exploration is highest for moderate values of δ, we depict the Explore regime as though it occurs between Go and NoGo, as in Fig. 5.15.

We have shown in Sect. 5.2.2 that increasing ϵ_s increases correlation among STN neurons. We have also suggested the intuitive idea that higher correlations among STN neurons should result in weaker exploration. Analogous to Figs. 5.12 and 5.16 shows the Go/Explore/NoGo or GEN profiles, and also as a function of δ for $\epsilon_s = 0.001$, 0.05, and 0.95 (Fig. 5.16a, b, c). Note the progressive reduction in the Explore regime with increasing ϵ_s (Fig. 5.16d).

In this section, we demonstrated the emergence of the Explore regime and showed how the output of the BG model engaged in binary action selection can be depicted by the GEN profile (Fig. 5.15). We also showed the effect of the dopamine level, δ, and STN dynamics as controlled by ϵ_s, on the GEN profile (Fig. 5.16).

We now extend the above binary action selection model to the full n-armed bandit problem. To achieve such an extension, we add concepts from RL like the Critic, Critic training, and TD error, to the model described in the preceding section.

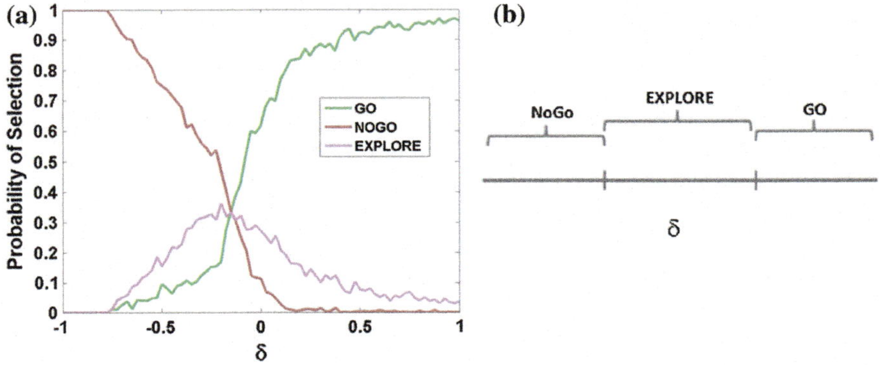

Fig. 5.15 **a** Probability of selection of Go/Explore/NoGo; **b** Schematic: Go occurs for higher range of δ, Explore for intermediate range, and NoGo for lower range

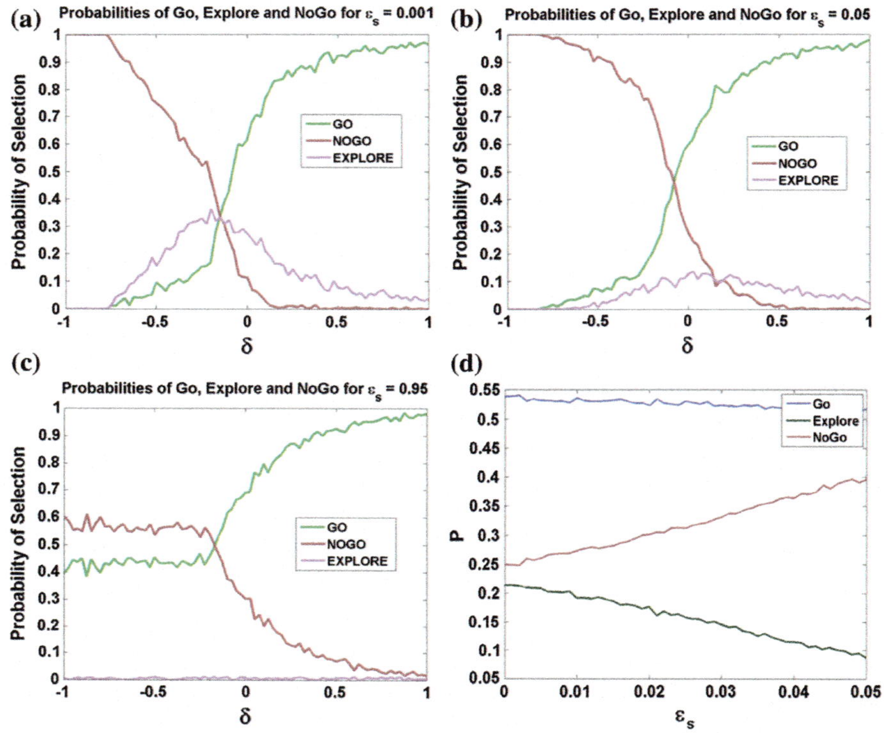

Fig. 5.16 Probability of selection of Go/Explore/NoGo regimes as: **a** a function of δ for $\epsilon_s = 0.001$; **b** a function of δ for $\epsilon_s = 0.05$; **c** a function of δ for $\epsilon_s = 0.95$; **d** a function of ϵ_s

5.3.2 Modeling the N-*Armed Bandit Problem*

The n-armed bandit problem consists of n slot machines each of which delivers a fixed reward (deterministically or probabilistically) when selected. The agent has to determine which machines to play, and how many times, so as to maximize total reward received. The n-armed bandit problem is a natural extension of the binary action selection problem described in the previous section. Therefore, we now extend the BG model of Sect. 5.3.1 and apply it to the n-armed bandit problem.

In the previous section, the BG model had to select between two inputs based on their magnitudes interpreted as saliencies. But now the BG model has n outputs, representing n actions, one of which has to be selected. Thus, the binary output vector has a single output that equals 1, while the rest are zeros. The input to BG model is also a n-dimensional binary (0/1) vector, where only a single component equals 1. Whereas the output vector represents the action to be selected, or the next action, the input vector represents the previous action.

We now describe the computations in each module of the BG model.

5.3.2.1 Computations in the Striatum

Let,

x: binary vector that denotes the cortical input that selects an arm. $x_i = 1$, if the ith arm is selected, $x_j = 0$ for $j \sim= i$.
w: cortico-striatal weights. Each cortical input is connected to a striatal neuron via a single cortico-striatal connection.

Value is computed in the striatum as,

$$V = \sum_{i=1}^{n} w_i x_i \tag{5.3.2.1}$$

Striatum has D1R- and D2R-expressing neurons, respectively. Striatal neurons expressing D1R project via the direct pathway to GPi, while striatal neurons expressing D2R project to the GPe. Since D1Rs are activated at higher DA levels, and D2R at lower levels, the effect of DA on striatal D1R and D2R-expressing cells is modeled as follows,

$$V_i^{\mathrm{StrD1}} = \lambda^{\mathrm{D1}} w_i x_i \tag{5.3.2.2}$$

and

$$V_i^{StrD2} = \lambda^{D2} w_i x_i \tag{5.3.2.3}$$

where

V_i^{StrD1} is the output of the ith striatal D1 neuron
V_i^{StrD2} is the output of the ith striatal D2 neuron.

The gain terms, $\lambda^{\mathrm{D1}}$and λ^{D2}, are similar to those defined in Eqs. (5.3.2.2, 5.3.2.3) and are defined as a function of dopamine (δ_v) as follows:

$$\lambda^{\mathrm{D1}} = \left(\frac{1}{1 + \mathrm{e}^{\kappa_1(\delta_V - \theta_{\mathrm{D1}})}}\right) \tag{5.3.2.4}$$

and

$$\lambda^{\mathrm{D2}} = \left(\frac{1}{1 + \mathrm{e}^{\kappa_2(\delta_V - \theta_2)}}\right) \tag{5.3.2.5}$$

$\lambda^{\mathrm{D1}}$and λ^{D2} are gain functions that model the effect of dopamine (δ_V) on the D1R- and D2R-expressing neurons in striatum, respectively. Since λ^{D1} and λ^{D2} are increasing and decreasing functions of δ_V, respectively, we have $\kappa_1 < 0$ and $\kappa_2 > 0$.

Note that we denote dopamine by δ_v in Eqs. (5.3.2.4, 5.3.2.5) above and not by the usual δ, which denotes TD error. We introduce δ_v as a novel quantity, as the temporal gradient of value function, V, expressed as,

$$\delta_V(t) = V(t) - V(t-1) \tag{5.3.2.6}$$

This new dopamine-related variable, δ_v, has a crucial role in exploration, an idea that will be elaborated upon shortly.

5.3.2.2 Computations in STN–GPe System

Dynamics of the STN–GPe system in the n-armed bandit model is identical to Eqs. (5.2.2.5–5.2.2.7) described earlier. The outputs of the D2R-expressing neurons in the striatum are presented as input to GPe. Thus, V_i^{D2} in Eq. (5.2.2.7) is set to be equal to V_i^{StrD2} in Eq. (5.3.2.3) above. Output of the STN, U^{STN}, is presented as input to the GPi.

5.3.2.3 Computations in GPi

Information flowing in from the D1R-expressing neurons of the striatum and STN is combined by the GPi, analogous to Eq. (5.2.3.1), as follows.

$$U_i^{\mathrm{GPi}} = -\lambda^{\mathrm{D1}} V_i^{\mathrm{StrD1}} + w^{\mathrm{STN\_GPi}} \lambda^{\mathrm{D2}} U_i^{\mathrm{STN}} + c \tag{5.3.2.7}$$

where the gain functions λ^{D1} and λ^{D2} are as in Eqs. (5.2.3.2, 5.2.3.3), and c is a constant. In the n-armed bandit model, we focus only on the arm selected and average reward obtained; reaction time is not a focus of the present model. Therefore, we simplify the action selection process in this case and omit the integration step as in Eq. (5.2.4.2). Action selected in the thalamus, as represented by the vector, x, is computed as follows:

$$\begin{aligned} x(i) &= 1, \quad \text{if } -U_j^{\mathrm{GPi}} > -U_i^{\mathrm{GPi}} \quad \forall j \neq i \\ x(i) &= 0, \quad i \neq j \end{aligned} \tag{5.3.2.8}$$

Ideally, action must be selected based on the thalamic neuron with the highest activation. Since GPi neurons have inhibitory projections into thalamus, if we assume one-to-one connections between GPi and thalamus, the thalamic neuron with the highest activation will be the one that receives input from the GPi neuron with the smallest activation. Therefore, in Eqs. (5.3.2.7, 5.3.2.8) above, we do not explicitly model the thalamic neuron, but select the action based on the GPi neuron with the smallest activation. Furthermore, note that the NoGo regime is disallowed in the above mechanism of action selection, since some action must be selected in every cycle. The selected action must be fed back to the striatum for the next cycle. If NoGo is allowed, then the same input needs to be presented repeatedly until some action is selected in the output. To avoid this trivial difficulty, NoGo is disallowed in the present version of the BG model.

Note that the action thus selected, x, at the output of the BG, may be different from the original arm selected, represented by the input vector, x. To distinguish the two, we denote the arm selected in Eq. (5.3.2.8) above as x^{post}. The new arm selected, x^{post}, is fed back to the striatum, and the cycle continues.

5.3.2.4 Reward and Learning

The jth arm selected as per x^{post} results in reward r_j. Value corresponding to x^{post} is computed as,

$$V_j^{\mathrm{post}} = \sum_{i=1}^{n} w_i x_i^{\mathrm{post}} \tag{5.3.2.9}$$

Instantaneous error, δ, is defined as,

$$\delta = r_j - V_j^{\mathrm{post}} \tag{5.3.2.10}$$

Note that δ in Eq. (5.3.2.10) above may be identified with temporal difference (TD) error (5.3.2.11) in RL for discount factor, $\gamma = 0$. Here, 'V' denotes the value function, 't' the time, and 'r' the reward.

$$\delta = r(t) + \gamma V(t+1) - V(t) \tag{5.3.2.11}$$

This δ is used to update the cortico-striatal connections as,

$$\Delta w_i = \eta \delta x_i^{\mathrm{post}} \tag{5.3.2.12}$$

We now describe the rationale of the operation of the BG model just described. Figure 5.1 shows a schematic of the signal flow in the model. The D1 neurons of striatum compute value, using standard RL (Eq. 5.3.2.9). Note that the weights used to compute the value function (Eq. 5.3.2.12) are the same as those used to compute the outputs of D1 (Eq. 5.3.2.2) and D2 neurons (Eq. 5.3.2.3). Therefore, the weight, w_i, associated with ith D1 (or D2) neuron, is the expected reward or value associated with ith action. If a certain weight, w_i, is high, then the magnitude of V_i^{StrD1} is likely to be high, if the corresponding input equals 1 (i.e. $x_i = 1$) (Eq. 5.3.2.2). In addition, if δ_v is a large positive number, the contribution of DP to GPi will dominate that of STN, and therefore, the ith action is likely to be selected at the output. Thus for high dopamine levels (δ_v), the output of GPi is dominated by

DP. In such a situation, the action selected will have a tendency to remain the same as the previous section.

On the contrary, for large negative values of δ_v, D1 neurons have small activation, and D2 neurons are more active. Furthermore, in Eq. (5.3.2.7), the response of GPi neurons is dominated more by the output of STN than the output of D1 neurons. Due to the complex oscillatory activity of STN, the action selected is likely to be random, unrelated to the input to the BG.

Let us now consider what makes δ_v positive or negative. When the value obtained in a cycle is greater than the value of previous cycle, δ_v is positive. Assuming the value estimates are accurate, it means that a more rewarding arm is selected in this cycle than the previous cycle. Thus when the values obtained in two successive cycles show an increasing trend, there is a low probability of changing the action selected. On the other hand, when successive values show a decreasing trend, there is a significant chance that the next action is selected randomly.

Let us now apply the above model to a numerical n-armed bandit problem.

5.3.2.5 *N*-Armed Bandit—A Simulation Study

Equations (5.3.2.1–5.3.2.8) describe the BG model applied to the n-armed bandit. We now simulate a large number of n-armed bandit problems (n = 5), on the lines described in Chap. 2 of Sutton and Barto (1998). The simulations are repeated for different values of model parameters. Results obtained by averaging over 500 instances are reported. Each instance is simulated for 1000 steps. The rewards are generated by the following distribution: $r_i = i/n + A * v$, where r_i is the reward of the ith arm, v is a random variable uniformly distributed over [0, 1], and $A = 0.3$.

The first variable considered is ϵ_s, which represents the weight of STN lateral connections (Eq. 5.2.2.5). Results described in Sect. 5.2.2 show that increasing ϵ_s increases the correlation among neural activities within STN and between STN and GPe. We now show that ϵ_s serves as a kind of an inverse of the exploration parameter, ϵ, used in ϵ-greedy methods (Sutton, & Barto, 1998). Increasing ϵ_s corresponds to weaker exploration. Figure 5.17a shows variation of average reward with iterations for different values of ϵ_s. Figure 5.17d shows the dependence of steady-state reward (obtained by averaging the reward profiles from Fig. 5.17a over the interval [900–1000] iterations) on ϵ_s. Note that highest rewards are obtained at an intermediate level of ϵ_s, beyond which average reward drops rapidly. The η used in Eq. (5.3.2.12) is 0.01; $\kappa_1 = -0.05$ and $\kappa_2 = 5$ in Eqs. (5.3.2.4) and (5.3.2.5), respectively, with θ_{D1} and $\theta_{D2} = 0$. $w_{sg} = w_{gs} = 0.91$; $w^{STN\_GPi} = 0.3$ in the Eqs. (5.2.2.4), (5.2.2.5), and (5.3.2.7), respectively. With the change in time dt = 0.01: $\tau_s = 0.1$ (Eq. 5.2.2.5); $\tau_g = 0.3$ (Eq. 5.2.2.7); $\eta = 0.01$ (Eq. 5.3.2.12); The resulting trend is similar to what is observed in standard approaches to n-armed bandit problems like the ϵ-greedy method.

By way of confirmation, we applied the ϵ-greedy method to the above set of n-armed bandit problems. Figure 5.17b shows average reward profiles (each of them averaged over 500 instances). Figure 5.17d shows that the highest reward is

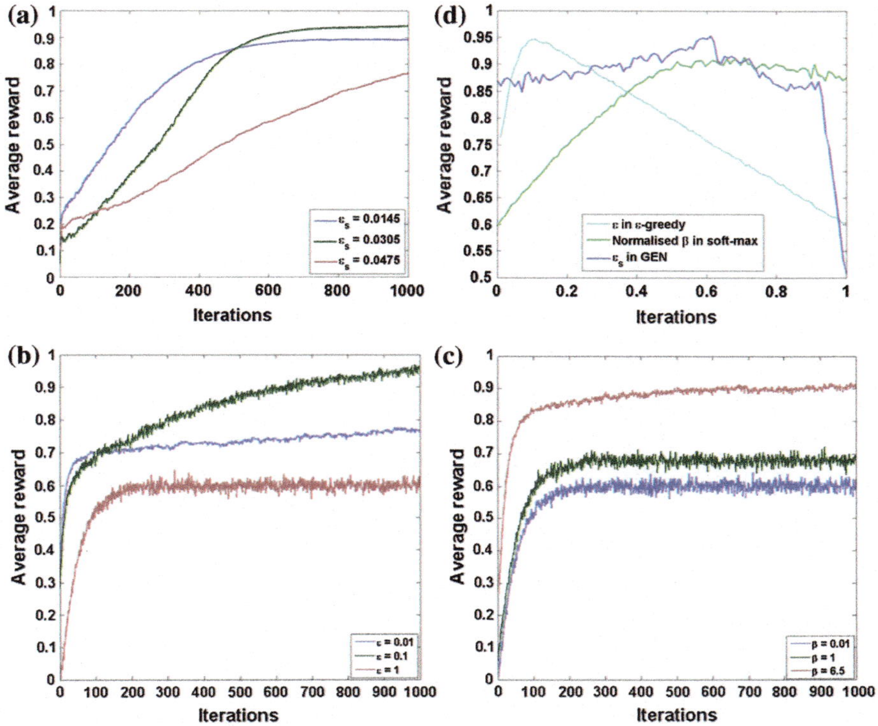

Fig. 5.17 Five-arm bandit task averaged over 500 instances: The mean reward versus iterations obtained for **a** different ϵ_s using GEN policy, **b** different ϵ values using ϵ-greedy policy, **c** different β-values using softmax policy, and **d** steady-state reward (averaged from 900 to 1000 iterations) versus ϵ of ϵ-greedy, β in range [0 10] of softmax normalized to [0 1], ϵ_s of GEN policy

obtained at $\epsilon = 0.07$, with average reward falling rapidly for higher values. We also applied softmax policy (Sutton, & Barto, 1998) to the same n-armed problem, and the average rewards over iterations for various β-values are given in Fig. 5.17b, with the steady-state average reward values plotted against β of range [0, 10] normalized to [0, 1] (Fig. 5.17d).

5.3.3 Climbing Value Gradient Using δ_V

In the previous section, we have shown how the proposed BG network model can be applied to the n-armed bandit problem. A novel feature of the model is the introduction of an additional dopamine-related variable called δ_v in addition to the classical TD error, δ. The TD error was used, as in classical RL-based BG models, for value training (Eqs. 5.3.2.9, 5.3.2.10, 5.3.2.12). But the new δ_v is used to switch between DP and IP since it controls the gain function of D1 and D2 MSNs in striatum (Eqs. 5.3.2.2–5.3.2.5). The quantity δ_v is simply value gradient (Eq. 5.3.2.6). As noted earlier, it is different from TD error in that there is no reward term, and $\gamma = 1$ (Eq. 5.3.2.11). From Eqs. (5.3.2.1–5.3.2.8), we note that, for positive values of δ_v, since DP is selected, there is a tendency to repeat the previous action, and for moderate or negative values of δ_v, since IP is selected, there is a tendency to select a random action. Due to this effect of δ_v on the value, Eqs. (5.3.2.7–5.3.2.8) appear to perform some sort of hill-climbing on the value function. In fact, the aforementioned effect of δ_v on value is strongly reminiscent of simulated annealing, a form of stochastic optimization (Kirkpatrick, Jr. & Vecchi, 1983). A primitive form of simulated annealing can be expressed as follows:

Let,

- $E(x)$ be the cost function that must be maximized
- $x(t) = i$ is the current state
- j = one of the neighboring states of i

$$\Delta E = E(t+1) - E(t) \tag{5.3.3.1}$$

The decision whether to switch from the state 'i' to state 'j' is made by the following rule:

$$\begin{array}{lll} \textit{If } E(i) < E(j) & & \\ X(t+1) = j; & & \text{(a)} \\ \text{else} & & \\ X(t+1) = j; & \text{with probability } P = \exp(-(E(j) - E(i))/T); & \text{(b)} \\ X(t+1) = i; & \text{with probability } 1 - P & \text{(c)} \end{array} \tag{5.3.3.2}$$

Here, (a) denotes the switch to state 'j' with probability 1, if $E(j)$ is larger than $E(i)$; (b) describes the switch to state 'j' with probability, P, if $E(i)$ is larger than $E(j)$; and (c) represents the stay in the previous state with probability $1 - P$.

ΔE in the above case of simulated annealing is analogous to δ_v, and E is analogous to the value function, in the BG context.

In this section, we focus on the mechanism of hill-climbing of value function, driven by δ_v. Unlike the previous sections, where we considered discrete action spaces, we show how the above mechanism can perform hill-climbing in continuous action spaces. However, in the present section we do not consider the full RL problem with continuous state and action spaces.

From the binary action selection problem of Sect. 5.3.1, we saw that the BG model exhibits three behaviors depending on δ_v: (1) Go regime for large positive δ_v, (2) Explore regime for intermediate values of δ_v, and (3) NoGo regime for large negative values of δ_v. In other words, repeat the previous action for large positive

δ_v; select a random action for moderate δ_v; and take no action for large negative δ_v. These regimes inspire a simple mechanism for hill-climbing as follows:

Let

- $V(x)$ = value function
- x = n-dimensional state vector ($x \in S$, where $S \subset R^n$)
- $\delta_v = V(t) - V(t-1)$
- $\Delta x = x(t) - x(t-1)$

Δx is updated by the following equations:

$$\begin{array}{lll} \text{if}(\delta_v > D_{\text{hi}}) & & \\ \quad \Delta x(t+1) = \Delta x(t) & \text{-``}Go\text{''} & \text{(a)} \\ \text{else if}(\delta_v > D_{\text{lo}} \wedge \delta_v(t) \le D_{\text{hi}}) & & \\ \quad \Delta x(t+1) = \phi & \text{-``Explore''} & \text{(b)} \\ \textit{else}\ (\delta_v \le D_{\text{lo}}) & & \\ \quad \Delta x(t+1) = 0 & \text{-``NoGo''} & \text{(c)} \end{array} \tag{5.3.3.3}$$

The three cases of Eq. (5.3.3.3) represent the Go, Explore, and NoGo regimes, respectively, and may be interpreted as follows: (a) Go regime repeats the last update in state, (b) Explore regime updates 'x' in a random direction, and (c) NoGo regime does not update state. $D_{\text{hi}} > 0$, $D_{\text{lo}} < 0$, ϕ is a random vector whose each component, ϕ_i, is given as,

$$\phi_i = G(0,1)\exp\left(-\delta_v^2/\sigma^2\right) \tag{5.3.3.4}$$

where $G(0, 1)$ is a Gaussian random variable with mean 0 and standard deviation 1.

From purely algorithmic point of view, the last Eq. (5.3.3.3c) seems wasteful since there is no state update in that case. Whenever the NoGo regime is selected, since there is no state update, we have $\delta_v = 0$. In the next iteration, therefore, Explore regime is selected since $D_{\text{hi}} > 0 > D_{\text{lo}}$. To make the computation more efficient, Eq. (5.3.3.3c) can be altered slightly as follows:

$$\begin{array}{lll} \text{if}\ (\delta_v \le D_{\text{lo}}) & & \\ \quad \Delta x(t+1) = -\Delta x(t) & \text{``NoGo''} & \text{(c)} \end{array}$$

Now, in NoGo regime, the state x is updated in an opposite direction compared to the previous update. Thus, the Eq. (5.3.3.3a, b, c) may be expressed in modified form as follows:

$$\begin{aligned}
&\text{if}(\delta_v > D_{\text{hi}}) \\
&\qquad \Delta x(t+1) = \Delta x(t) \qquad -\text{"Go"} \qquad \text{(a)} \\
&\text{else if}(\delta_v > D_{\text{lo}} \wedge \delta_v(t) \leq D_{\text{hi}}) \\
&\qquad \Delta x(t+1) = \phi \qquad -\text{"Explore"} \qquad \text{(b)} \\
&\text{else } (\delta_v \leq D_{\text{lo}}) \\
&\qquad \Delta x(t+1) = -\Delta x(t) \qquad -\text{"NoGo"} \qquad \text{(c)}
\end{aligned} \tag{5.3.3.5}$$

In Eq. (5.3.3.5) above, the Go, Explore, and NoGo regimes are depicted as discrete, disjoint regimes demarcated by thresholds—D_{hi} and D_{lo}. But the regimes as observed in the binary action selection simulations of Sect. 5.3.1 are not disjoint, with multiple regimes occurring for a given δ_v. In this section, we combine the Eq. (5.3.3.5a, b, c) in a single update equation so that the three regimes smoothly overlap.

Let us begin by reformulating the three regimes of Eq. (5.3.3.5a, b, c) in a single equation as follows:

$$\begin{aligned}
\Delta x(t+1) = {} & \text{step}(\delta_V(t) - D_{\text{hi}})\Delta x(t) \\
& + \phi \text{pulse}(\delta_V(t), D_{\text{hi}}, D_{\text{lo}}) \\
& - \text{step}(D_{\text{lo}} - \delta_V(t))\Delta x(t)
\end{aligned} \tag{5.3.3.6}$$

where step(.) is the step function or the Heaviside function defined as,

$$\begin{aligned}
\text{step}(x) &= 1, \quad \text{for } x \geq 0 \\
&= 0, \quad \text{elsewhere}
\end{aligned} \tag{5.3.3.7}$$

and pulse(x, a, b) is defined as,

$$\begin{aligned}
\text{pulse}(x) &= 1, \quad \text{for } b \leq x \leq a \\
&= 0, \quad \text{elsewhere}
\end{aligned} \tag{5.3.3.8}$$

The functions step(x) and pulse(x) may be approximated for $\kappa_3 > 0$ and $\kappa_4 < 0$ by their continuous versions as sigmoid and Gaussian functions, respectively. The update rule is thereby rewritten as,

$$\begin{aligned}
\Delta x(t+1) = {} & \log\,\text{sig}(\kappa_3(\delta_V(t) - D_{\text{hi}}))\Delta x(t) \\
& + \psi \exp(-\delta_V^2(t)/\sigma_E^2) \\
& - \log\,\text{sig}(\kappa_4(\delta_V(t) - D_{\text{lo}}))\Delta x(t)
\end{aligned} \tag{5.3.3.9}$$

The expansion of log sig is provided in Eq. 5.3.3.10.

$$\log\,\text{sig}(x) = \frac{1}{1 + \exp(-x)} \tag{5.3.3.10}$$

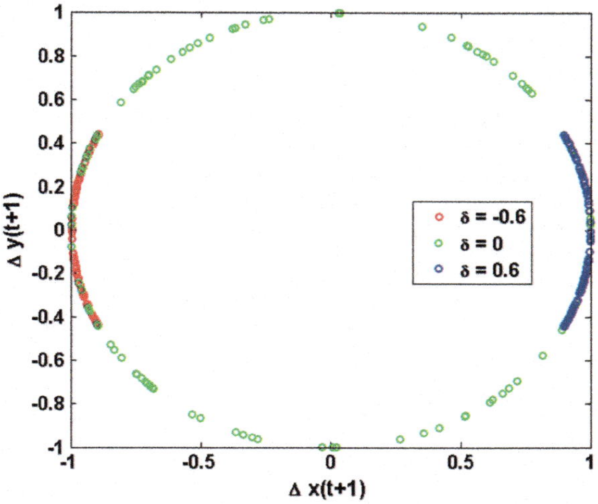

Fig. 5.18 An illustration of the map of Eq. (5.3.3.9) taking $\kappa_3 = -\kappa_4 = 1$; $\sigma = 0.5$. $\Delta x(t) = (1, 0)$

Thus, Eq. (5.3.3.9) describes a map between the current state update, $\Delta x(t)$, to the next state update, $\Delta x(t + 1)$. For large positive δ_V, $\Delta x(t + 1)$ is nearly in the same direction as $\Delta x(t)$; for large negative δ_V, $\Delta x(t + 1)$ is nearly in the same direction as $-\Delta x(t)$; for intermediate values of δ_V, $\Delta x(t + 1)$ is random. Note that the map of Eq. (5.3.3.9) is stochastic due the middle term on the right-hand side. The map of Eq. (5.3.3.9) is depicted in Fig. 5.18, for two-dimensional input space. $\Delta x(t) = [1, 0]$.

The parameters of Eq. (5.3.3.9) are as follows: $\kappa_3 = -\kappa_4 = 1$; $\sigma = 0.5$. The distribution of $\Delta x(t)$ for $\delta_V = 0.6$, 0, and -0.6 is shown by blue, green, and red circles, respectively.

Normalized $\Delta X(t + 1)$ is shown on the unit circle. The distribution of $\Delta x(t)$ for $\delta_V = 0.6$, 0, and -0.6 is shown by blue, green, and red circles, respectively.

A Numerical Example:

We now apply the GEN algorithm (Eq. 5.3.3.9a, b, c) to a simple optimization problem and study the effect of GEN parameters like D_{hi}, D_{lo}, and σ on the number of iterations required to find the maximum.

The function to be maximized is $E(x) = -\|x - x_T\|^2$ where $n = 10$ and $x \in R^n$. D_{hi} and D_{lo} are linked as $D_{hi} = -D_{lo} = \psi$. The value of σ, which arises in calculation of the random variable ϕ (in Eq. 5.3.3.9b), is also linked to ψ as follows: $\sigma = 0.2 * \psi$. The idea behind linking ψ with σ in the above fashion is that a large difference between the thresholds D_{hi} and D_{lo} implies a broad Explore regime and therefore consistent with a larger standard deviation of Gaussian in Eq. (5.3.3.6).

Now that the three GEN parameters (D_{hi}, D_{lo}, and σ) are expressed in terms of ψ, it is easy to sweep the ψ-space for the ψ-value for which one obtains the most efficient search. Figure 5.19 shows the number of steps necessary to find the

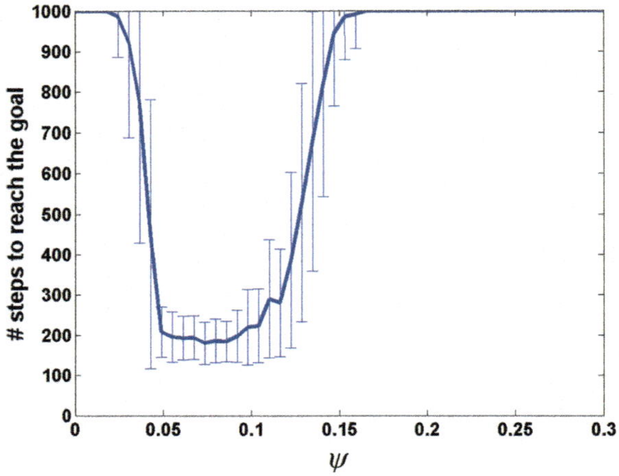

Fig. 5.19 Number of steps taken to reach the maximum of $E(x)$ as a function of ψ averaged over 100 instances (x is a 10-dimensional vector). The goal x_T is taken as the origin. A random vector of length 5 is taken as an initial value for x. The search is stopped when $(x - x_T)^2 \leq 0.1$. Parameters in Eq. (5.3.3.9) are set as: $\kappa_3 = -\kappa_4 = 10$. The maximum step length allowed is 0.2; that is, the output of Eq. (5.3.3.9) is multiplied by 0.2, before the state update

maximum of the cost function, $E(x)$, as a function of ψ. The upper limit for the number of steps is 1000, beyond which the search is aborted (averaged over 100 instances). Note that fastest searches were obtained for intermediate values of ψ: The search was abortive for both very large exploration and very limited exploration.

5.4 Discussion

In this chapter, we extend the perspective of the BG system as an RL engine, by hypothesizing that the indirect pathway does not merely inhibit action selection, but, by virtue of its complex dynamics, provide the exploratory drive necessary for the system to learn by reinforcement. Thus, our perspective takes a significant departure from the classical Go-NoGo interpretation of the functional anatomy of BG pathways (Albin, Young, & Penney, 1989). This simplified binary view, although useful, cannot explain how the BG circuit can perform its greater role as a complete RL engine. Experiments by Schultz and colleagues (Hollerman & Schultz, 1998; Schultz et al., 1997) that identify mesencephalic dopamine signals with TD error had inspired an extensive effort to use concepts of RL to describe BG function. A lot of attention was directed to the role of ventral striatum in value computation, which is natural since value computation occupies a central place in RL formalism. Optimal action policy can be implemented by climbing the value gradient, a process known as exploitation. However, exploration, the complementary process to exploitation, does not seem to have been given sufficient attention by BG models. Even in the standard RL literature, exploration is typically represented simply by a noise term, or by introduction of stochasticity in action selection, and not explicitly by a module in the manner of an Actor or a Critic. The Go/NoGo picture of BG can explain discrete action selection using RL mechanisms, but not in continuous state and action spaces.

The Go/NoGo theory of the BG function is supported by a simplistic interpretation of the functional neurochemistry of the two BG pathways (Albin et al., 1989; Contreras-Vidal & Stelmach, 1995). There are two inhibitory stages along the direct pathway, making the pathway effectively excitatory or disinhibitory; there are three inhibitory stages on the indirect pathway making it effectively inhibitory. These arguments hide an important anatomical detail, viz. presence of excitatory feedback connections from STN to GPe. Presence of feedback connections in IP makes the simple 'overall inhibitory' argument null and void and also compels us to study the consequences of the rich dynamics of the excitatory–inhibitory loop of neuronal populations (Brunel, 2000). The STN–GPe loop has been dubbed as the 'pacemaker' of BG considering its role in generating pathological oscillations associated with Parkinsonian tremor (Hurtado et al., 1999; Terman et al., 2002). The fact that neurons in this system exhibit uncorrelated firing patterns in normal conditions, and highly correlated and synchronized firing under dopamine-deficient pathological conditions, seems to offer an important clue to the possible role of this circuit in the overall BG function.

We hypothesize that, by virtue of its complex dynamics, the STN–GPe system is in the best position to serve as an explorer, thereby supplying the missing piece in the RL machinery of BG. The aim of this chapter is to develop this theme and demonstrate using a series of models the role of IP in exploration.

We begin with a simple network model of BG in Sect. 5.2. The model has representations for the striatum, GPe, STN, and GPi. The striatum has D1- and D2-expressing MSNs that project to GPi and GPe, respectively. A single pair of STN and GPe neurons is shown to produce limit cycle oscillations. When two pools of STN and GPe neurons are connected, the network produced complex oscillations. It was shown that correlations within and across STN and GPe can be controlled by varying the lateral connections in STN. At this stage of model formulation, there is no RL or any other form of learning. There are no cortico-striatal connections. There is a parameter, δ, that represents striatal dopamine, but not linked to TD error.

The network defined in Sect. 5.2 is applied to binary action selection problem in Sect. 5.3. Two inputs are presented to the network. Magnitude of the inputs represents their saliency. Go, Explore, and NoGo regimes are first defined here with respect to the action selected by the network. Go corresponds to selection of the more salient input, Explore corresponds to that of the less salient input, and NoGo corresponds to non-selection of any input. The effect of δ parameter on the three regimes is studied. It was observed that the Go regime is predominant for larger values of δ, the Explore regime for intermediate values, and the NoGo regime for smallest values. Thus, the simple BG model of Sect. 5.2, while confirming the classical Go/NoGo depiction of BG, goes beyond and suggests the presence of an Explore regime between the two older regimes. It is also shown that the Explore regime arises from the dynamics of STN–GPe system: Strengthening the lateral connections in STN, a change that increases correlations within STN, attenuates Explore regime.

The network model of Sect. 5.2.2 is next placed within the larger framework of RL, and the expanded model is applied to n-armed bandit problem in Sect. 5.3.2. Two dopamine-related signals are distinguished in this model: (1) TD error denoted by δ, and (2) value gradient denoted by δ_V. Whereas TD error is used for value training, or training of cortico-striatal connections, as in classical Actor–Critic models of BG (Joel et al., 2002), the value gradient is used for switching between DP and IP in the striatum. The quantity δ_V drives action selection in such a way that the network progressively selects arms with higher expected payoff. In other words, δ_V is used for hill-climbing over the value function.

Hill-climbing of the value profile, and the role played by the quantity δ_V therein, is the focus of Sect. 5.3.3. In this section, the Go/Explore/NoGo regimes observed empirically in Sect. 5.3.1, are formulated into an algorithm, and dubbed the GEN policy, that represents a mechanism for climbing the value gradient. Later, in Sect. 5.3.3, the GEN policy is presented, not just as a policy for climbing the value profile, but as a general algorithm for optimization. The GEN algorithm is applied to optimization of a quadratic function in 10 dimensions. The number of steps taken to reach the maximum is taken as a measure of efficiency of the algorithm. Simulations in Sect. 3.3 show that the number of steps depends critically on the width of the Explore regime: If the width of the regime is too narrow, exploration is inadequate, whereas if the regime is too wide, the search does not settle down. The simulation demonstrates the need for an Explore regime that is optimal for a given problem. The GEN algorithm which was formulated as three separate equations in Eq. (5.3.3.5) is reformulated as a single update equation (Eq. 5.3.3.9).

The GEN algorithm is comparable to a standard optimization alvrithm like simulated annealing, where a given change in state is accepted stochastically depending on the change in the cost function brought about by the change in state. However, a more formal comparison of the two algorithms, paying attention to the analogous roles of the temperature parameter, T, in simulated annealing and ψ parameter (Sect. 3.3) that determines the width of Explore regime, would provide useful insights into GEN algorithm.

Thus, we begin with a dynamic network model of BG in Sect. 2.2 and progressively characterize it and reduce it to a lumped model—the GEN policy.

In this chapter, we have shown the presence of three regimes—Go, Explore, and NoGo—using a simplified network model of BG. Recently, this Go-Explore-NoGo model was extended to a BG network model consisting of Izhikevich neurons (Mandali & Chakravarthy, 2016; Mandali, Rengaswamy, Chakravarthy, & Moustafa, 2015). Even in the spiking version of the BG model, it was possible to recover and identify the three regimes. It would be a logical next step to search for the three regimes in a more realistic, biophysical model of BG. In the next few chapters, we use the Go-Explore-NoGo regime to simulate several BG functions, including reaching and gait (Muralidharan et al., 2017; Muralidharan, Balasubramani, Chakravarthy, Lewis, & Moustafa, 2014).

Appendix

The system of equations for single oscillator is given by,

$$\frac{\mathrm{d}x}{\mathrm{d}t} = -x + v - s + I \tag{i}$$

$$v = \tan h(\lambda x) \tag{ii}$$

$$\frac{\mathrm{d}s}{\mathrm{d}t} = -s + v \tag{iii}$$

Differentiating (i)

$$\frac{\mathrm{d}^2x}{\mathrm{d}^2t} = -\frac{\mathrm{d}x}{\mathrm{d}t} + \lambda \sec h^2(\lambda x)\frac{\mathrm{d}x}{\mathrm{d}t} - \frac{\mathrm{d}s}{\mathrm{d}t} \tag{iv}$$

Substituting (ii) and (iii) in (iv)

$$\frac{\mathrm{d}^2x}{\mathrm{d}^2t} = -\frac{\mathrm{d}x}{\mathrm{d}t} + \lambda \sec h^2(\lambda x)\frac{\mathrm{d}x}{\mathrm{d}t} - (-s + \tan h(\lambda x)) \tag{v}$$

Substituting (i) and (ii) in (v)

$$\frac{\mathrm{d}^2x}{\mathrm{d}^2t} = -\frac{\mathrm{d}x}{\mathrm{d}t} + \lambda \sec h^2(\lambda x)\frac{\mathrm{d}x}{\mathrm{d}t} - \left(\frac{\mathrm{d}x}{\mathrm{d}t} + x - v - I + \tan h(\lambda x)\right)$$

on rearranging

$$\frac{\mathrm{d}^2x}{\mathrm{d}^2t} + \frac{\mathrm{d}x}{\mathrm{d}t}\left(2 - \lambda \sec h^2(\lambda x)\right) + (x - I) = 0 \tag{vi}$$

is similar to Lienard's equation $\frac{\mathrm{d}^2x}{\mathrm{d}^2t} + \frac{\mathrm{d}x}{\mathrm{d}t}f(x) + g(x) = 0$ where $f(x) = 2 - \lambda \sec h^2(\lambda x)$, and $g(x) = x - I$.

Checking for the Lienard's conditions, let us assume $I = 0$.

1. Both $f(x)$ and $g(x)$ are continuously differentiable for all x.
2. $g(-x) = -g(x)$ for all x (i.e., $g(x)$ is an odd function).

3. $g(x) > 0$ for $x > 0$.
4. For all x(i.e., $f(x)$ is an even function); The odd function $F(x) = \int_0^x f(u)du$ = $2x$ − tan $h(\lambda x)$ has exactly one positive zero at $x = x_o$, is negative for $0<x<x_o$, is positive and non-decreasing for $x > x_o$, and $F(x) \to \infty$ as $x \to \infty$ (one can estimate x_o from graph of $F(x)$). So the system has a unique stable limit cycle surrounding the origin in the phase plane.

References

Albin, R. L., Young, A. B., & Penney, J. B. (1989). The functional anatomy of basal ganglia disorders. *Trends in Neurosciences, 12*(10), 366–375.

Baunez, C., Humby, T., Eagle, D. M., Ryan, L. J., Dunnett, S. B., & Robbins, T. W. (2001). Effects of STN lesions on simple vs choice reaction time tasks in the rat: Preserved motor readiness, but impaired response selection. *European Journal of Neuroscience, 13*(8), 1609–1616.

Bergman, H., Wichmann, T., Karmon, B., & DeLong, M. (1994). The primate subthalamic nucleus. II. Neuronal activity in the MPTP model of Parkinsonism. *Journal of Neurophysiology, 72*(2), 507–520.

Brown, P., Oliviero, A., Mazzone, P., Insola, A., Tonali, P., & Di Lazzaro, V. (2001). Dopamine dependency of oscillations between subthalamic nucleus and pallidum in Parkinson's disease. *The Journal of Neuroscience, 21*(3), 1033–1038.

Brunel, N. (2000). Dynamics of Sparsely Connected Networks of Excitatory and Inhibitory Spiking Neurons. *Journal of Computational Neuroscience 8*, 183–208.

Chakravarthy, V. S., Joseph, D., & Bapi, R. S. (2010). What do the basal ganglia do? A modeling perspective. *Biological Cybernetics, 103*(3), 237–253. https://doi.org/10.1007/s00422-010-0401-y.

Contreras-Vidal, J., & Stelmach, G. E. (1995). Effects of Parkinsonism on motor control. *Life Sciences, 58*(3), 165–176.

Crick, F. (1984). Function of the thalamic reticular complex: The searchlight hypothesis. *Proceedings of the National Academy of Sciences, 81*(14), 4586–4590.

Daw, N. D., O'Doherty, J. P., Dayan, P., Seymour, B., & Dolan, R. J. (2006). Cortical substrates for exploratory decisions in humans. *Nature, 441*(7095), 876–879.

Flores-Hernandez, J., Cepeda, C., Hernandez-Echeagaray, E., Calvert, C. R., Jokel, E. S., Fienberg, A. A., … Levine, M. S. (2002). Dopamine enhancement of NMDA currents in dissociated medium-sized striatal neurons: Role of D1 receptors and DARPP-32. *Journal of Neurophysiol, 88*(6), 3010–3020.

Frank, M. J. (2005). Dynamic dopamine modulation in the basal ganglia: a neurocomputational account of cognitive deficits in medicated and nonmedicated Parkinsonism. *Journal of Cognitive Neuroscience, 17*(1), 51–72. https://doi.org/10.1162/0898929052880093.

Gangadhar, G., Joseph, D., & Chakravarthy, V. S. (2008). Understanding Parkinsonian handwriting through a computational model of basal ganglia. *Neural Computation, 20*(10), 2491–2525.

Gillies, A., Willshaw, D., Atherton, J., & Arbuthnott, G. (2002). Functional interactions within the subthalamic nucleus. In *The basal ganglia VII* (pp. 359–368). Boston: Springer.

Grabli, D., McCairn, K., Hirsch, E. C., Agid, Y., Féger, J., François, C., et al. (2004). Behavioural disorders induced by external globus pallidus dysfunction in primates: I. Behavioural study. *Brain, 127*(9), 2039–2054.

Gurney, K., Prescott, T. J., & Redgrave, P. (2001). A computational model of action selection in the basal ganglia. I. A new functional anatomy. *Biological Cybernetics, 84*(6), 401–410.

Hollerman, J. R., & Schultz, W. (1998). Dopamine neurons report an error in the temporal prediction of reward during learning. *Nature Neuroscience, 1*(4), 304–309.

Houk, J. C., Davis, J. L., & Beiser, D. G. (1995). *Models of information processing in the basal ganglia*. Cambridge: The MIT press.

Humphries, M., & Gurney, K. (2002). The role of intra-thalamic and thalamocortical circuits in action selection. *Network: Computation in Neural Systems, 13*(1), 131–156.

Humphries, M. D., & Prescott, T. J. (2010). The ventral basal ganglia, a selection mechanism at the crossroads of space, strategy, and reward. *Progress in Neurobiology, 90*(4), 385–417. https://doi.org/10.1016/j.pneurobio.2009.11.003.

Hurtado, J. M., Gray, C. M., Tamas, L. B., & Sigvardt, K. A. (1999). Dynamics of tremor-related oscillations in the human globus pallidus: a single case study. *Proceedings of the National Academy of Sciences, 96*(4), 1674–1679.

Joel, D., Niv, Y., & Ruppin, E. (2002). Actor-critic models of the basal ganglia: New anatomical and computational perspectives. *Neural Networks, 15*(4–6), 535–547.

Kalva, S. K., Rengaswamy, M., Chakravarthy, V. S., & Gupte, N. (2012). On the neural substrates for exploratory dynamics in basal ganglia: A model. *Neural Networks, 32,* 65–73. https://doi.org/10.1016/j.neunet.2012.02.031.

Kandel, E. R., Schwartz, J. H., & Jessell, T. M. (2000). *Principles of neural science* (Vol. 4). New York: McGraw-Hill.

Kirkpatrick, S., Gelatt, C. D., Jr., & Vecchi, M. P. (1983). Optimization by simulated annealing. *Science, 220*(4598), 671–680.

Kliem, M. A., Maidment, N. T., Ackerson, L. C., Chen, S., Smith, Y., & Wichmann, T. (2007). Activation of nigral and pallidal dopamine D1-like receptors modulates basal ganglia outflow in monkeys. *Journal of Neurophysiology, 98*(3), 1489–1500.

Knutson, B., Adams, C. M., Fong, G. W., & Hommer, D. (2001). Anticipation of increasing monetary reward selectively recruits nucleus accumbens. *Journal of Neuroscience, 21*(16), 159.

Kravitz, A. V., Freeze, B. S., Parker, P. R., Kay, K., Thwin, M. T., Deisseroth, K., et al. (2010). Regulation of Parkinsonian motor behaviours by optogenetic control of basal ganglia circuitry. *Nature, 466*(7306), 622–626.

Krishnan, R., Ratnadurai, S., Subramanian, D., Chakravarthy, V. S., & Rengaswamy, M. (2011). Modeling the role of basal ganglia in saccade generation: is the indirect pathway the explorer? *Neural Networks, 24*(8), 801–813. https://doi.org/10.1016/j.neunet.2011.06.002.

Magdoom, K. N., Subramanian, D., Chakravarthy, V. S., Ravindran, B., Amari, S., & Meenakshisundaram, N. (2011). Modeling basal ganglia for understanding Parkinsonian reaching movements. *Neural Computation, 23*(2), 477–516. https://doi.org/10.1162/NECO_a_00073.

Magill, P., Bolam, J., & Bevan, M. (2001). Dopamine regulates the impact of the cerebral cortex on the subthalamic nucleus–globus pallidus network. *Neuroscience, 106*(2), 313–330.

Mandali, A., & Chakravarthy, V. S. (2016). Probing the role of medication, DBS electrode position, and antidromic activation on impulsivity using a computational model of basal ganglia. *Frontiers in Human Neuroscience, 10,* 450.

Mandali, A., Rengaswamy, M., Chakravarthy, S., & Moustafa, A. A. (2015). A spiking Basal Ganglia model of synchrony, exploration and decision making. *Frontiers in Neuroscience, 9,* 191.

Michmizos, K. P., & Nikita, K. S. (2011). *Local field potential driven Izhikevich model predicts a subthalamic nucleus neuron activity.* Paper presented at the Engineering in Medicine and Biology Society, EMBC, 2011 Annual International Conference of the IEEE.

Montague, P. R., Dayan, P., & Sejnowski, T. J. (1996). A framework for mesencephalic dopamine systems based on predictive Hebbian learning. *Journal of Neuroscience, 16*(5), 1936–1947.

Muralidharan, V., Balasubramani, P. P., Chakravarthy, V. S., Gilat, M., Lewis, S. J., & Moustafa, A. A. (2017). A neurocomputational model of the effect of cognitive load on freezing of gait in Parkinson's disease. *Frontiers in Human Neuroscience, 10,* 649. https://doi.org/10.3389/fnhum.2016.00649.

Muralidharan, V., Balasubramani, P. P., Chakravarthy, V. S., Lewis, S. J., & Moustafa, A. A. (2014). A computational model of altered gait patterns in Parkinson's disease patients negotiating narrow doorways. *Frontiers in Computational Neuroscience, 7,* 190. https://doi.org/10.3389/fncom.2013.00190.

O'Doherty, J., Dayan, P., Schultz, J., Deichmann, R., Friston, K., & Dolan, R. J. (2004). Dissociable roles of ventral and dorsal striatum in instrumental conditioning. *Science, 304* (5669), 452–454.

Plenz, D., & Kital, S. T. (1999). A basal ganglia pacemaker formed by the subthalamic nucleus and external globus pallidus. *Nature, 400*(6745), 677–682.

Pragathi Priyadharsini, B., Ravindran, B., & Srinivasa Chakravarthy, V. (2012). Understanding the role of serotonin in basal ganglia through a unified model. In A. P. Villa, W. Duch, P. Érdi, F. Masulli, & G. Palm (Eds.), *Artificial Neural Networks and Machine Learning—ICANN 2012* (Vol. 7552, pp. 467–473). Berlin: Springer.

Redgrave, P., Prescott, T. J., & Gurney, K. (1999). The basal ganglia: A vertebrate solution to the selection problem? *Neuroscience, 89*(4), 1009–1023.

Rushworth, M. F., & Behrens, T. E. (2008). Choice, uncertainty and value in prefrontal and cingulate cortex. *Nature Neuroscience, 11*(4), 389–397.

Schultz, W., Dayan, P., & Montague, P. R. (1997). A neural substrate of prediction and reward. *Science, 275*(5306), 1593–1599.

Sridharan, D., Prashanth, P. S., & Chakravarthy, V. S. (2006). The role of the basal ganglia in exploration in a neural model based on reinforcement learning. *International Journal of Neural Systems, 16*(2), 111–124.

Stein, P. S., Grillner, S., Selverston, A. I., & Stuart, D. G. (1997). *Neurons, networks, and behavior*. Cambridge, MA: MIT Press.

Steiner, H., & Tseng, K. Y. (2010). *Handbook of basal ganglia structure and function: A decade of progress* (Vol. 20). Access Online via Elsevier.

Sukumar, D., Rengaswamy, M., & Chakravarthy, V. S. (2012). Modeling the contributions of Basal ganglia and Hippocampus to spatial navigation using reinforcement learning. *PLoS ONE, 7*(10), e47467. https://doi.org/10.1371/journal.pone.0047467.

Sutton, R., & Barto, A. (1998). *Reinforcement learning: An introduction. Adaptive computations and machine learning*. MIT Press/Bradford.

Terman, D., Rubin, J., Yew, A., & Wilson, C. (2002). Activity patterns in a model for the subthalamopallidal network of the basal ganglia. *The Journal of Neuroscience, 22*(7), 2963–2976.

Willshaw, D., & Li, Z. (2002). Subthalamic–pallidal interactions are critical in determining normal and abnormal functioning of the basal ganglia. *Proceedings of the Royal Society of London, Series B: Biological Sciences, 269*(1491), 545–551.

Yoshida, W., & Ishii, S. (2006). Resolution of uncertainty in prefrontal cortex. *Neuron, 50*(5), 781–789.

Chapter 6
Synchronization and Exploration in Basal Ganglia—A Spiking Network Model

Alekhya Mandali and V. Srinivasa Chakravarthy

Abstract Making an optimal decision could be to either 'Explore' or 'exploit' or 'not to take any action,' and basal ganglia (BG) are considered to be a key neural substrate in decision making. In earlier chapters, we had hypothesized earlier that the indirect pathway (IP) of the BG could be the subcortical substrate for exploration. Here, we build a spiking network model to relate exploration to synchrony levels in the BG (which are a neural marker for tremor in Parkinson's disease). Key BG nuclei such as the subthalamic nucleus (STN), Globus Pallidus externus (GPe), and Globus Pallidus internus (GPi) were modeled as Izhikevich spiking neurons, whereas the striatal output was modeled as Poisson spikes. We have applied reinforcement learning framework with the dopamine signal representing the reward prediction error used for cortico-striatal weight update. We apply the model to two decision-making tasks: a binary action selection task and an n-armed bandit task. The model shows that exploration levels could be controlled by STN's lateral connection strength which also influenced the synchrony levels in the STN–GPe circuit. An increase in STN's lateral strength led to a decrease in exploration which can be thought as the possible explanation for reduced exploratory levels in Parkinson's patients.

6.1 Introduction

Chakravarthy, Joseph, and Bapi (2010) suggested that STN–GPe loop, a coupled excitatory–inhibitory network in the IP, might be the substrate for exploration (Chakravarthy et al., 2010). It is well known that coupled excitatory–inhibitory pools of neurons can exhibit rich dynamic behavior like oscillations and chaos (Borisyuk, Borisyuk, Khibnik, & Roose, 1995; Sinha, 1999). This hypothesis has inspired models simulating various BG functions ranging from action selection in continuous spaces (Krishnan, Ratnadurai, Subramanian, Chakravarthy, & Rengaswamy, 2011), reaching movements (Magdoom et al., 2011), spatial navigation (Sukumar, Rengaswamy, & Chakravarthy, 2012), precision grip (Gupta, Balasubramani, & Chakravarthy, 2013), and gait (Muralidharan, Balasubramani, Chakravarthy, Lewis, & Moustafa, 2013) in normal and Parkinsonian conditions.

V. S. Chakravarthy and A. Moustafa, *Computational Neuroscience Models of the Basal Ganglia*, Cognitive Science and Technology,
https://doi.org/10.1007/978-981-10-8494-2_6

Using a network of rate-coding neurons, Kalva, Rengaswamy, Chakravarthy, and Gupte (2012) showed that exploration emerges out of the chaotic dynamics of the STN–GPe system (Kalva et al., 2012). Most rate-coded models, by design, fail to capture dynamic phenomena like synchronization found in more realistic spiking neuron models (Bevan, Magill, Terman, Bolam, & Wilson, 2002; Park, Worth, & Rubchinsky, 2010; Park, Worth, & Rubchinsky, 2011). Synchronization within BG nuclei had gained attention since the discovery that STN, GPe, and GPi neurons show high levels of synchrony in Parkinsonian conditions (Bergman, Wichmann, Karmon, & DeLong, 1994; Bevan et al., 2002; Hammond, Bergman, & Brown, 2007; Tachibana, Iwamuro, Kita, Takada, & Nambu, 2011; Weinberger & Dostrovsky, 2011). This oscillatory activity was found to be present in two frequency bands, one around the tremor frequency [2–4 Hz] and another in [10–30 Hz] frequency (Weinberger & Dostrovsky, 2011). Park et al. (2011) report the presence of intermittent synchrony between STN neurons and its local field potentials (LFP), recorded using multiunit activity electrodes from PD patients undergoing deep brain stimulation (DBS) surgery (Park et al., 2011) which is absent in healthy controls.

One of the key objectives of the current study is to use a 2D spiking neuron model to understand and correlate STN–GPe's synchrony levels to exploration. As the second objective, we apply the above-mentioned model to the n-armed bandit problem of Daw, O'Doherty, Dayan, Seymour, and Dolan (2006) and Bourdaud, Chavarriaga, Galán, and del R Millan (2008) (Bourdaud et al., 2008; Daw et al., 2006) with the specific aim of studying the contributions of STN–GPe dynamics to exploration. The proposed model shares some aspects of classical RL-based approach to BG modeling. For example, dopamine signal is compared to reward prediction error (Schultz, 1998). Furthermore, DA is allowed to control cortico-striatal plasticity (Reynolds and Wickens 2002), modulate the gains of striatal neurons (Hadipour-Niktarash, Rommelfanger, Masilamoni, Smith, & Wichmann, 2012; Kliem, Maidment, Ackerson, Chen, Smith, & Wichmann, 2007), and influence the dynamics of STN–GPe by modulating the connections (Fan, Baufreton, Surmeier, Chan, & Bevan, 2012; Kreiss, Mastropietro, Rawji, & Walters, 1997).

6.2 Methods

6.2.1 Spiking Neuron Model of the Basal Ganglia

The network model of BG (Mandali, Rengaswamy, Chakravarthy, & Moustafa, 2015) described earlier was used to simulate the binary action selection and n-arm bandit task. For details of the model and its related equations, refer to earlier sections. The details of the tasks and the related measures are explained below.

6.2.2 *Binary Action Selection Task*

The first task we simulated was the simple binary action selection similar to Humphries, Stewart, and Gurney (2006), where two competing stimuli were presented to the model (Humphries et al., 2006). The input firing frequency is thought to represent 'saliency,' with higher frequencies representing higher salience (Humphries et al., 2006). The response of striatal output to cortical input falls in the range of a few tens of Hz (Sharott, Doig, Mallet, & Magill, 2012). Therefore, the frequencies that represent the 2 actions were assumed to be around 4 Hz (stimulus #1) and 8 Hz (stimulus #2). Spontaneous output firing rate of the striatal neurons (without input) is assumed to be around 1 Hz (Plenz & Kitai, 1998; Sharott et al., 2012). Selection of higher salient stimulus among the available choices could be considered as 'exploitation' while selecting the less salient one as 'exploration' (Sutton & Barto, 1998). So, the action selected is defined as 'Go' if stimulus #2 (more salient) is selected, 'Explore' if stimulus #1 (less salient) is selected, and 'NoGo' if none of them is selected.

The inputs were given spatially such that the neurons in the upper half of the lattice receive stimulus #1 and lower half the other (Fig. 6.1). The striatal outputs from D1 and D2 neurons of the striatum are given as input to GPi and GPe modules, respectively, with the projection pattern as shown in Fig. 6.1. Poisson spike trains corresponding to stimulus #1 were presented as input to neurons (1–1250) and were fully correlated among themselves. Similarly, Poisson spike trains corresponding to stimulus #2 were presented as input to neurons (1251–2500) and were fully correlated among themselves. Stimulus #1 and #2 are presented for an interval of 100 ms between 100 and 200 ms; at other times, uncorrelated spike trains at 1 Hz are presented to all the striatal neurons.

6.2.3 *The N-Armed Bandit Task*

We now describe the four-armed bandit task (Bourdaud et al., 2008; Daw et al., 2006) used to study exploratory and exploitatory behavior. In this experimental task, subjects were presented with four arms where one among them is to be selected in every trial for a total of 300 trials. The reward/payoff for each of these slots was obtained from a Gaussian distribution whose mean changes from trial to trial with payoff ranging from 0 to 100. The payoff, $r_{i,k}$ associated with the ith machine at the kth trial, was drawn from a Gaussian distribution of mean $\mu_{i,k}$ and standard deviation (SD) σ_0. The payoff was rounded to the nearest integer, in the range [0, 100]. At each trial, the mean is diffused according to a decaying Gaussian random walk. The trial was defined as an 'exploitatory' trial if highest reward giving arm was selected else defined as an 'exploratory' trial.

The payoffs generated by the slot machines are computed as follows,

$$\mu_{i,k+1} = \lambda_m \mu_{i,k} + (1 - \lambda_m)\theta_m + \mathrm{e} \tag{6.1}$$

$$r'_{i,k} \approx N(\mu_{i,k}, \sigma_0^2) \tag{6.2}$$

$$r_{i,k} = \mathrm{round}(r'_{i,k}) \tag{6.3}$$

where

$\mu_{i,k}$ is the mean of the Gaussian distribution with standard deviation (σ_0) for ith machine during k^{th} trial. λ_m and θ_m control the random walk of mean ($\mu_{i,k}$), and e $\sim N(0, \sigma_d^2)$ is obtained from Gaussian distribution of mean 0 and standard deviation σ_d. $r_{i,k}$ and $r'_{i,k}$ are the payoffs before and after rounding to nearest integer, respectively. The initial value of mean payoff, $\mu_{i,0}$, is set to a value of 50. All the values for the parameters λ_m, θ_m, σ_d, σ_0 were adapted from (Bourdaud et al., 2008).

To make an optimal decision, the subjects need to keep track of rewards associated with each of the four arms. The subject's decision to either Explore or exploit would depend on this internal representation which would closely resemble the actual payoff that is being obtained. It is quite difficult to identify whether the subject made an exploratory decision or an exploitative one just by observing the EEG and selected slot data. A subject-specific model is required to classify their decisions and identify the strategy (Bourdaud et al., 2008; Daw et al., 2006). Keeping this in mind, Bourdaud et al. (2008) used a 'behavioral model' that uses the softmax principle of RL to fit the selection pattern of human subjects. The parameter 'β' of the behavioral model was adjusted such that the final selection pattern matches that of individual subjects in the experiment (given below). The parameter 'β' which controls the exploration level in the behavioral model is tuned to match % exploitation obtained for each of the eight subjects (one subject's data were discarded because of artifacts); two out of the eight subjects had similar exploration levels. Hence, a total of six subjects' data are taken into account to check the performance of the proposed spiking BG model.

6.2.3.1 Behavioral Model (Adapted from Bourdaud et al. (2008))

The behavioral model labels each trial as corresponding to either an exploratory or exploitative decision. The model assumes that the user estimates the mean payoff of each machine using a Bayesian linear Gaussian rule (i.e., a Kalman filter). Using these estimations, he/she selects a machine according to a softmax rule. All the subjects are assumed to share the same model for tracking the payoff means, and thus, parameters are computed using the entire available data. The parameters of the model (for both mean tracking and machine selection) are estimated by maximizing the model likelihood with respect to the subject's choices.

At any given trial, the behavioral model provides the mean payoff for all machines considering previous observations (i.e., the payoff obtained at previous trials). Comparison between the model's estimated payoffs for all machines is used to label that trial as either exploration or exploitation. Those trials in which the user selects the machine with the highest estimated mean are labeled as corresponding to exploitative decisions.

The subject strategy for tracking the payoff of each machine is modeled by a Kalman filter, whose parameters are assumed to remain constant over trials. Once the *j*th machine is selected, at the *k*th trial, the estimated payoff distribution is updated from its preselection values $\left(\widehat{\mu}_{j,k}^{\text{pre}}, \left(\widehat{\sigma}_{j,k}^{\text{pre}}\right)^2\right)$ to its post-selection values $\left(\widehat{\mu}_{j,k}^{\text{post}}, \left(\widehat{\sigma}_{j,k}^{\text{post}}\right)^2\right)$ as follows

$$\widehat{\mu}_{j,k}^{\text{post}} = \widehat{\mu}_{j,k}^{\text{post}} + K_k\left(r_k - \widehat{\mu}_{j,k}^{\text{pre}}\right) \tag{6.4}$$

$$\left(\widehat{\sigma}_{j,k}^{\text{post}}\right)^2 = (1 - K_k)\left(\widehat{\sigma}_{j,k}^{\text{pre}}\right)^2 \tag{6.5}$$

where

$$(K_k) = \frac{\left(\widehat{\sigma}_{j,k}^{\text{pre}}\right)^2}{\left(\widehat{\sigma}_{j,k}^{\text{pre}}\right)^2 + (\widehat{\sigma}_0)^2} \tag{6.6}$$

The mean estimation for the remaining machines does not change as result of the choice since the user cannot observe the payoff of these machines. That is,

$$\forall i \neq j$$

$$\widehat{\mu}_{j,k}^{\text{post}} = \widehat{\mu}_{j,k}^{\text{pre}} \tag{6.7}$$

$$\widehat{\sigma}_{j,k}^{\text{post}} = \widehat{\sigma}_{j,k}^{\text{pre}} \tag{6.8}$$

Then, the estimations are also evolved according to the diffusion rule:

$$\widehat{\mu}_{j,k+1}^{\text{pre}} = \widehat{\lambda}\widehat{\mu}_{j,k}^{\text{post}} + (1 - \widehat{\lambda})\widehat{\theta} \tag{6.9}$$

$$\left(\mu_{j,k+1}^{\prime\text{pre}}\right)^2 = \widehat{\lambda}^2\left(\sigma_{j,k}^{\prime\text{post}}\right)^2 + \sigma_d^2 \tag{6.10}$$

The choice of subjects is modeled by a softmax rule; i.e., at each trial *k*, the probability of choosing the machine is

Table 6.1 Estimation of parameters of the behavioral model (Bourdaud et al., 2008)

	λ	θ	σ_d	σ_0
Real values	0.9836	50	2.8	4
Estimated values	0.92	51.37	8.12	N/A
Subject	1 2	3 4	5 6	7 8
β	0.37	0.19	0.19	0.29
	0.28	0.21	0.29	0.23

$$P_{i,k} = \frac{\exp\left(\beta \widehat{\mu}_{i,k}^{\text{pre}}\right)}{\sum_j \exp\left(\beta \widehat{\mu}_{j,k}^{\text{pre}}\right)} \tag{6.11}$$

where 'β' is a scaling parameter. Higher values of β drive the system to exploitative behavior and vice versa. The parameters of the behavioral model $\left(\sigma_0, \widehat{\theta}, \widehat{\lambda}, \widehat{\sigma}_d\right)$ are estimated by maximizing the log likelihood under the following constraints. To speed up convergence, estimated parameters $\left(\sigma, \widehat{\mu}_{j,0}^{\text{pre}} \,\&\, \widehat{\sigma}_{j,0}^{\text{pre}}\right)$ are initialized to the parameters of the original model $(\sigma_0, \mu_{j,0} \,\&\, \sigma_{j,0})$, respectively. Fixing the last two parameters does not significantly affect the estimation of the others, because their influence vanishes quickly within a few trials. Table 6.1 shows the estimated values of the model, which are consistent with the real values of the machines.

6.2.3.2 Strategy for Slot Machine Selection

To simulate the experiment, we utilized the concepts of RL and combined the dynamics of BG model to select an optimally rewarding slot in each trial. Experimental data show that BG receives reward-related information in the form of dopaminergic input to striatum (Chakravarthy et al., 2010; Niv, 2009). Cortico-striatal plasticity changes due to dopamine (Reynolds & Wickens, 2002) were incorporated in the model by allowing DA signals to modulate the Hebb-like plasticity of cortico-striatal synapses (Surmeier, Ding, Day, Wang, & Shen, 2007).

The architecture of the proposed network model is depicted in Fig. 6.1. The output of striatum (both D1 and D2 parts) was divided equally into four quadrants which receive input from corresponding stimulus. The stimuli are associated with 2 weights $\left(w_{i,0}^{\text{D1}}, w_{i,0}^{\text{D2}}\right)$ initialized with equal value of 50 which represent the cortico-striatal weights of D1 and D2 MSNs in the striatum. Each of the cortico-striatal weights represents the saliency (in terms of striatal spike rate) for that corresponding arm. These output spikes generated from each of the D1 and D2 striatum project to GPi and GPe, respectively. The final selection of an arm is made as in Sect. 6.2.4. The reward $r_{i,k}$ received for the selected slot was sampled from Gaussian distribution with mean $\mu_{i,k}$ and SD (σ_0) (Eq. 6.3).

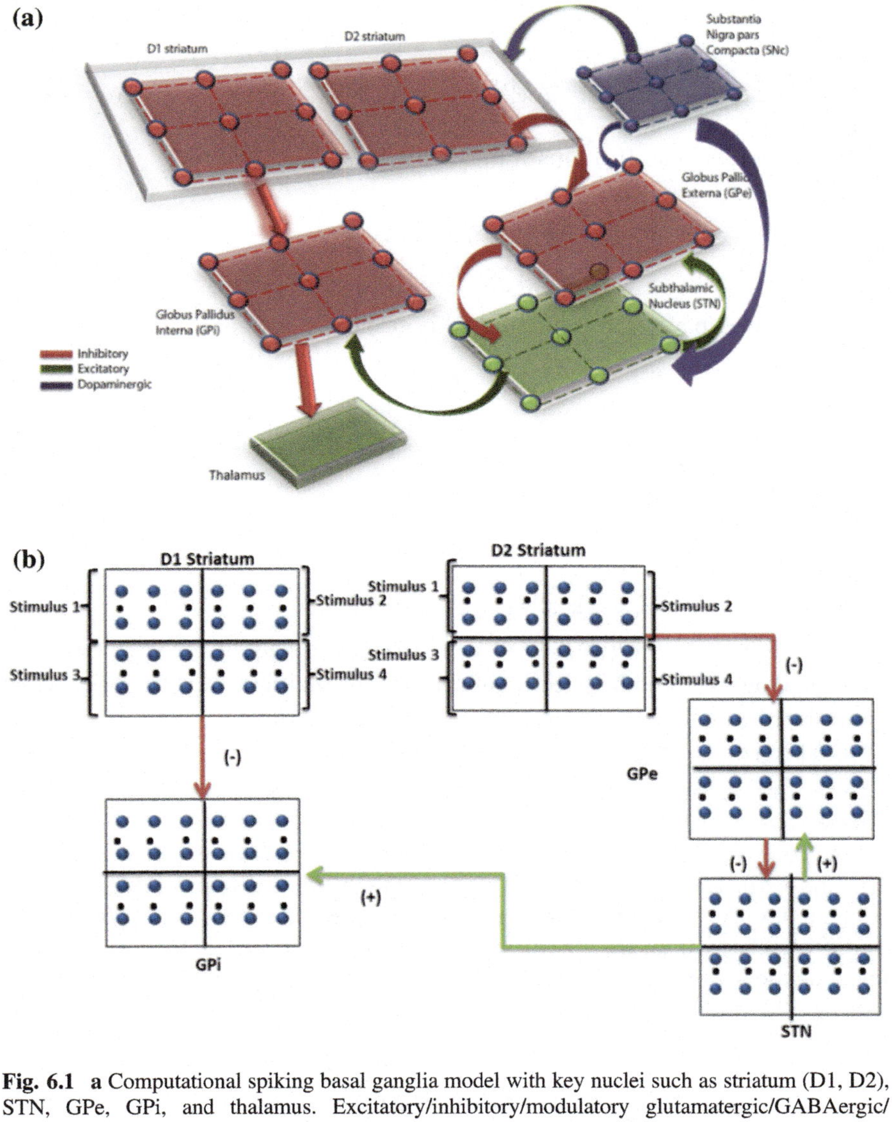

Fig. 6.1 **a** Computational spiking basal ganglia model with key nuclei such as striatum (D1, D2), STN, GPe, GPi, and thalamus. Excitatory/inhibitory/modulatory glutamatergic/GABAergic/dopaminergic projections are shown by green/red/violet arrows. **b** The BG model and the regions within each nuclei corresponding to the four decks are indicated

Utilizing the reward obtained for the input 'i' and trial 'k', the expected value of the slots, inputs to D1 and D2 striatum are updated using the following equations,

$$\Delta w^{\mathrm{D1}}_{i,k+1} = \eta \delta_k x^{\mathrm{inp}}_{i,k} \tag{6.12}$$

$$\Delta w^{\mathrm{D2}}_{i,k+1} = -\eta \delta_k x^{\mathrm{inp}}_{i,k} \tag{6.13}$$

The expected value (V_k) for *k*th trial is calculated as

$$V_k = \sum_{i=1}^{4} w^{\mathrm{D1}}_{i,k} * x^{\mathrm{inp}}_{i,k} \tag{6.14}$$

The received payoff (Re_k) for *k*th trial is calculated as

$$\mathrm{Re}_k = \sum_{i=1}^{4} r_{i,k} * x^{\mathrm{inp}}_{i,k} \tag{6.15}$$

The error (δ) for *k*th trial is defined as

$$\delta_k = \mathrm{Re}_k - V_k \tag{6.16}$$

where $w^{\mathrm{D1}}_{i,k}$ are the cortico-striatal weights of D1 striatum for *i*th machine in *k*th trial, $w^{\mathrm{D2}}_{i,k}$ are the cortico-striatal weights of D2 striatum for *i*th machine for *k*th trial, $r_{i,k}$ is the reward obtained for the selected *i*th machine for *k*th trial, $x^{\mathrm{inp}}_{i,k}$ is the binary input vector representing the four slot machines, e.g., if the first slot machine is selected $x^{\mathrm{inp}}_{i,k}$ = [1 0 0 0], η (=0.3) is the learning rate of D1 and D2 striatal MSNs, Re_k is the received payoff for selected slot for *k*th trial, and V_k is the expected value for selected slot for *k*th trial.

The cortico-striatal weights are updated (Eqs. 6.12 and 6.13) using the error term 'δ' (Eq. 6.16). The reward-related information in the form of dopaminergic input to striatum has been correlated to the error (δ) (Chakravarthy et al., 2010; Niv, 2009). The δ calculated from Eq. (6.16) has both positive and negative values with no upper and lower boundaries but the working DA range in the model was limited to small positive values (0.1–0.9). Hence, a mapping from δ to DA is defined as follows:

$$\mathrm{DA} = \mathrm{sig}(\lambda * \delta_k) \tag{6.17}$$

where

DA is the dopamine signal within range of 0.1–0.9, λ is the slope of sigmoid (=0.2), δ_k is the error obtained for *k*th trial (Eq. 6.16), and sig () is the sigmoid function.

6.2.4 *Measures*

6.2.4.1 Synchronization

The phenomenon of neural synchrony has attracted the attention of many computational and experimental neuroscientists in the recent decades (Hauptmann & Tass, 2007; Kumar, Cardanobile, Rotter, & Aertsen, 2011; Park et al., 2011; Pinsky & Rinzel, 1995; Plenz & Kital, 1999). It is believed that partial synchrony helps in the generation of various EEG rhythms such as alpha and beta (Izhikevich, 2007). Studying synchrony in neural networks has been gaining importance due to its presence in normal functioning (coordinated movement of the limbs) and in pathological states (e.g., synchronized activity of CA3 neurons in the hippocampus during an epileptic seizure) (Pinsky & Rinzel, 1995). Plenz and Kital (1998) proposed that STN–GPe might act as a pacemaker (Plenz & Kital, 1999), a source for generating oscillations in pathological conditions such as Parkinson's disease. Park et al. (2011) report the presence of intermittent synchrony between STN neurons and its local field potentials (LFP), recorded using multiunit activity electrodes from PD patients undergoing DBS surgery (Park et al., 2011). They also calculated the duration of synchronized and desynchronized events in neuronal activity by estimating transition rates, which were obtained with the help of first return maps plotted using phase of neurons (Park et al., 2010, 2011). To observe how dopamine changes synchrony in STN–GPe, we calculated the phases of individual neurons as defined in (Pinsky & Rinzel, 1995).

The phase of *j*th neuron was calculated as follows:

$$\emptyset_j(t) = 2 * \pi * \frac{\left(T_{j,k} - t_{j,k}\right)}{\left(t_{j,k+1} - t_{j,k}\right)} \tag{6.18}$$

$$R^{\text{sync}}(t) * \mathrm{e}^{i\theta(t)} = \frac{1}{N}\sum_{j=1}^{N} \mathrm{e}^{i\emptyset_j(t)} \tag{6.19}$$

where

$t_{j,k}$ and $t_{j,k+1}$ are the onset times of *k*th and $k + 1$th spike of the *j*th neuron $T_{j,k} \in \left[t_{j,k}, t_{j,k+1}\right]$, $\emptyset_j(t)$ = phase of *j*th neuron at time '*t*', R^{sync} is the synchronization measure $0 \le R^{\text{sync}} \le 1$, θ = average phase of neurons, N = total number of neurons in the network.

6.2.5 *Action Selection Using the Race Model*

Action selection is modulated by BG output nucleus GPi which projects back to the cortex via the thalamus. We have used the race model (Vickers, 1970) for the final action selection where an action is selected when temporally integrated neuronal activity of the output neurons crosses a threshold (Frank, 2006; Frank, Samanta, Moustafa, & Sherman, 2007; Humphries, Khamassi, & Gurney, 2012).

The dynamics of the thalamic neurons is as follows:

$$\frac{dz_k(t)}{dt} = -z_k(t) + f_{\mathrm{Gpik}}(t) \tag{6.20}$$

$$\begin{aligned} f'_{\mathrm{Gpik}} &= \frac{1}{(N*N)/k}\sum_{t=1}^{T}\left(\sum_{i=1}^{N}\sum_{j=1}^{N/k} S_{ij}^{\mathrm{GPik}}(t)\right) \\ f_{\mathrm{GPik}} &= \frac{f_{\mathrm{GPi}}^{\max} - f'_{\mathrm{Gpik}}}{f_{\mathrm{GPi}}^{\max}} \end{aligned} \tag{6.21}$$

where

z_k (t) = integrating variable for kth stimulus, f_{GPik} (t) = normalized and reversed average firing frequency of GPi neurons receiving kth stimulus from striatum, $f_{\mathrm{GPi}}^{\max}$ = highest firing rate among the GPi neurons, S_{ij}^{Gpik} = neuronal spikes of GPi neurons receiving kth stimulus, N = number of neurons in a single row/column of GPi array (=50), and T = duration of simulation.

The first neuron (z_k) among k stimuli to cross the threshold (=0.15) represents the action selected. All the variables representing neuron activity are reset immediately after each action selection.

6.3 Results

We start with results of neural dynamics (STN–GPe) as a function of DA and then present with decision-making results.

6.3.1 *Neural Dynamics*

Pathological oscillations of STN and GP have been associated with various PD symptoms (Brown, 2003; Plenz & Kital, 1999). Correlated neural firing patterns in STN and GPi can be seen in both experimental conditions of dopamine depletion and in Parkinsonian conditions. In the present model, we show increased synchronized behavior under conditions of reduced dopamine, resembling the situation in dopamine-deficient conditions of Parkinson's disease. The effect of DA on the synchronization of STN and GPe neurons was studied by estimating the values of $R_{\mathrm{STN}}^{\mathrm{sync}}$, $R_{\mathrm{GPe}}^{\mathrm{sync}} R_{\mathrm{STNGPe}}^{\mathrm{sync}}$ for increasing values of DA (0.1–0.9).

The three 'R^{sync}' (Eq. 6.19) values showed a decrease in amplitude with an increase in DA level (Fig. 6.2a–c). Under low DA conditions, GPe activity follows STN activity (Plenz & Kital, 1999), thus forming a pacemaker kind of circuit, which could be the source of STN–GPe oscillations Fig. 6.2d. One of the suspected reasons of bursting activity in STN is the decreased inhibition from GPe neurons (Plenz & Kital, 1999) at low DA levels. This feature is captured by the model since GPe firing rates are smaller for lower DA levels. The STN neurons showed oscillations around the frequency of 10 Hz at low DA but were absent at high DA level (Kang & Lowery, 2013).

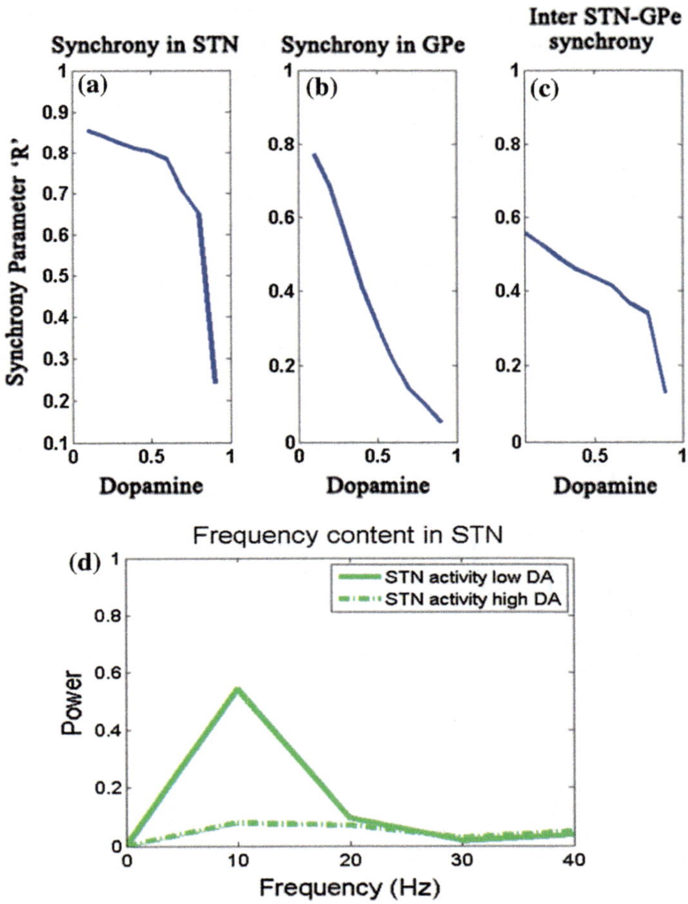

Fig. 6.2 Change in the three synchronization values R_{STN}^{sync} (**a**), R_{GPe}^{sync} (**b**) and R_{STNGPe}^{sync} (**c**) oscillatory activity in STN neurons (**d**) frequency content with the value of DA (0.1–0.9). Simulations show reduced synchronization within STN and GPe networks, and also between STN and GPe networks, as DA is increased

6.3.2 Decision Making

After the model's performance was quantified at neural level, we studied the role of BG in decision making using two tasks especially in explorative and exploitative dynamics. This work is in continuation to our earlier hypothesis that the source for exploration comes from STN–GPe dynamics (Kalva et al., 2012). The first task was a simple binary action selection similar to Humphries et al., (2006), where two competing stimuli were presented to the model. The input firing frequency is thought to represent 'saliency,' with higher frequencies representing higher salience. Selection of stimulus with the higher salience between the two available choices could be considered as 'exploitation' while selecting the less salient one as 'exploration' (Sutton & Barto, 1998). So the action selected is defined as 'Go' if stimulus #2 (more salient) is selected, 'Explore' if stimulus #1 (less salient) is selected, and 'NoGo' if none of them is selected. Simulations were run for 100 trials, and the percentage of actions selected under each regime (Go, Explore, and NoGo) was calculated for dopamine levels ranging from low (0.1) to high (0.9) (Fig. 6.3). We may note that the probability of NoGo, where no action is selected, decreases with increase in dopamine; probability of Go increases with dopamine; the peak of exploration is found at intermediate levels of dopamine (Fig. 6.3). The range of DA where a peak in exploration was observed is the same where STN and GPe network showed chaotic activity.

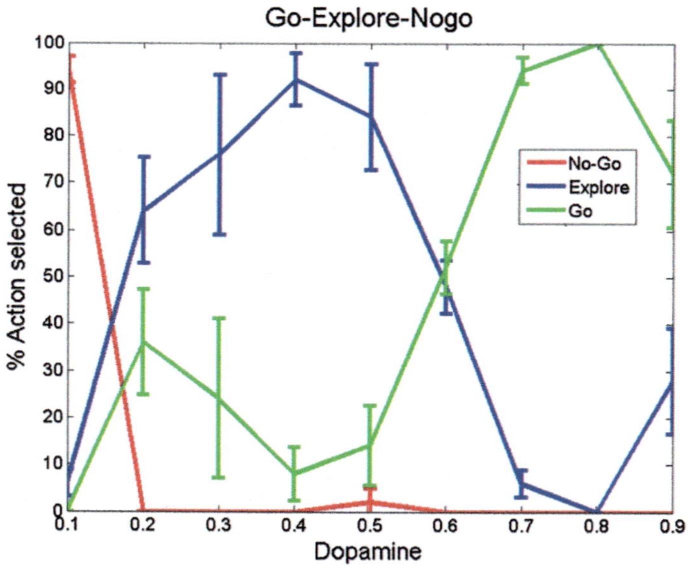

Fig. 6.3 Percentage of action selection observed in the Go, NoGo, and Explore regimes averaged over 200 trials with DP and IP weight values at $w_{STN \rightarrow GPi} = 1.15$ & $w_{Str \rightarrow GPi} = 0.8$. We ran the simulation for 100 trials and segmented into 4 bins (25 trials each). We then calculated the variance of each regime across all DA levels

The second task was a four-armed bandit task (Bourdaud et al., 2008; Daw et al., 2006) which is similar to a real-world decision-making scenario. In this task, the subjects are presented with four arms where one among them is to be selected in every trial for a total of 300 trials. The reward/payoff for each of these slots was obtained from a Gaussian distribution whose mean changes from trial to trial with payoff ranging from 0 to 100. The model's performance (% exploitation) was compared with behavioral model, which represents the experimental data in the n-armed bandit task (Fig. 6.4). The parameter 'β' of the behavioral model which controls the Exploit–Explore balance was adjusted to match the performance of individual subjects in the experiment. Exploration in the model can be obtained by either increasing the IP weight (influence from STN) or decreasing DP weight (influence from striatum).

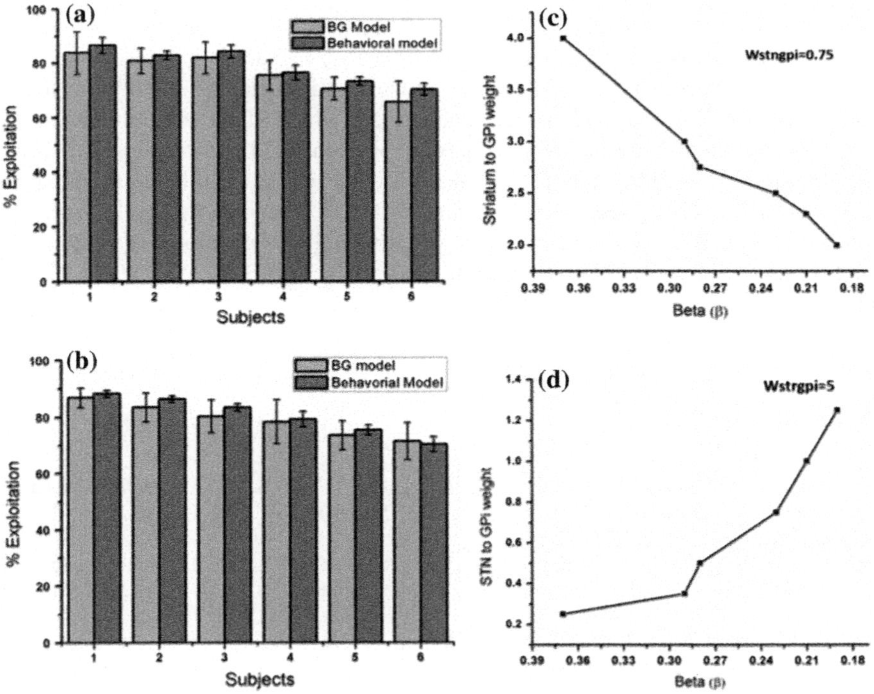

Fig. 6.4 Compares the performance of BG model with the behavioral model. **a** The percentage exploitation obtained for each of the six subjects from BG and behavioral model. The relationship between betas (β) of the behavioral model and DP weights ($w_{Str \to GPi}$) with a constant $w_{STN \to GPi}$ value (=0.75) used to attain (**a**) are shown in (**c**). **b** The relationship between betas (β) of the behavioral model and IP weights ($w_{STN \to GPi}$) of BG model with a constant $w_{Str \to GPi}$ value of (=5) used to attain (**b**) are shown in (**d**). *Y*-axis represents percentage exploitation, and *X*-axis represents a subject which is a specific beta value (β) in behavioral model and the IP or DP weight in the BG model

6.4 Discussion

The synchrony results tally with the general observation from electrophysiology that at higher levels of dopamine, the STN–GPe system shows desynchronized activity and under dopamine-deficient conditions of PD exhibits synchronized bursts (Bergman et al., 1994; Gillies, Willshaw, Gillies, & Willshaw, 1998; Park et al., 2011). We observed that STN activity showed oscillatory activity with a frequency (=10 Hz) which falls under the beta frequency range observed in experimental PD study (Weinberger & Dostrovsky, 2011). One of the aims of the present work is also to show that the complex dynamics of STN–GPe system contributes to exploration. To this end, we first simulated the binary action selection task [similar to Humphries et al., (2006)] where saliency was coded in the firing rate. The selection of higher one was defined as 'exploitation/Go' and lesser one as 'exploration/Explore' and not selecting any of the inputs as 'NoGo'. The model showed NoGo at low DA levels (0.1–0.3) and Go at high DA levels (0.7–0.9) consistent with the classical picture of BG function. Along with this, a peak in 'Explore' at intermediate levels of DA (0.4–0.6) was also observed (Fig. 6.3). To check whether any other module in the network is influencing exploration in the system, we removed the STN to GPi connection (which effectively eliminated the IP). This omission rendered the system to display only Go and NoGo regimes (no exploration, results not included). We then moved to simulating the n-armed bandit task, where the performance of model was compared with experimental result. The results obtained from BG model closely match with the behavioral model (Fig. 6.4) reinforcing the idea that STN–GPe could be a source for exploration at subcortical level.

References

Bergman, H., Wichmann, T., Karmon, B., & DeLong, M. (1994). The primate subthalamic nucleus. II. Neuronal activity in the MPTP model of parkinsonism. *Journal of Neurophysiology, 72*(2), 507–520.

Bevan, M. D., Magill, P. J., Terman, D., Bolam, J. P., & Wilson, C. J. (2002). Move to the rhythm: Oscillations in the subthalamic nucleus–external globus pallidus network. *Trends in Neurosciences, 25*(10), 525–531.

Borisyuk, G. N., Borisyuk, R. M., Khibnik, A. I., & Roose, D. (1995). Dynamics and bifurcations of two coupled neural oscillators with different connection types. *Bulletin of Mathematical Biology, 57*(6), 809–840.

Bourdaud, N., Chavarriaga, R., Galán, F., & del R Millan, J. (2008). Characterizing the EEG correlates of exploratory behavior. *IEEE Transactions on Neural Systems and Rehabilitation Engineering, 6*(6), 549–556.

Brown, P. (2003). Oscillatory nature of human basal ganglia activity: Relationship to the pathophysiology of Parkinson's disease. *Movement Disorders, 18*(4), 357–363.

Chakravarthy, V., Joseph, D., & Bapi, R. S. (2010). What do the basal ganglia do? A modeling perspective. *Biological Cybernetics, 103*(3), 237–253.

Daw, N. D., O'Doherty, J. P., Dayan, P., Seymour, B., & Dolan, R. J. (2006). Cortical substrates for exploratory decisions in humans. *Nature, 441*(7095), 876–879.

Fan, K. Y., Baufreton, J., Surmeier, D. J., Chan, C. S., & Bevan, M. D. (2012). Proliferation of external globus pallidus-subthalamic nucleus synapses following degeneration of midbrain dopamine neurons. *The Journal of Neuroscience, 32*(40), 13718–13728.

Frank, M. J. (2006). Hold your horses: A dynamic computational role for the subthalamic nucleus in decision making. *Neural Networks, 19*(8), 1120–1136.

Frank, M. J., Samanta, J., Moustafa, A. A., & Sherman, S. J. (2007). Hold your horses: Impulsivity, deep brain stimulation, and medication in parkinsonism. *Science, 318*(5854), 1309–1312.

Gillies, A., Willshaw, D., Gillies, A., & Willshaw, D. (1998). A massively connected subthalamic nucleus leads to the generation of widespread pulses. *Proceedings of the Royal Society of London, Series B: Biological Sciences, 265*(1410), 2101–2109.

Gupta, A., Balasubramani, P. P., & Chakravarthy, V. S. (2013). Computational model of precision grip in Parkinson's disease: A utility based approach. *Frontiers in Computational Neuroscience, 7*.

Hadipour-Niktarash, A., Rommelfanger, K. S., Masilamoni, G. J., Smith, Y., & Wichmann, T. (2012). Extrastriatal D2-like receptors modulate basal ganglia pathways in normal and parkinsonian monkeys. *Journal of Neurophysiology, 107*(5), 1500–1512.

Hammond, C., Bergman, H., & Brown, P. (2007). Pathological synchronization in Parkinson's disease: Networks, models and treatments. *Trends in Neurosciences, 30*(7), 357–364.

Hauptmann, C., & Tass, P. A. (2007). Therapeutic rewiring by means of desynchronizing brain stimulation. *Biosystems, 89*(1), 173–181.

Humphries, M. D., Khamassi, M., & Gurney, K. (2012). Dopaminergic control of the exploration-exploitation trade-off via the basal ganglia. *Frontiers in Neuroscience, 6*.

Humphries, M. D., Stewart, R. D., & Gurney, K. N. (2006). A physiologically plausible model of action selection and oscillatory activity in the basal ganglia. *The Journal of Neuroscience, 26* (50), 12921–12942.

Izhikevich, E. M. (2007). *Dynamical systems in neuroscience*. Cambridge: The MIT press.

Kalva, S. K., Rengaswamy, M., Chakravarthy, V. S., & Gupte, N. (2012). On the neural substrates for exploratory dynamics in basal ganglia: A model. *Neural Networks, 32,* 65–73. https://doi.org/10.1016/j.neunet.2012.02.031.

Kang, G., & Lowery, M. M. (2013). Interaction of oscillations, and their suppression via deep brain stimulation, in a model of the cortico-basal ganglia network. *IEEE Transactions on Neural Systems and Rehabilitation Engineering, 21*(2), 244–253.

Kliem, M. A., Maidment, N. T., Ackerson, L. C., Chen, S., Smith, Y., & Wichmann, T. (2007). Activation of nigral and pallidal dopamine D1-like receptors modulates basal ganglia outflow in monkeys. *Journal of Neurophysiology, 98*(3), 1489–1500.

Kreiss, D. S., Mastropietro, C. W., Rawji, S. S., & Walters, J. R. (1997). The response of subthalamic nucleus neurons to dopamine receptor stimulation in a rodent model of Parkinson's disease. *The Journal of Neuroscience, 17*(17), 6807–6819.

Krishnan, R., Ratnadurai, S., Subramanian, D., Chakravarthy, V. S., & Rengaswamy, M. (2011). Modeling the role of basal ganglia in saccade generation: Is the indirect pathway the explorer? *Neural Networks, 24*(8), 801–813.

Kumar, A., Cardanobile, S., Rotter, S., & Aertsen, A. (2011). The role of inhibition in generating and controlling Parkinson's disease oscillations in the basal ganglia. *Frontiers in Systems Neuroscience, 5*.

Magdoom, K., Subramanian, D., Chakravarthy, V. S., Ravindran, B., Amari, S.-I., & Meenakshisundaram, N. (2011). Modeling basal ganglia for understanding Parkinsonian reaching movements. *Neural Computation, 23*(2), 477–516.

Mandali, A., Rengaswamy, M., Chakravarthy, S., & Moustafa, A. A. (2015). A spiking Basal Ganglia model of synchrony, exploration and decision making. *Frontiers in Neuroscience, 9,* 191.

Muralidharan, V., Balasubramani, P. P., Chakravarthy, V. S., Lewis, S. J., & Moustafa, A. A. (2013). A computational model of altered gait patterns in parkinson's disease patients negotiating narrow doorways. *Frontiers in Computational Neuroscience, 7.*

Niv, Y. (2009). Reinforcement learning in the brain. *Journal of Mathematical Psychology, 53*(3), 139–154.

Park, C., Worth, R. M., & Rubchinsky, L. L. (2010). Fine temporal structure of beta oscillations synchronization in subthalamic nucleus in Parkinson's disease. *Journal of Neurophysiology, 103*(5), 2707–2716.

Park, C., Worth, R. M., & Rubchinsky, L. L. (2011). Neural dynamics in parkinsonian brain: the boundary between synchronized and nonsynchronized dynamics. *Physical Review E, 83*(4), 042901.

Pinsky, P. F., & Rinzel, J. (1995). Synchrony measures for biological neural networks. *Biological Cybernetics, 73*(2), 129–137.

Plenz, D., & Kitai, S. T. (1998). Up and down states in striatal medium spiny neurons simultaneously recorded with spontaneous activity in fast-spiking interneurons studied in cortex–striatum–substantia nigra organotypic cultures. *The Journal of Neuroscience, 18*(1), 266–283.

Plenz, D., & Kital, S. T. (1999). A basal ganglia pacemaker formed by the subthalamic nucleus and external globus pallidus. *Nature, 400*(6745), 677–682.

Reynolds, J. N. J., & Wickens, J. R. (2002). Dopamine-dependent plasticity of corticostriatal synapses. *Neural Networks, 15*(4), 507–521.

Schultz, W. (1998). Predictive reward signal of dopamine neurons. *Journal of Neurophysiology, 80*(1), 1–27.

Sharott, A., Doig, N. M., Mallet, N., & Magill, P. J. (2012). Relationships between the firing of identified striatal interneurons and spontaneous and driven cortical activities in vivo. *The Journal of Neuroscience, 32*(38), 13221–13236.

Sinha, S. (1999). Noise-free stochastic resonance in simple chaotic systems. *Physica A: Statistical Mechanics and its Applications, 270*(1), 204–214.

Sukumar, D., Rengaswamy, M., & Chakravarthy, V. S. (2012). Modeling the contributions of Basal ganglia and Hippocampus to spatial navigation using reinforcement learning. *PLoS ONE, 7*(10), e47467.

Surmeier, D. J., Ding, J., Day, M., Wang, Z., & Shen, W. (2007). D1 and D2 dopamine-receptor modulation of striatal glutamatergic signaling in striatal medium spiny neurons. *Trends in Neurosciences, 30*(5), 228–235.

Sutton, R. S., & Barto, A. G. (1998). *Reinforcement learning: An introduction* (Vol. 1). UK: Cambridge University Press.

Tachibana, Y., Iwamuro, H., Kita, H., Takada, M., & Nambu, A. (2011). Subthalamo-pallidal interactions underlying parkinsonian neuronal oscillations in the primate basal ganglia. *European Journal of Neuroscience, 34*(9), 1470–1484.

Vickers, D. (1970). Evidence for an accumulator model of psychophysical discrimination. *Ergonomics, 13*(1), 37–58.

Weinberger, M., & Dostrovsky, J. O. (2011). A basis for the pathological oscillations in basal ganglia: The crucial role of dopamine. *NeuroReport, 22*(4), 151.

Chapter 7
A Basal Ganglia Model of Freezing of Gait in Parkinson's Disease

Vignesh Muralidharan, Pragathi Priyadharsini Balasubramani, V. Srinivasa Chakravarthy and Ahmed A. Moustafa

Abstract Freezing of gait (FOG) is a mysterious clinical phenomenon seen in Parkinson's disease (PD) patients, a neurodegenerative disorder of the basal ganglia (BG), where there is cessation of locomotion under specific contexts. These contexts could include motor initiation, i.e., when starting movement, passing through narrow passages and corridors, while making a turn and as they are about to reach a destination. We have developed computational models of the BG which explains the freezing behavior seen in PD. The model uses reinforcement learning framework, incorporating Actor–Critic architecture, to aid learning of a virtual subject to navigate through these specific contexts. The model captures the velocity changes (slowing down) seen in PD freezers upon encountering a doorway, turns, and under the influence of cognitive load compared to PD non-freezers and healthy controls. The model throws interesting predictions about the pathology of freezing suggesting that dopamine, a key neurochemical deficient in PD, might not be the only reason for the occurrences of such freeze episodes. Other neuromodulators which are involved in action exploration and risk sensitivity influence these motor arrests. Finally, we have incorporated a network model of the BG to understand the network level parameters which influence contextual motor freezing.

7.1 Introduction

Parkinson's disease (PD) is a neurodegenerative disorder which arises due to a loss of dopaminergic neurons in the brain, specifically in a region called substantia nigra pars compacta (SNc), an integral part of a subcortical circuit called the basal ganglia (BG) (Hughes, Daniel, Kilford, & Lees, 1992). The BG system is involved in several crucial functions such as reward-based learning, action selection, motor preparation, and motor planning (Chakravarthy, Joseph, & Bapi, 2010). The BG circuit is part of a reward-processing system where the neuromodulator, dopamine, plays a significant role. Dopamine deficiency due to loss of SNc cells in PD manifests as a wide variety of motor, cognitive, and affective symptoms.

One of the defining features of PD pathology is the presence of gait disorders, where there is impairment in locomotion in a subset of the patients (Shine,

V. S. Chakravarthy and A. Moustafa, *Computational Neuroscience Models of the Basal Ganglia*, Cognitive Science and Technology,
https://doi.org/10.1007/978-981-10-8494-2_7

Moustafa, Matar, Frank, & Lewis, 2013). Parkinsonian gait is often characterized by shuffling steps, also called festinating gait. Several spatiotemporal changes are observed in PD gait including reduced walking speeds, lower step lengths, increased cadence (steps/time), and higher double-support times (time during both feet are touching the ground) (Hughes et al., 1990; Morris, Iansek, Matyas, & Summers, 1998). In addition to these features, a more debilitating aspect of PD gait is known as freezing of gait (FOG) (Giladi et al., 2001). It is an episodic phenomenon where there is a sudden and paroxysmal cessation of locomotion, often triggered by certain environmental contexts which include approaching narrow doorways/passages, turns, and also during movement-related scenarios such as movement initiation and dual-tasking (Schaafsma et al., 2003). One consequence of freezing is the increased rate of falling and an overall decrease in quality of life (Latt, Lord, Morris, & Fung, 2009). There have been several hypotheses which try to explain the factors contributing to freezing behavior, ranging from a defect in lower level areas such as spinal mechanisms leading to improper gait execution (Chee, Murphy, Danoudis, Georgiou-Karistianis, & Iansek, 2009) to improper communication among different cortico-subcortical (mainly cortico-basal ganglia) networks ultimately leading to motor arrests (Lewis & Barker, 2009).

The contextual triggers for freezing which usually include turning environments, and narrow doorways, among others suggest that there might be an inherent problem of evaluating space underlying FOG. We have modeled the cortico-BG network to understand the influence of the specific contexts that lead to FOG (Muralidharan et al., 2017; Muralidharan, Balasubramani, Chakravarthy, Lewis, & Moustafa, 2014). Thus, modeling freezing is approached as a problem where the environment being navigated by an agent (simulated subject) is successfully negotiated by building a function which computes the 'value of space.' The meaning of this 'value of space' may be explained as follows. If a region of space in the proximity of an animal has high value, it means that moving into that space is beneficial, safe, or rewarding to that animal. A spatial region with low value means that it is probably unsafe or unrewarding. Thus, the notion of evaluation of space, which is explored in this chapter in the context of Parkinsonian gait, is proposed as a key guiding principle underlying spatial navigation. This value of space is a notion borrowed from reinforcement learning literature where it is called the value function, which estimates the expected reward from a given state (Sutton & Barto, 1998). Thus by building a value function over space, the motor machinery would be able to use this information to guide the agent successfully through a given environment. The different scenarios considered for modeling include understanding the influence of doorways and narrow passages on gait, studying the role of cognition in freezing, influence of turning on gait. Importantly, freezing is known not to be limited just to gait movements and could be a global movement deficit affecting hand movement and speech (Nieuwboer et al., 2009; Park et al., 2009, 2014; Ricciardi et al., 2016). Accordingly, we also developed a network model of the cortico-BG network to study freezing of arm movements.

7.2 Motivation, Objective, and Scope

In this chapter, we aim to achieve the following objectives:

- Build a cortico-BG-spinal model, taking into account the different levels of motor control, to understand the influence of environment contexts like doorways, narrow passages, and turns on the gait of PD subjects.
- Understand the role of cognition on FOG, by considering the interaction among multiple cortico-BG loops (motor and cognitive loops) and elucidate the motor arrests seen during dual-tasking.
- Extend the existing approach to a network model of the cortico-BG circuitry and study the network parameters that lead to freezing.
- Elucidate the neurobiological correlates of key model parameters, to gain better insights into the mechanism of FOG.

7.3 Methods and Results

7.3.1 The Influence of Doorways on FOG

The proposed model (Fig. 7.1) has two stages of control: (1) the higher level of control representing the cortico-basal ganglia system and (2) the spinal level central pattern generators (CPGs) that translate the higher level gait commands such as velocity into gait rhythm. The BG model is essentially simulated using the Actor–Critic architecture, with the difference that the Actor is modeled by the Go/Explore/NoGo (GEN) model (Chakravarthy et al., 2010; Kalva, Rengaswamy, Chakravarthy, & Gupte, 2012; Magdoom et al., 2011). The spinal CPGs are modeled by networks of Hopf oscillators (Righetti, Buchli, & Ijspeert, 2006). Our model is used to simulate the results of two PD gait studies (Almeida & Lebold, 2010; Cowie, Limousin, Peters, & Day, 2010). The model simulates the approach of a subject to a doorway and computes the velocity profile along the track leading to the doorway. The agent repeatedly approaches a doorway, walking along a short track. The agent aims at passing through the doorway without bumping into the sides of the doorway. Due to the well-known trade-off between accuracy and speed in motor function (Bradshaw & Sparrow, 2000; Duarte & Latash, 2007; MacKay, 1982), rapid approaches to the doorway are more likely to result in a collision. Therefore, in our model, the agent learns to reduce its speed in the vicinity of the doorway, which it does using RL mechanisms.

The Critic in the network approximates the value, $V(t)$, which is a function of the view vector, ϕ, as

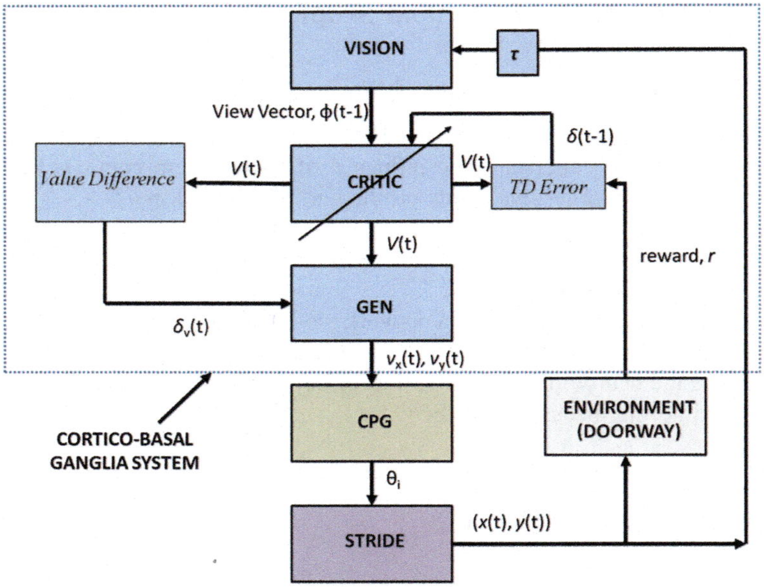

Fig. 7.1 Block diagram detailing the cortico-basal ganglia system and the central pattern generator module used in our study. The arrow on the critic represents the module training. The figure also projects the cortico-BG system, CPG, and locomotor apparatus in the shades of blue, brown, and violet, respectively

$$V(t) = \tanh\left(\sum W_{\mathrm{i}}(t)\phi_{\mathrm{i}}(t)\right) \tag{7.1}$$

where W are the weights updated using the temporal difference (TD) error $\delta = r(t) + \gamma V(t) - V(t-1)$ which is a correlate of dopamine signaling (Schultz, 2010). Here $r(t)$ is the reward obtained at time t, and γ is the discount factor. The Actor/GEN policy performs stochastic hill-climbing over the value function using gradient information also called value difference ($\delta_{\mathrm{V}} = V(t) - V(t-1)$)

$$\Delta X(t) = A_{\mathrm{G}}\mathrm{sig}(\lambda_{\mathrm{G}}\delta_{\mathrm{V}})\,\Delta X(t-1) + A_{\mathrm{E}}\chi\exp(-\delta_{\mathrm{V}}^2/\sigma_{\mathrm{E}}^2) - A_{\mathrm{N}}\mathrm{sig}(\lambda_{\mathrm{N}}\delta_{\mathrm{V}})\,\Delta X(t-1) \tag{7.2}$$

ΔX represents the velocity of the agent which the GEN updates using δ_{V}. Here δ_{V} which is also thought to represent a form of dopamine signal, similar to the TD error (Schultz, 2010), can switch between different regimes of action selection, that is, Go, Explore, and NoGo which is thought to be implemented by the BG circuitry. sig is a nonlinear sigmoid function. $A_{\mathrm{G/N}}$ and $\lambda_{\mathrm{G/N}}$ represent the gain and sensitivity of Go/NoGo. σ_{E} is the exploratory parameter.

Using these concepts, we simulated performance in two experimental studies, that investigated the gait pattern of PD patients as they approach a doorway (Almeida & Lebold, 2010; Cowie et al., 2010). The task setup in these experiments involved three subject groups: healthy controls, PD freezers, and PD non-freezers, to pass through a doorway of a specific width while monitoring their gait parameters as they performed the task. The study of Cowie et al. shows a sharp dip in velocity as a PD patient approaches the doorway, a dip that becomes sharper in the case of narrower doorways; this effect was more pronounced in PD patients (ON and OFF freezers with and without dopaminergic medication) than in healthy controls. Almeida and Lebold (2010) consider a similar setup but compare the gait patterns of PD freezers with non-freezers in terms of step lengths and its variability.

The velocity profile obtained from the model of Cowie et al. (2010) for controls and the PD condition is as shown in Fig. 7.2a, b, respectively. In controls, there seems to be a reduction in velocity on approaching the doorway which is

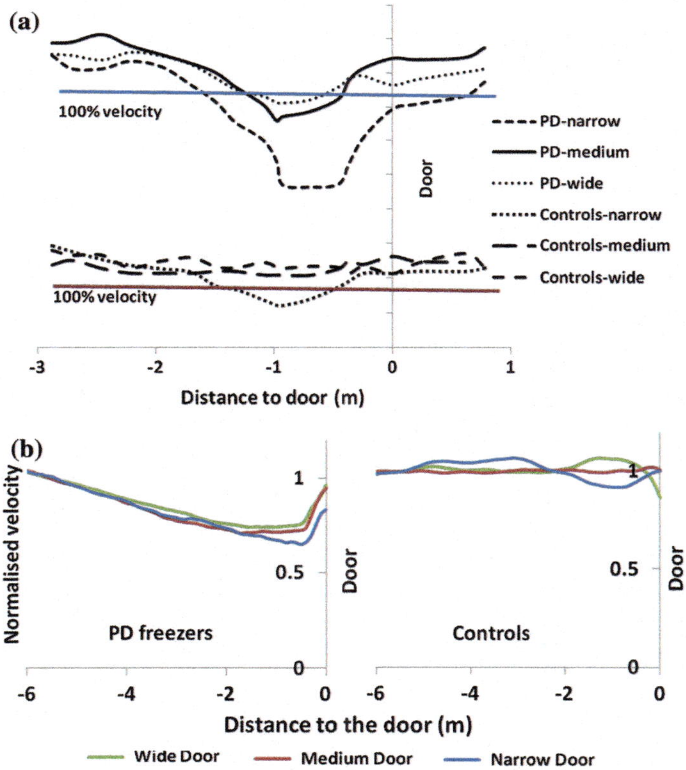

Fig. 7.2 Normalized velocity profile for controls and PD freezers in **a** experiment (Cowie et al., 2010) and **b** simulation under different doorway conditions. 100% velocity in the experimental results represents the velocity profile under a no-door condition. In simulation results, the velocity profiles are normalized by an average velocity far before (5–6 m) from the doorway

exaggerated in PD conditions, thus capturing the experimental results. The velocity near the doorway is normalized by the average velocity calculated far before the doorway (5–6 m). Additionally, our simulation results show a certain door-size-dependent scaling of velocity in case of PD subjects.

The simulated stride length profile for controls, PD ON, and PD OFF, under different doorway sizes is shown in Fig. 7.3b, and that of the experiments (Cowie et al., 2010) in Fig. 7.3a. The average stride length of controls is higher than that of the PD patients. In the model, we also found that simulated PD ON condition has higher mean velocities than PD OFF, in agreement with experimental data. Our simulation results also show that there is a significant difference in stride lengths between the wide/medium door and the narrow door conditions in both PD ON and PD OFF states. Almeida and Lebold in their study show differences in gait patterns between PD ON—freezers and non-freezers. The experiments conducted in the ON condition (Fig. 7.3c) report that the PD freezers produce significantly lower step lengths, compared to non-freezers and controls. This reduction in step lengths is further amplified in the case of reduced door sizes and the model captures this effect (Fig. 7.3d).

The simulations lead to the conclusion that dopamine reduction, modeled here by clamping the temporal difference error (δ), alone cannot replicate the gait patterns seen in the experiments (Almeida & Lebold, 2010; Cowie et al., 2010) and the involvement of several other factors including exploration in the GEN policy represented by the parameter σ_E and discount factor (γ) is necessary to produce the observed effect of freezing.

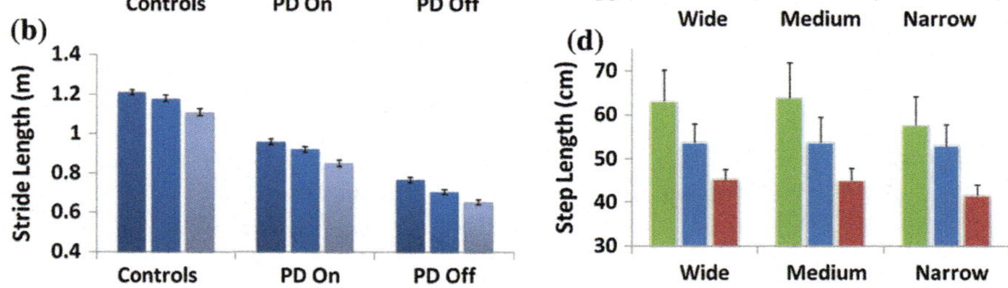

Fig. 7.3 Mean stride lengths and standard errors for controls, PD ON and PD OFF under different doorway conditions in **a** experiments (Cowie et al., 2010) and **b** simulations, reported with $p < 0.005$. Mean and standard deviation of step length profiles for PD freezers and non-freezers under wide, medium, and narrow door conditions in **c** experiments (Almeida and Lebold, 2010), and **d** simulations. PD freezers show significantly reduced step lengths compared to non-freezers ($p < 0.05$) and control ($p < 0.005$) under all door conditions ($N = 50$)

7.3.2 The Role of Cognition in FOG

Experimental data show that perceptual cues can either exacerbate or ameliorate FOG in PD patients. For example, simple visual stimuli like stripes on the floor can alleviate freezing, whereas complex stimuli like narrow doorways can trigger it. Competitive interactions among the cortico-basal ganglia loops are thought to be a major factor for triggering freezing (Lewis & Barker, 2009). We model the behavior of PD subjects, freezers and non-freezers on a virtual reality (VR) gait paradigm. These behavioral experiments used a modified version of the Stroop task (Treisman & Fearnley, 1969) where there is an association of a color-word stimulus to a specific motor action (i.e., to walk or to stop) while subjects navigate a series of doorways. These VR tasks, which assess gait performance quantified in terms of step latency (time between two consecutive steps), require effective interaction between the cortico-basal ganglia circuits (Matar, Shine, Naismith, & Lewis, 2013; Shine et al., 2013).

We developed a cortico-basal ganglia model that simulates the interaction between the motor and cognitive loops (using a 'Motor Module' and 'Cognitive Module') essential to understand the influence of cognition on gait. Both the Cognitive and Motor Modules of the proposed BG model are based on the Actor–Critic architecture, each having its respective Critic and Actor. Evidence from the two modules is combined to produce the final output. These two modules build their respective evidences based on different sensory stimuli—the Motor Module based on visual appearance of the doorway and the Cognitive Module based on the word cue. The first evidence (EI) involves the Cognitive Module identifying the salience of a word cue upon its presentation (Fig. 7.4). The second evidence (EII) involving the Motor Module takes the visual appearance of the doorway as input and computes the direction of the step as well as the latency associated with it as outputs. EII is computed at every time step. In this case the GEN policy, which is the Actor, adopts hill-climbing over the utility (a different learning framework) landscape to calculate the velocity of the agent. The evidences of the two modules are combined subsequently to compute the step latency.

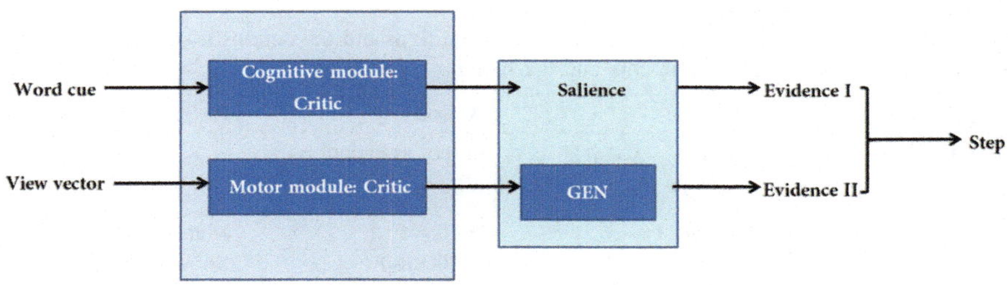

Fig. 7.4 A schematic of the model. There are two modules (cognitive and motor) each computing its evidence. The two evidences are combined subsequently to compute step latency

The Utility formulation is ideal to study cognitive load as it combines the value function (Q), that is expected reward, with the risk function (h), which represents reward variance (Bell, 1995; d'Acremont, Lu, Li, Van der Linden, & Bechara, 2009). In recent work, it has been shown that the utility function formulation can be effectively used to model the interactions between dopamine and serotonin in BG (Balasubramani, Chakravarthy, Ravindran, & Moustafa, 2014, 2015). The utility function is given as

$$U(t) = Q(t) - \alpha\, \mathrm{sign}(Q(t))\sqrt{h(t)} \quad (7.3)$$

where α controls the risk sensitivity. The term α was correlated to the functioning of serotonin (5HT) in the BG and the sign(Q) term in Eq. (7.3) represents the non-linear risk sensitivity (Balasubramani, Chakravarthy, Ravindran, & Moustafa, 2014).

The Effect of Conflict The model simulates the result of Matar and colleagues (Matar et al., 2013) to understand the effect of cognitive cues on motor activity (See Table 7.1 for the cues used). Modal latency, which is basically the preferred step latency, in Fig. 7.5a shows no change in the latency among controls, PD non-freezers, and PD freezers, similar to experimental results. This suggests that the preferred walking speeds for all subject groups are nearly similar and freezing is an intermittent effect. The experimental results (Fig. 7.5b) show that cues like GREEN (green) which have an implicit salience for 'walk' response, evoke little or no change in the step latency for PD freezers. The RED (red) cue which has an implicit salience to 'stop' response seems to increase the step latency of the PD freezers. The model replicates this effect where the BLUE (blue) and RED (red) cues produced maximum footstep latency (MFSL) in the freezers in comparison with controls and non-freezers (Fig. 7.5c).

The Effect of Cognitive Load The effect of cognitive load on motor responses is a result of the ability of the subject to map cues to appropriate actions, depending on the nature of the cues. In this respect, simple cues are easily associated with their corresponding actions—walking or stopping. This is different for complex cues as

Table 7.1 List of cues used in the virtual reality paradigm and the actions associated with their appearance as used in the experiments and the model

Type	Cues	Actions	References
Simple	WALK, STOP, WALK, STOP	Direct associations	
Complex	Congruent: RED, GREEN, BLUE	Walk Walk/stop	Matar et al. (2013) Shine et al. (2013)
	Incongruent: RED, GREEN, BLUE, RED, GREEN, BLUE	Stop Stop/walk	Matar et al. (2013) Shine et al. (2013)

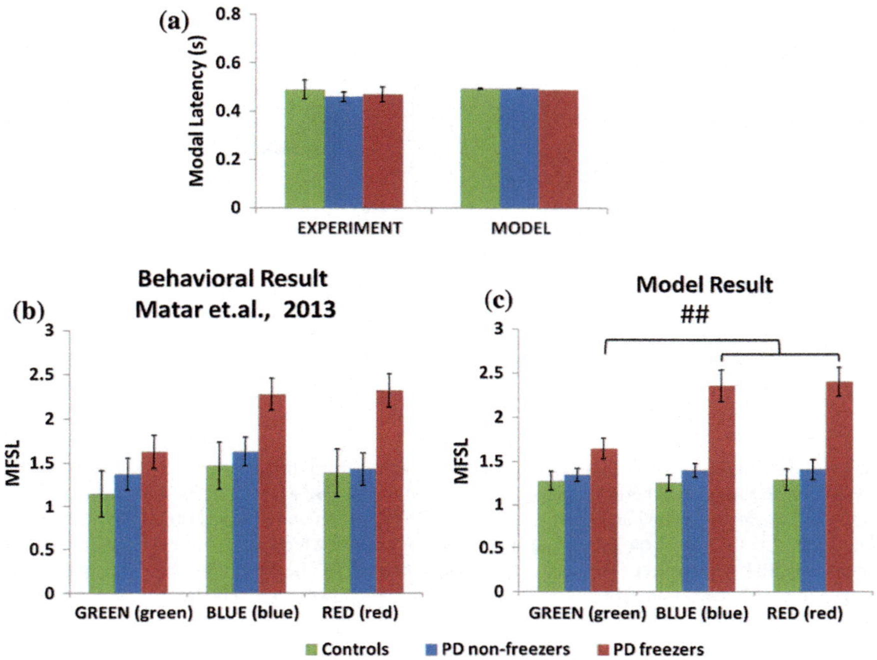

Fig. 7.5 Experimental (Matar et al., 2013) and modeling data of modal latency (**a**) observed in controls, PD non-freezers and PD freezers. The maximum-scaled footstep latency (MFSL) exhibited on the presentation of the congruent cues as seen in the Matar et al. experiment (**b**) and the model (**c**). It illustrates that PD freezers show increased latencies on the high-conflict cues like RED (red) compared to the low-conflict case GREEN (green). (##—$p < 0.005$)

mapping to actions is not straightforward. In the Shine et al. (2013) experiment, non-freezers and freezers PD patients were presented with cues both in the OFF and the ON medication conditions. The trials were also counterbalanced among the patients such that a congruent cue is associated with 'walk' and incongruent to 'stop' and vice versa. According to the experiments, which were conducted on both PD freezers and non-freezers, the outcome of loading is evident from the number of motor arrests observed. The experiments were conducted with patients ON and OFF their dopamine medications. The PD freezers (OFF) showed the highest number of motor arrests (Fig. 7.6c), with the tendency of freezing about 2.7 times more than the non-freezers. Though PD freezers were generally more likely to suffer a motor arrest (both OFF and ON) compared to the non-freezers, the high load situation triggered more arrests in the PD freezers. Similar to the previous experiment, there were also no significant differences in the modal latency between the PD freezers and the non-freezers.

In the model, a similar strategy is imposed and the trials including the low and high load cues are extracted. The number of motor arrests is estimated using the distribution of the step latency (Fig. 7.6a, b). A motor arrest is any event with step

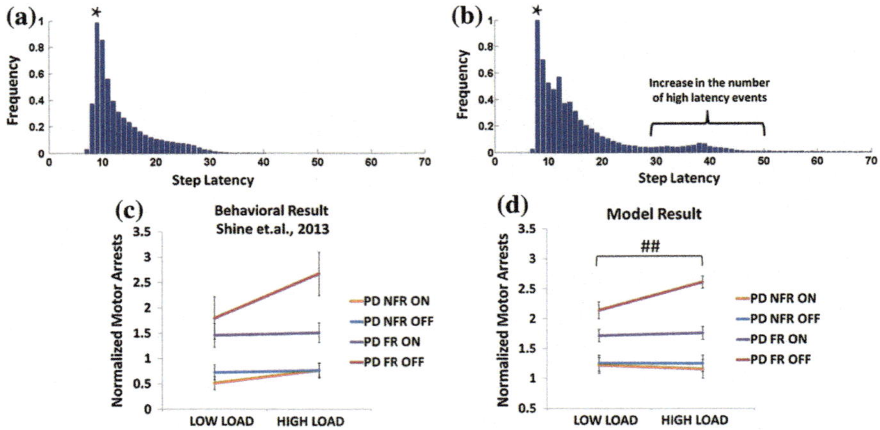

Fig. 7.6 Frequency distributions of step latency observed from the model for the PD non-freezers (**a**) and the PD freezers where * represents the modal points (**b**). Motor arrests seen in PD freezers and non-freezers under low and high levels of cognitive load in experiments (**c**) and the model (**d**). The PD freezers (OFF) show a large number of motor arrests, which comes down under medicated conditions. PD non-freezers show no significant changes in the both the loads as well as the medication. (Abbreviation NFR: Non-freezer, FR: Freezer) (##—$p < 0.005$)

latency that is twice the modal (preferred) step latency of the subject. The PD freezers seem to have higher frequency of higher step latency events especially in the regions of 30–50 in Fig. 7.6b. The modeling results are similar to experimental results (Fig. 7.6d), where under high load scenario, the PD freezers OFF medication show maximum motor arrests, which is comparatively less in the low load case.

The involvement of the Cognitive Module in motor arrests can be ascertained by analyzing the utility of the Cognitive and Motor Modules at the time of a motor arrest (Fig. 7.7a). It is clear that the average values of the utility of walking $\left(U_{\mathrm{w}}^{\mathrm{cog}}\right)$ of the Cognitive Module are much lower than utility of the Motor Module (U^{mot}) in the case of PD freezers compared to non-freezers. This strengthens the claim that there is a shift to more cognition-based decision during a freeze episode and understanding the role of these areas would lead to further insights into the phenomenon.

The contribution of risk sensitivities from each module (α^{mot} and α^{cog}) to the motor arrests also reveals several trends (Fig. 7.7b). Although lower α^{mot} values are used to simulate PD freezer conditions and are the optimal range for accounting for the experimental data, the trends suggest that higher α^{mot} would lead to a high number of motor arrests. The role of α^{cog} to elicit motor arrests seems to be more effective under conditions of low α^{mot} values where there is an increase in the number of motor arrests as α^{cog} increases. The model thus predicts an increase of α^{cog} to differentiate a control from a non-freezer but an increase in α^{mot} to increase the number of motor arrests. Medications are also found to play a role in decreasing the number of motor arrests, and in particular, a case of low α^{mot} is shown to better respond to medications (Fig. 7.7c).

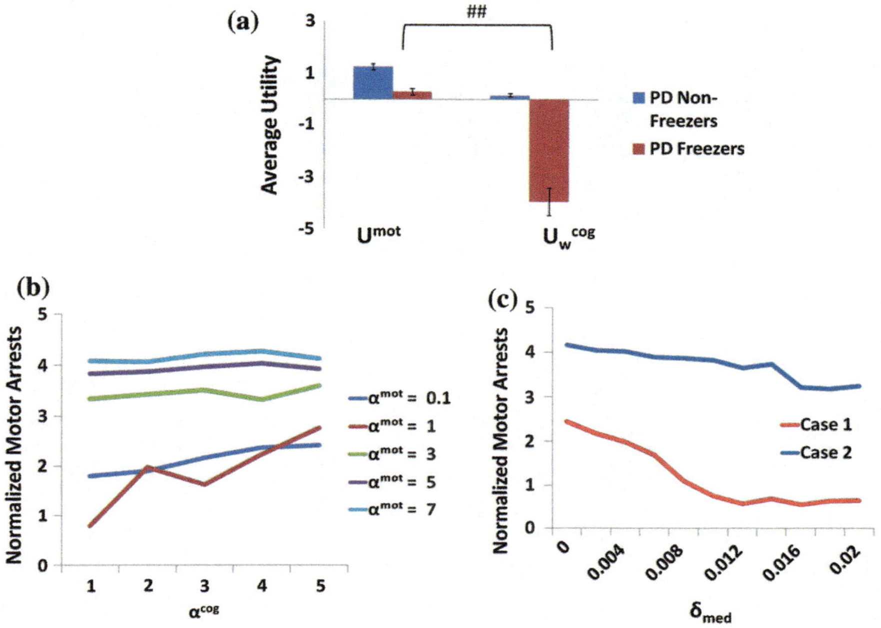

Fig. 7.7 Average utility (**a**) for both the PD non-freezers and the PD freezers during events of motor arrests triggered upon the presentation of a word cue. (##—$p < 0.005$). The normalized motor arrests in the model (**b**) seen in PD freezers for different values of α^{mot} and α^{cog}. **c** The trend for the motor arrest as a function of the medication factor (δ_{med}) for two cases (Case 1 $\alpha^{mot} = 0.1$; $\alpha^{cog} = 7$ and Case 2 $\alpha^{mot} = 7$; $\alpha^{cog} = 7$)

7.3.3 Influence of Turning on FOG

Turning is another scenario which induces freezing in PD (Giladi et al., 2001; Schaafsma et al., 2003). As in the case of doorways, an environment with a turn is considered for simulating the motion of an agent. Our model simulates performance in the VR turning task where healthy controls and PD subjects (both freezers and non-freezers) execute locomotion in a virtual reality environment as described in the previous section, with the introduction of turns in the task (Gilat et al., 2015). The subjects involved in the study, including healthy controls, PD non-freezers, and PD freezers, performed alternate pressing of foot pedals led to forward motion in the VR task. There were an average of 23 turns in the task, and the subjects did not perform a different motor pattern to navigate the turn. So as the subjects pedaled toward a turn the change of the visual scene indicated the execution of a turn. Step latency again was taken as the behavioral output of the subject.

We adopt a similar strategy as the previous models, and the model architecture is similar to the motor-cognitive model, wherein the Motor Module step latency is estimated. The deviation from the previous models is the absence of the cognitive

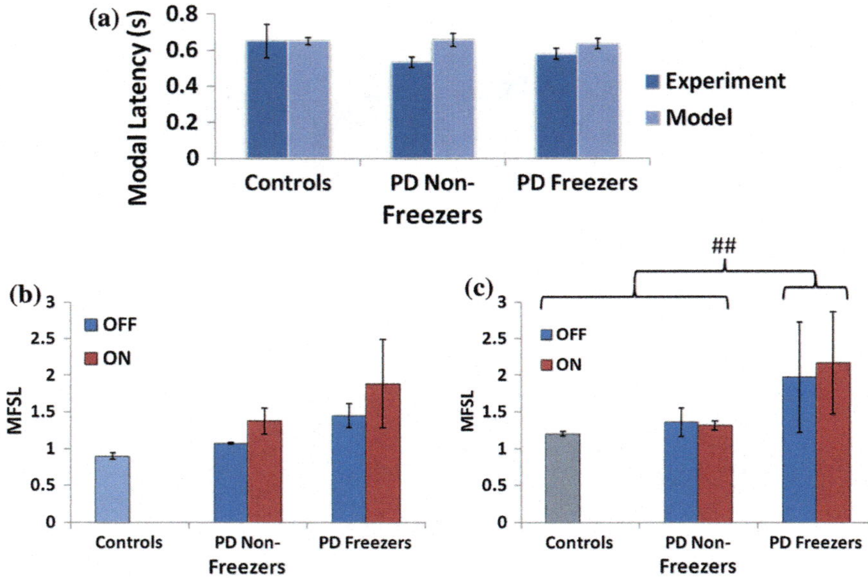

Fig. 7.8 **a** Modal latency of controls, PD non-freezers and PD freezers in the experiments and the model. The experimental (**b**) and model (**c**) maximum footstep latency (MFSL) of the three subject groups while encountering a turn with PD freezers showing the higher MFSL in both the ON and OFF cases. (##—$p < 0.005$)

loop (as word cues were not considered in the experiments), and the risk term is not considered. On simulating the performance of healthy controls, PD non-freezers and PD freezers, we see that modal latency estimated for the subjects groups reveals no significant differences (Fig. 7.8a). In the model, we consider the latency values at all other points except for the turn. The maximum footstep latency (MFSL) is again estimated in the experiment as the largest latency exhibited whenever the subject encounters a turn, and the same method is adopted in the model as well. In Fig. 7.8b, we see that the PD freezers show higher MFSL compared to controls and PD non-freezers which is also captured by the model (Fig. 7.8c). There was no significant change in the MFSL for PD subjects in the ON and OFF medicated states, and the same is seen in the model as well.

7.3.4 Freezing in Other Modalities

Freezing is known to be triggered not only during gait movements but can occur in other motor movements as well. It is seen that freezing occurs in upper limb movements, also referred to as freezing of upper limb (FOUL) (Vercruysse et al., 2013). We utilized an approach to model the areas involved in general motor control and developed a network level model which performs arm reaching movements. The objective was to study freezing of arm and understand the network parameters which give rise to this behavior. Although it is seen that FOUL occurs mostly in bimanual tasks, we here aim to understand how the velocity profiles of the arm changes while moving through a passage, similar to that of navigating through a doorway as all our modeling strategy has been concentrated on building value function over space. Accordingly, we simulate a task where an arm moves through a passage of different sizes and understand whether the same slowing down of hand would happen close to the passage.

We consider two cases: a narrow (0.2 m) and a wide passage (0.4 m) and a target which is placed beyond this passage that the arm tries to reach (Fig. 7.9a, b). The reward given for a successful reach and punishment is delivered if the arm hits the barriers. The velocity profile of the arm is tracked in the model for both controls and PD conditions. We observe that the velocity while passing through the passage for PD in both cases (narrow and wide) is lower than the controls (Fig. 7.9c–e) exemplifying the same effect seen in doorways and turns. Average velocity at the passage indicates that the variability in PD in the case of a narrow doorway is higher than the controls, which also conforms to the experimental results of higher gait variability in freezers close to a doorway (Almeida & Lebold, 2010).

7.4 Conclusions

We used cortico-basal ganglia models to understand FOG in PD patients and studied the influence of environment (doorways and turns), the effect of cognitive cues on motor freezing and even arm freezing through passages. To our knowledge, there are no prior computational models explaining the FOG in PD. We attempt to capture this by carefully considering the impact of different levels of control on gait. The cortico-BG module uses RL concepts for learning the environment in which the agent is placed (for e.g., navigating through doorway). The BG dynamics is modeled through GEN that has been tested in many of our earlier studies (Kalva et al., 2012; Magdoom et al., 2011; Sridharan, Prashanth, & Chakravarthy, 2006). This module outputs a higher level control parameter such as velocity of gait to be passed on to the next level of control.

From all of our modeling studies, we found that just by modulating δ, the dopamine correlates representing the TD error did not capture the results seen in the behavioral studies. Since dopamine deficiency is generally considered the crucial factor, the 'star of the show' (LeWitt, 2012), responsible for PD-related impairment, RL-based computational models of BG function typically propose TD error (a dopamine correlate) as the key variable that controls normal and pathological function. We saw that three other parameters σ, which controls the extent of exploration in the actor, γ, which discounts future rewards, and α, which controls risk sensitivity were also crucial in explaining freezing.

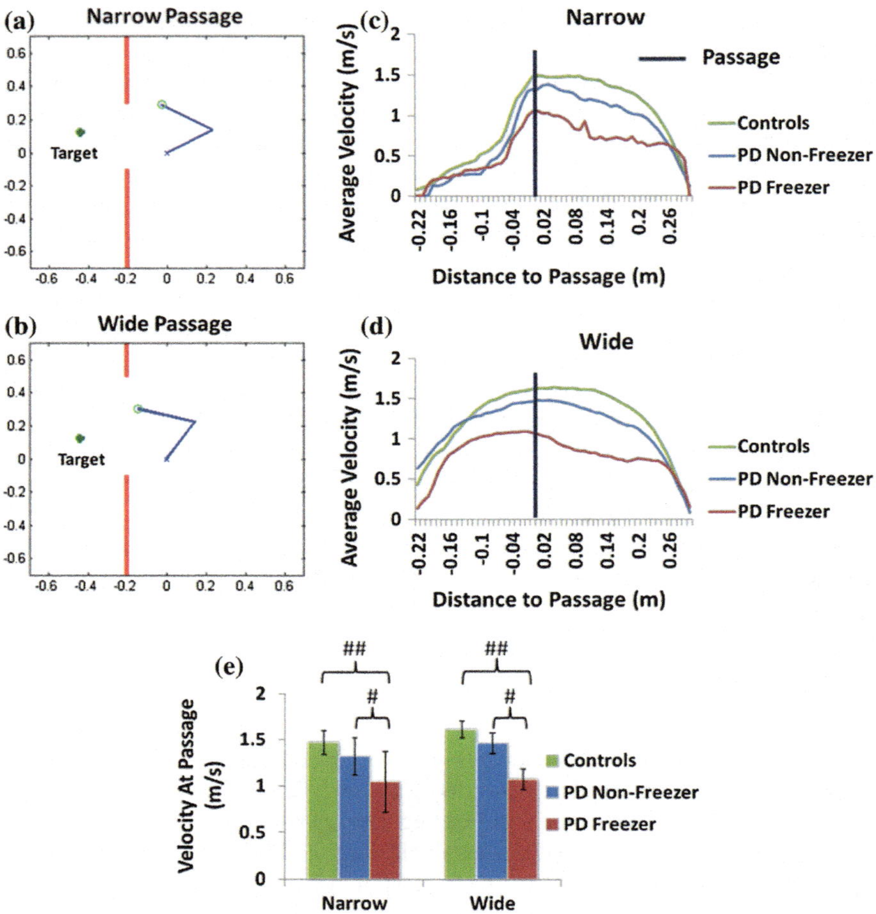

Fig. 7.9 Arm along with the **a** narrow and the **b** wide passages. The averaged velocity profile of the arm while reaching the target through the **c** narrow and **d** wide passage (represented by the black line in plots). **e** The velocity at the passage between controls, and PD shows increased variability in PD when passing through it. (#—$p < 0.05$, ##—$p < 0.005$)

There was an attempt to accommodate the function of different neuromodulators—dopamine, serotonin, and norepinephrine—in an unified theoretical framework based on RL (Doya, 2002). According to this view, dopamine represents TD error, norepinephrine represents exploration denoted by the temperature parameter, β, and serotonin represents discount parameter, γ. Specifically, within the BG circuitry, it was suggested that Globus Pallidus (GP) is the substrate for exploration (Doya, 2002). The Globus Pallidus is also known to have high levels of norepinephrine (Russell, Allin, Lamm, & Taljaard, 1992). The chaotic dynamics of STN–GPe system qualifies to serve as a source of exploratory drive, an idea that has been investigated extensively using computational models (Kalva et al., 2012; Sridharan et al., 2006). In the model, the exploration parameter, σ, denotes the extent of exploration and therefore may be described as a neural correlate for norepinephrine in the BG. Similarly, serotonin has been linked to the discount factor, γ, or the timescale of reward integration, with larger values of γ corresponding to higher levels of serotonin (Tanaka et al., 2007). Low levels of serotonin were associated with impulsivity, a behavior that may be thought to be a result of short-term reward seeking (Rogers, 2011). Based on the arguments just described, we adjust both γ and σ that represent serotonin and norepinephrine, respectively, in addition to δ_{lim} and δ_{med} that are related to dopamine levels, in the present model to capture PD-related gait changes.

Additionally, when cognitive load comes into play, the parameters α^{mot} and α^{cog} which correspond to risk sensitivity in each module seem to bring about the behavioral differences between controls, non-freezers, and freezers. Modeling efforts suggest that risk sensitivity correlates with the levels of serotonin in the motor areas (Balasubramani, Chakravarthy, Ravindran, & Moustafa, 2014, 2015). The concentration of serotonin and its derivatives in PD have been shown to be lower in the cerebrospinal fluid, with a strong correlation with freezing of gait (Tohgi, Abe, Takahashi, Takahashi, & Hamato, 1993). Therefore, besides dopamine, the modeling efforts suggest the need to understand the role of other neuromodulators in PD when it comes to FOG.

References

Almeida, Q. J., & Lebold, C. A. (2010). Freezing of gait in Parkinson's disease: A perceptual cause for a motor impairment? *Journal of Neurology, Neurosurgery and Psychiatry, 81*(5), 513–518.

Balasubramani, P. P., Chakravarthy, V. S., Ravindran, B., & Moustafa, A. A. (2014). An extended reinforcement learning model of basal ganglia to understand the contributions of serotonin and dopamine in risk-based decision making, reward prediction, and punishment learning.

Balasubramani, P. P., Chakravarthy, V. S., Ravindran, B., & Moustafa, A. A. (2015). A network model of basal ganglia for understanding the roles of dopamine and serotonin in reward-punishment-risk based decision making. *Frontiers in Computational Neuroscience, 9,* 76.

Bell, D. E. (1995). Risk, return, and utility. *Management Science, 41*(1), 23–30.

Bradshaw, E. J., & Sparrow, W. (2000). The speed-accuracy trade-off in human gait control when running towards targets. *Journal of Applied Biomechanics, 16*(4), 331–341.

Chakravarthy, V. S., Joseph, D., & Bapi, R. S. (2010). What do the basal ganglia do? A modeling perspective. *Biological Cybernetics, 103*(3), 237–253.

Chee, R., Murphy, A., Danoudis, M., Georgiou-Karistianis, N., & Iansek, R. (2009). Gait freezing in Parkinson's disease and the stride length sequence effect interaction. *Brain, 132*(8), 2151–2160.

Cowie, D., Limousin, P., Peters, A., & Day, B. L. (2010). Insights into the neural control of locomotion from walking through doorways in Parkinson's disease. *Neuropsychologia, 48*(9), 2750–2757.

d'Acremont, M., Lu, Z.-L., Li, X., Van der Linden, M., & Bechara, A. (2009). Neural correlates of risk prediction error during reinforcement learning in humans. *Neuroimage, 47*(4), 1929–1939.

Doya, K. (2002). Metalearning and neuromodulation. *Neural Networks, 15*(4), 495–506.

Duarte, M., & Latash, M. L. (2007). Effects of postural task requirements on the speed–accuracy trade-off. *Experimental Brain Research, 180*(3), 457–467.

Giladi, N., McDermott, M., Fahn, S., Przedborski, S., Jankovic, J., Stern, M.,... Group, P. S. (2001). Freezing of gait in PD prospective assessment in the DATATOP cohort. *Neurology, 56*(12), 1712–1721.

Gilat, M., Shine, J. M., Walton, C. C., O'Callaghan, C., Hall, J. M., & Lewis, S. J. (2015). Brain activation underlying turning in Parkinson's disease patients with and without freezing of gait: A virtual reality fMRI study. *npj Parkinson's Disease, 1,* 15020.

Hughes, A. J., Daniel, S. E., Kilford, L., & Lees, A. J. (1992). Accuracy of clinical diagnosis of idiopathic Parkinson's disease: A clinico-pathological study of 100 cases. *Journal of Neurology, Neurosurgery and Psychiatry, 55*(3), 181–184.

Hughes, J., Bowes, S., Leeman, A., O'Neill, C., Deshmukh, A., Nicholson, P., et al. (1990). Parkinsonian abnormality of foot strike: A phenomenon of ageing and/or one responsive to levodopa therapy? *British Journal of Clinical Pharmacology, 29*(2), 179–186.

Kalva, S. K., Rengaswamy, M., Chakravarthy, V. S., & Gupte, N. (2012). On the neural substrates for exploratory dynamics in basal ganglia: A model. *Neural Networks, 32,* 65–73.

Latt, M. D., Lord, S. R., Morris, J. G., & Fung, V. S. (2009). Clinical and physiological assessments for elucidating falls risk in Parkinson's disease. *Movement Disorders, 24*(9), 1280–1289.

Lewis, S. J., & Barker, R. A. (2009). A pathophysiological model of freezing of gait in Parkinson's disease. *Parkinsonism & Related Disorders, 15*(5), 333–338.

LeWitt, P. A. (2012). Norepinephrine: The next therapeutics frontier for Parkinson's disease. *Translational Neurodegeneration, 1*(1), 4.

MacKay, D. G. (1982). The problems of flexibility, fluency, and speed–accuracy trade-off in skilled behavior. *Psychological Review, 89*(5), 483.

Magdoom, K., Subramanian, D., Chakravarthy, V. S., Ravindran, B., Amari, S.-I., & Meenakshisundaram, N. (2011). Modeling basal ganglia for understanding Parkinsonian reaching movements. *Neural Computation, 23*(2), 477–516.

Matar, E., Shine, J. M., Naismith, S. L., & Lewis, S. J. (2013). Using virtual reality to explore the role of conflict resolution and environmental salience in Freezing of Gait in Parkinson's disease. *Parkinsonism & Related Disorders, 19*(11), 937–942.

Morris, M., Iansek, R., Matyas, T., & Summers, J. (1998). Abnormalities in the stride length-cadence relation in parkinsonian gait. *Movement Disorders, 13*(1), 61–69.

Muralidharan, V., Balasubramani, P. P., Chakravarthy, V. S., Gilat, M., Lewis, S. J., & Moustafa, A. A. (2017). A neurocomputational model of the effect of cognitive load on freezing of gait in Parkinson's disease. *Frontiers in Human Neuroscience, 10,* 649. https://doi.org/10.3389/fnhum.2016.00649.

Muralidharan, V., Balasubramani, P. P., Chakravarthy, V. S., Lewis, S. J., & Moustafa, A. A. (2014). A computational model of altered gait patterns in parkinson's disease patients negotiating narrow doorways. *Frontiers in Computational Neuroscience, 7,* 190. https://doi.org/10.3389/fncom.2013.00190.

Nieuwboer, A., Vercruysse, S., Feys, P., Levin, O., Spildooren, J., & Swinnen, S. (2009). Upper limb movement interruptions are correlated to freezing of gait in Parkinson's disease. *European Journal of Neuroscience, 29*(7), 1422–1430.

Park, H. K., Kim, J. S., Im, K. C., Oh, S. J., Kim, M. J., Lee, J. H., et al. (2009). Functional brain imaging in pure akinesia with gait freezing: [18F] FDG PET and [18F] FP-CIT PET analyses. *Movement Disorders, 24*(2), 237–245. https://doi.org/10.1002/mds.22347.

Park, H. K., Yoo, J. Y., Kwon, M., Lee, J. H., Lee, S. J., Kim, S. R., et al. (2014). Gait freezing and speech disturbance in Parkinson's disease. *Neurological Sciences, 35*(3), 357–363. https://doi.org/10.1007/s10072-013-1519-1.

Ricciardi, L., Ebreo, M., Graziosi, A., Barbuto, M., Sorbera, C., Morgante, L., et al. (2016). Speech and gait in Parkinson's disease: When rhythm matters. *Parkinsonism & Related Disorders, 32,* 42–47. https://doi.org/10.1016/j.parkreldis.2016.08.013.

Righetti, L., Buchli, J., & Ijspeert, A. J. (2006). Dynamic hebbian learning in adaptive frequency oscillators. *Physica D: Nonlinear Phenomena, 216*(2), 269–281.

Rogers, R. D. (2011). The roles of dopamine and serotonin in decision making: Evidence from pharmacological experiments in humans. *Neuropsychopharmacology, 36*(1), 114–132.

Russell, V., Allin, R., Lamm, M., & Taljaard, J. (1992). Regional distribution of monoamines and dopamine D 1-and D 2-receptors in the striatum of the rat. *Neurochemical Research, 17*(4), 387–395.

Schaafsma, J., Balash, Y., Gurevich, T., Bartels, A., Hausdorff, J. M., & Giladi, N. (2003). Characterization of freezing of gait subtypes and the response of each to levodopa in Parkinson's disease. *European Journal of Neurology, 10*(4), 391–398.

Schultz, W. (2010). Dopamine signals for reward value and risk: Basic and recent data. *Behavioral and Brain Functions, 6*(1), 24.

Shine, J. M., Matar, E., Ward, P. B., Frank, M. J., Moustafa, A. A., Pearson, M.,... Lewis, S. J. (2013). Freezing of gait in Parkinson's disease is associated with functional decoupling between the cognitive control network and the basal ganglia. *Brain*, https://doi.org/10.1093/brain/awt272.

Shine, J. M., Moustafa, A. A., Matar, E., Frank, M. J., & Lewis, S. J. (2013b). The role of frontostriatal impairment in freezing of gait in Parkinson's disease. *Frontiers in Systems Neuroscience, 7,* 61. https://doi.org/10.3389/fnsys.2013.00061.

Sridharan, D., Prashanth, P., & Chakravarthy, V. (2006). The role of the basal ganglia in exploration in a neural model based on reinforcement learning. *International Journal of Neural Systems, 16*(02), 111–124.

Sutton, R. S., & Barto, A. G. (1998). *Reinforcement learning: An introduction* (Vol. 1). MIT press Cambridge: Cambridge.

Tanaka, S. C., Schweighofer, N., Asahi, S., Shishida, K., Okamoto, Y., Yamawaki, S., et al. (2007). Serotonin differentially regulates short-and long-term prediction of rewards in the ventral and dorsal striatum. *PLoS ONE, 2*(12), e1333.

Tohgi, H., Abe, T., Takahashi, S., Takahashi, J., & Hamato, H. (1993). Concentrations of serotonin and its related substances in the cerebrospinal fluid of parkinsonian patients and their relations to the severity of symptoms. *Neuroscience Letters, 150*(1), 71–74.

Treisman, A., & Fearnley, S. (1969). The Stroop test: Selective attention to colours and words. *Nature, 222*(5192), 437–439.

Vercruysse, S., Spildooren, J., Heremans, E., Wenderoth, N., Swinnen, S., Vandenberghe, W., & Nieuwboer, A. (2013). The neural correlates of upper limb motor blocks in Parkinson's disease and their relation to freezing of gait. *Cereb Cortex,* https://doi.org/10.1093/cercor/bht170.

Chapter 8
Modeling Precision Grip Force in Controls and Parkinson's Disease Patients

Ankur Gupta and V. Srinivasa Chakravarthy

Abstract Precision grip (PG) is the ability to hold an object between forefinger and thumb. Lifting objects in PG require delicate finger grip force (GF) control. Healthy controls modulate GF depending on size, weight, surface curvature, and friction. The difference between the actual GF generated and the minimum GF required to prevent the object from slipping is known as safety margin (SM). Published results suggest that OFF-medicated Parkinson's disease (PD) patients generated average SM identical to that of controls with increased SM variance. PD patients on medication demonstrated higher average SM with SM variance identical to that of controls. Previously known computational models provide an insight on how the GF is generated and controlled but are unsuitable for modeling the GF in PD patients. In this chapter, we present a Go/Explore/NoGo (GEN) algorithm in a utility-based decision-making framework to explain the SM generated by healthy controls and PD patients both during ON and OFF medication. The study suggests that PD GF is a result of dopamine-level-dependent suboptimal decision-making-based force selection and the suitability of the GEN algorithm to model decision-making tasks.

Development of an opposable thumb in primates has been seen as a very crucial evolutionary step (Almecija, Moya-Sola, & Alba, 2010). It bestowed upon the organism the ability to develop complex tools and effectively use it as early as 2.5 million years ago (Panger, Brooks, Richmond, & Wood, 2002) to obtain/extract food, execute offensive/defensive strategies, and attract peers (Van Schaik, Deaner, & Merrill, 1999). Evidence also suggests that the manufacture and use of tool shaped the brain development in primates, specifically the motor system (Calvin, 1982).

In daily life, we interact and manipulate multiple objects, many of which require grasping. Due to difference in the object's shape, size, and the task requirement, the movement strategy changes. Movements are broadly classified into two groups—prehensile and non-prehensile movements. Prehension is derived from Latin word *prehendere* that means 'to grasp.' Hence, prehensile movements are the ones in which the object is secured by hand/fingers wrapping around the object. Napier (1956) further classified the prehensile movements based on the grasping strategy

V. S. Chakravarthy and A. Moustafa, *Computational Neuroscience Models of the Basal Ganglia*, Cognitive Science and Technology,
https://doi.org/10.1007/978-981-10-8494-2_8

and task objectivity, mainly power and precision grip. Power grip is the grasping strategy (formed by enveloping the object in the palm) employed when task requires large inaccurate forces. The precision grip (PG) is the prehensile strategy employed to generate small yet accurate forces with enhanced control. The grip is classified as a PG when the object is held between the opposable thumb and a finger. PG-lift tasks are noninvasive tools to understand the motor system integration and their effect on behavior. Thus, PG has been employed in multiple studies ranging from Huntington's disease (Gordon, Quinn, Reilmann, & Marder, 2000), dystonia (Nowak & Hermsdörfer, 2005), sensory impairments (Witney, Wing, Thonnard, & Smith, 2004) to Parkinson's disease (PD) (Ingvarsson, Gordon, & Forssberg, 1997; Fellows, Noth, & Schwarz, 1998).

The precision grip lift task (PGLT) is an experimental task requiring the object to be grasped between index finger and thumb and lifting it to the experimenter specified height. The index finger and thumb (referred to as fingers for simplicity in the remaining text) generate two forces on the object: one for gripping the object (referred as grip force, GF) and one for lifting (referred as lift force or load force, LF). Both the GF and LF interact with the object through the boundary of finger–object friction. The GF couples the fingers to the object, and a sufficiently high GF is required to prevent the object from slipping (critical force, CF), where slip is defined as relative motion between the finger and the object. Studies have also shown that the GF generated is dependent on size (Gordon, Forssberg, Johansson, & Westling, 1991; Gordon, Forssberg, Johansson, Eliasson, & Westling, 1992), weight (Johansson and Westling, 1988a, 1988b, Forssberg et al., 1992), surface curvature (Jenmalm, Goodwin, & Johansson, 1998), and friction (Johansson & Westling, 1984; Westling & Johansson, 1984; Forssberg, Eliasson, Kinoshita, Westling, & Johansson, 1995; Saels, Thonnard, Detrembleur, & Smith, 1999). LF, when coupled with a GH higher than critical slipping force, lifts the object. It might appear that the GF and LF are independent of each other; however, there is a very intimate coordination between the two forces in PGLT. In case of low LF, the fingers would fail to lift the object from the table on which the object rests and a very high LF would cause the object to overshoot the target position. In addition, if the optimal LF is generated before the GF reaches CF, the fingers would slip past the object, thereby causing the task to end in failure. One of the strategies employed to prevent slip is to sense the slip through the mechanoreceptors and increase the GF in an ongoing task. This is followed by generation of the updated GF for the new trial (Wolpert, Ghahramani, & Jordan, 1995; Wolpert & Flanagan, 2001; Davidson & Wolpert, 2003; Flanagan, Vetter, Johansson, & Wolpert, 2003; Nowak, Glasauer, & Hermsdörfer, 2004; Schmitz, Jenmalm, Ehrsson, & Forssberg, 2005).

The grip force profile (Fig. 8.1) shows an initial peak known as peak grip force (PGF) and a steady-state or stable grip force (SGF) that is defined as the average grip force generated between 3.5 and 4.5 s after the lift onset. The minimum amount of force required to prevent the object from slipping is known as slip force or critical force (F_{slip}). The difference between the SGF and F_{slip} is the safety margin (SM).

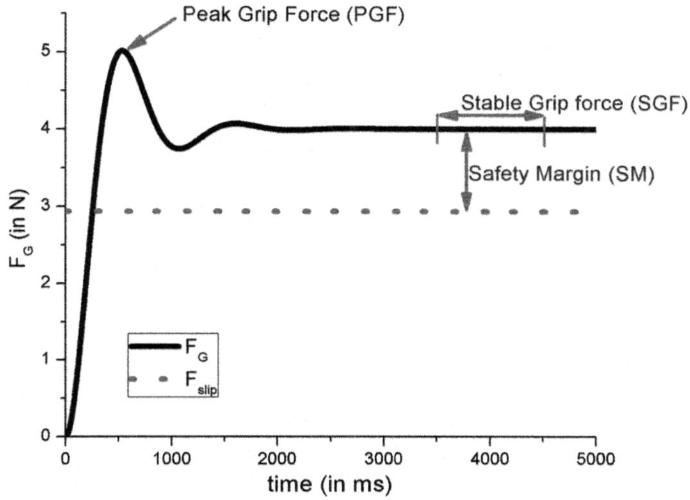

Fig. 8.1 An illustration of the grip force profile showing the peak grip force (PGF), stable grip force (SGF), and safety margin (SM). F_{slip} is the minimum amount of force required to prevent the object from slipping. SGF is computed as the average grip force generated between 3.5 and 4.5 s. Modified from Gupta et al. (2013a, 2013b)

SM is a crucial aspect of the GF profile. For a task like PGLT employing a low SM could lead to task failure if a small perturbation is encountered. However, a large SM could cause muscle fatigue. Therefore, subjects generate SM in a small range. Interestingly, PD patients are known to generate higher SM compared to healthy controls (Ingvarsson et al., 1997; Fellows et al., 1998). Understanding how the GF is generated and controlled in PD pathology could provide valuable insights to develop a framework and suggest robust algorithms for future research in motor control and disorders.

This chapter focuses on illustrating the neural control of precision grip force generation, models developed for grip force production, and a utility-based decision-making model for grip force production in healthy controls and PD patients.

8.1 Precision Grip Force Neural Control

The grip force production, similar to other motor control process, follows perception–action cycle paradigm (Warren, 2006; Newman-Norlund, van Schie, van Zuijlen, & Bekkering, 2007). Before the initial contact with the object, vision conveys approximate information about the weight, surface texture, surface curvature, and size of the object. This information is then used to determine the potential GF and LF that could be used for successful lifting the object. However, after a contact with the object and during the lifting process, this information is updated.

The neuromuscular spindle, Golgi tendon, and the mechanoreceptors in the fingertips (Ruffini and Pacinian corpuscles) convey information pertaining to muscle length, muscle tension, and joint configurations, respectively. This sensory information is then transmitted to the dorsal root ganglion in spinal cord. These signals then travel through medulla, pons, and midbrain to reach thalamus. Thalamus relays the information received from spinal cord and basal ganglia (BG) to primary somatosensory cortex (S-I) that contains four-key Brodmann's area (BA) 1, 2, 3a, and 3b. BA3a and BA3b receive the signals from thalamus and project to BA1 and BA2 (Strick, 1976). The neurons from BA 1, 2, 3a, and 3b project individually to the secondary somatosensory cortex (S-II). S-II innervates project to temporal lobe (for temporal memory) via insular cortex. S-I also projects to association areas in posterior parietal cortex (BA5 and BA7) (Kandel, Schwartz, & Jessell, 2000). BA5 and BA7 relay the information to premotor areas which in turn projects to supplementary motor area and primary motor cortex. The premotor areas are involved in motor preparation by sensory and motor planning. Transient virtual lesion study suggests that the ventral premotor area (PMv) is involved in correct finger placement on the object and sequential recruitment of fingers, whereas impaired dorsal premotor area (PMd) affects grasping and lifting (Davare, Andres, Cosnard, Thonnard, & Olivier, 2006). The primary motor cortex sends the muscle control signals to muscles via cortico-spinal tract (Muir & Lemon, 1983; Lemon, Johansson, & Westling, 1995). The overall neural control mechanism is illustrated in Fig. 8.2. Primary motor cortex also projects to basal ganglia, which is a key aspect of our model presented here.

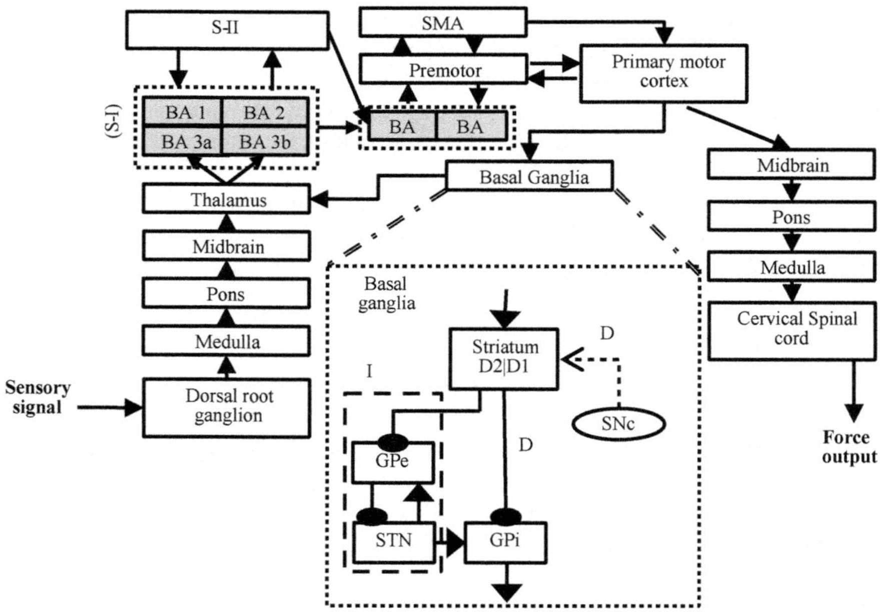

Fig. 8.2 Neural control of grip force generation

8.1.1 Role of BG in PGLT

An in-depth description of the basal ganglia anatomy and function is provided in Chaps. 2 and 3. This subsection contains information about the contribution of various neural substrates to specific GF components.

The GF generation requires GF planning and GF modulation. These two aspects are separately controlled by different neural components. Initial force selection requires identification of a constant or a variable force level or switching between the force levels based on the trial requirement. Vaillancourt, Yu, Mayka, and Corcos (2007) found that constant GF amplitude activated caudate, putamen, GPe, GPi, and STN, whereas, the force selection showed anterior BG, i.e., caudate, anterior putamen, and GPe activation. Anterior BG was also found to be responding to force switching and not for timings and in constant force tasks (Pope, Wing, Praamstra, & Miall, 2005). One of the crucial aspects of lifting the object is predicting the optimal GF for the trial. Putamen and caudate show scaled activations for GF predictability (Wasson, Prodoehl, Yu, Corcos, & Vaillancourt, 2007). Anterior BG was also shown to be involved in predictability (Boecker et al., 2005).

During the initial parameter selection, the rate of GF and LF and the amplitude of GF are determined. STN and GPi show scaled activations for GF rate, and GPe and putamen show increased activity for increased duration of the task (Vaillancourt, Mayka, Thulborn, & Corcos, 2004; Prodoehl, Yu, Wasson, Corcos, & Vaillancourt, 2008). Ehrsson, Fagergren, Johansson, and Forssberg (2003) found that the ipsilateral fronto-parietal area, contralateral fronto-parietal area, and subcortical motor structures are simultaneously activated in PGLT. During heavy object lifting, putamen shows increased activations whereas switching between the object weights shows contralateral putamen and thalamus activations (Kinoshita, Oku, Hashikawa, & Nishimura, 2000). GPi and STN are shown to be involved in force scaling, but GPe, putamen, and caudate do not show increased activations during the same task (Spraker, Yu, Corcos, & Vaillancourt, 2007).

Therefore, anterior BG (caudate, anterior putamen, and GPe) is involved in planning, that is, prediction and selection of initial GF, whereas posterior BG (posterior putamen, GPi, and STN) controls the dynamic aspects like GF scaling and rate of force control.

Various attempts have also been made to model the GF production to understand pathology or develop gripping systems with performance similar to that of humans.

8.2 Computational Models of Precision Grip

8.2.1 Kim and Inooka (1994)

Kim and Inooka (1994) developed a computational model for robotic hand force control. The model featured contact dynamics between the finger and the object, and thus, the GF generated was function of frictional force between the object and the finger. If the grip force was sufficiently high, there would be no relative acceleration between the finger and the object. However, in cases when the object slipped, the grip force was increased proportional to the difference between the acceleration of the finger and the object. This simple yet elegant model was successful in demonstrating the grip force control for no-slip lifting of object, but failed to explain the reasons (and model the same) for stable grip force generation (SGF).

8.2.2 de Gruijl, van der Smagt, and De Zeeuw (2009)

de Gruijl et al. (2009) proposed an olivocerebellar computational model for anticipatory GF control. The granule cells received information about various physical and kinematic parameters (arm position, acceleration, surface texture, slip force, and noise). The granule cell to Purkinje cell projections inhibited deep cerebellar nucleus and inferior olive. Deep cerebellar nucleus controlled the GF output. The difference between the experimentally obtained GF and simulated GF serves as the input to the granule cells. These inputs update the state information and weights of inferior olive through Purkinje cells. This study incorporated error correction mechanism to determine optimal GF; however, the model lacked a detailed finger–object interaction module that could be used to make more realistic GF predictions.

8.2.3 Ulloa, Bullock and Rhodes (2003)

Ulloa et al. developed a computational model that involved the cerebellar component as error corrector. The model comprised of multiple modules for arm transport, grip aperture, slip error, load error, Go signaling, and cerebellar error correction. The arm transport module moved the arm in position for grasping the object, whereas the grip aperture module generated the grip aperture required to grasp the object. The slip error component generated the slip information based on the motion between the fingers and the object, whereas the load error component generated error based on the actual load force required and the simulated load force. Both the slip error and the load error modules were dependent on the weight of the object and the friction between the object and fingers lifting it. Cerebellar component was involved in the reactive and anticipatory performance error correction based on the Go component's signal to initiate the movement. The model generated forces that were sensitive to object's weight and friction. One of the major limitations of this study was the inability of the model to generate realistic GF as observed in Johansson and Westling (1984, 1988a, 1988b), Forssberg, Eliasson, Kinoshita, Johansson, and Westling (1991), Forssberg et al. (1992), Johansson, Riso, Hager, and Backstrom (1992),

Johansson and Cole (1994), Forssberg, Eliasson, Kinoshita, Westling and Johansson (1995), Flanagan and Wing (1997), Flanagan, Bowman, and Johansson (2006) probably due to the absence of finger–object contact dynamics component.

8.2.4 *Fagergren, Ekeberg, and Forssberg (2000)*

Fagergren et al. (2000) modeled the GF profile as a second-order system to model active and reactive responses. The transfer function for active components (when the subjects increased the GF in steps) and reactive component (when the object was suddenly loaded) was combined to account for both active and reactive components. The GF developed was similar to the ones generated by healthy humans, but the lack of finger–object interaction model provided limited insight to the GF selection mechanism.

8.2.5 *Fagergren, Ekeberg, and Forssberg (2003)*

Fagergren et al. (2003) presented a modified model of Fagergren, Ekeberg et al. (2000) study that included a hand–object interaction component. The physical properties for weight (0.3 kg) and hand–object friction coefficient ($\mu_{\text{static, sandpaper}} = 1.21$, $\mu_{\text{static, silk}} = 0.35$) were borrowed from Johansson and Westling (1984). The authors assumed $\mu_{\text{dynamic}} = 0.8^* \mu_{\text{static}}$. Model generated neural GF and LF that in turn generated GF and LF with GF updated based on the sensory feedback. The model further described the GF [similar to Fagergren et al. (2000)] using the transfer function to propose the GF–LF delay was involved in slip prevention. This model though suggested how the LF–GF coordination prevented slips but in limited capacity explained how the GF was selected; thereby, the study was unsuitable to model PD condition.

8.2.6 *Kim, Nakazawa, and Inooka (2002)*

Handing over a small object in precision grip from one person to another requires dynamic regulation of finger forces. Kim et al. (2002) modeled the GF regulation in a smooth transfer task between the human and robotic fingers. The model featured a simple slip–GF relationship where the presence of slip increased the GF proportional to the slip and the GF developed was friction-dependent. The model was successful in smooth handling of the object but lacked an approach for the identification of forces produced in Parkinson's patients.

8.2.7 *Grip Force During Transient Friction Change (Gupta et al., 2013a, 2013b)*

Previous studies failed to present a consolidated model to explain the grip force variability that can be used to explain the abnormalities demonstrated in PD patients. The following model provide a more comprehensive GF production mechanism under friction changes in healthy subjects that serves as a backbone for the modeling the GF in PD patients and hence discussed in detail.

As discussed earlier, the GF and LF produced are tightly coupled to the size, weight, texture, surface curvature of the object, and the friction (between the object surface and interacting finger boundary). In a controlled experimental setup, the physical properties of the object are kept constant and the only sources of variation are the object–finger friction and neural control.

Therefore, a model of PGLT was developed to account for the GF and LF variability due to subjective friction differences in healthy subjects (Gupta et al., 2013a, 2013b).

Gupta et al. (2013a, 2013b) proposed a computational model to demonstrate the GF–LF variability in high friction (when the grip formed was using dry fingers) and low friction (when the fingers were wetted >2 min, as described by Johansson and Westling (1984) to prevent μ changes) conditions. The model (Fig. 8.3) comprised of two PID force controllers—F_G controller and F_L controller—that generate F_G and F_L, respectively. The F_G controller received slip information as the error input, where slip is defined as the relative motion between the finger and the object. Therefore, any velocity difference between the fingers and the object led to an increase in F_G that in turn tries to prevent the slip. The lift force controller receives difference between the target position and the actual object position as the input. The dynamics of the object–finger interaction is modeled in plant that generates the acceleration, velocity, and position profiles for both the finger and the object.

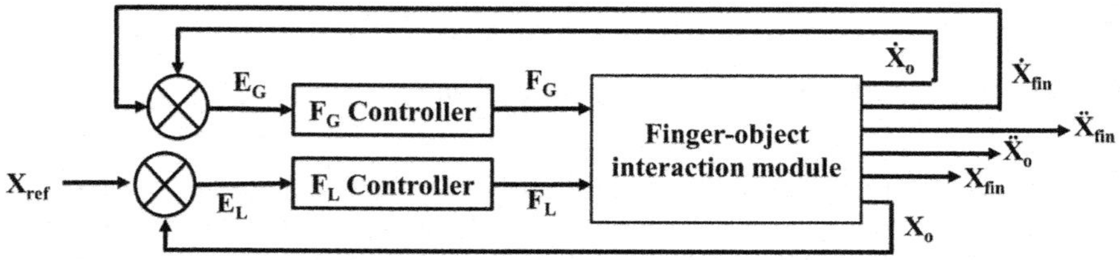

Fig. 8.3 An overview of the Gupta et al. (2013a, 2013b) model. The grip force controller (F_G controller) receives the slip information (as absolute difference between the finger velocity and object velocity) to generate a grip error (E_G) that is converted to the grip force (F_G). The lift force controller (F_L controller) receives the difference between the target position and the current position as the input and outputs the lift force (F_L). Finger–object interaction module based on the F_G and F_L generates the position, velocity, and accelerations for both finger (subscript fin) and object (subscript o). Modified from Gupta et al. (2013a, 2013b)

The F_L controller receives E_L as the input, that is, the difference between the target position (X_{ref}) and the current position (X_o).

$$E_L = X_{ref} - X_o \tag{8.1}$$

Based on E_L the PID controller generates a force

$$F_{L,PID} = K_{P,L}E_L + K_{I,L}\int_0^{\tau} E_L(\tau)d\tau + K_{D,L}\frac{dE_L}{dt} \tag{8.2}$$

To prevent discontinuities in the force production, a smoothing function was used with a smoothing factor $\tau_{F,L}$.

$$\tau_{F,L}\frac{dF_L}{dt} = -F_L + F_{L,PID} \tag{8.3}$$

Since the slip causes an increase in F_G produced, the error received by the controller is always positive; that is, the difference between the finger velocity (dX_{fin}/dt) and the object velocity (dX_o/dt) is always positive. Hence, E_G is defined as

$$E_G = \left(\frac{dX_{fin}}{dt} - \frac{dX_o}{dt}\right)^2 \tag{8.4}$$

The error was first smoothened to prevent large controller outputs that destabilized the system.

$$\tau_{E,G}\frac{dE_G}{dt} = -E_G + \left(\frac{dX_{fin}}{dt} - \frac{dX_o}{dt}\right)^2 \tag{8.5}$$

This smoothened error then served as the input to the PID controller.

$$F_{G,PID} = K_{P,G}E_G + K_{I,G}\int_0^{\tau} E_G(\tau)d\tau + K_{D,G}\frac{dE_G}{dt} \tag{8.6}$$

Similar to the F_L controller's output, the output of F_G controller was also smoothened.

$$\tau_{F,G}\frac{dF_G}{dt} = -F_G + F_{G,PID} \tag{8.7}$$

The object–finger interaction module (Fig. 8.4) comprises of an object (mass = M_o) gripped with two fingers (mass = M_{fin}). The fingers generate both the grip force (F_G) and lift force (F_L) that interacts to the object through a frictional

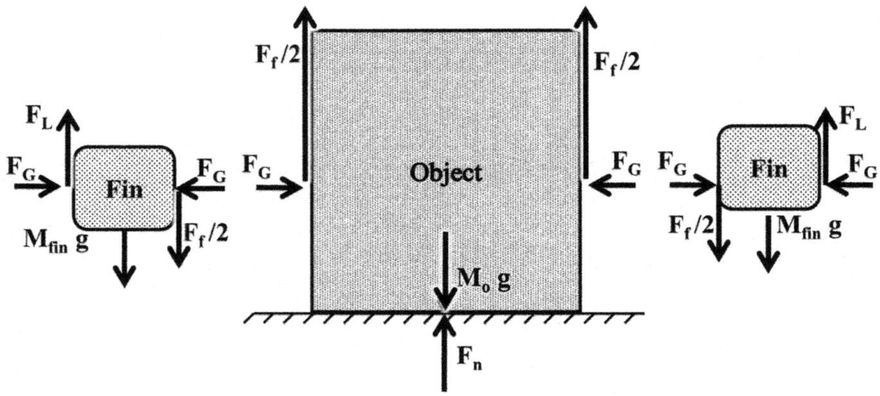

Fig. 8.4 Free body diagram showing the various forces acting on the object and the fingers. Fin = finger; F_G = grip force; F_L = lift force; F_n = normal force; F_f = frictional force, M_{fin} = mass of finger; M_o = mass of object; g = acceleration due to gravity. Modified from Gupta et al. (2013a, 2013b)

force (F_f). Since two fingers are employed in isometric force generation, each of the fingers generates half of the total frictional force on the object. When the object is resting on the table, a normal force (F_n) equivalent of the weight of the object (M_og) acts perpendicular to the surface of the table to prevent the net movement.

The net force acting on finger is a resultant of the F_L and F_f and force generated on finger as an action of gravity.

$$F_{fin} = F_L - F_f - M_{fin}g \tag{8.8}$$

The net force on the object is a result of F_f, normal force F_n, and force generated on object due to gravity.

$$F_o = F_f + F_n - M_o g \tag{8.9}$$

Since there are two fingers involved, the slip force (F_{slip}) is modeled as

$$F_{slip} = 2\mu F_G \tag{8.10}$$

When no-slip condition is observed, both the object and finger move with the same velocity.

$$F_{noslip} = \frac{M_o M_{fin}}{M_o + M_{fin}} \left(\frac{F_L}{M_{fin}} - \frac{F_n}{M_o} \right) \tag{8.11}$$

F_f computation is based on the numerical values of the F_{slip} and F_{noslip}.

$$F_{\mathrm{f}} = \begin{cases} F_{\mathrm{noslip}}, & \mathrm{if} F_{\mathrm{noslip}} < F_{\mathrm{slip}} \\ F_{\mathrm{slip}}, & \mathrm{otherwise} \end{cases} \quad (8.12)$$

And F_{n} is determined as

$$F_{\mathrm{n}} = \begin{cases} M_{\mathrm{o}}g, & \text{if } \mathrm{X_o} = 0 \text{ and } \mathrm{M_o g} > F_{\mathrm{f}} \\ 0, & \text{otherwise} \end{cases} \quad (8.13)$$

F_{fin} and F_{o} are determined using second law of motion.

$$F_{\mathrm{o}} = M_{\mathrm{o}} \frac{\mathrm{d}^2 X_{\mathrm{o}}}{\mathrm{d}t^2} \quad (8.14)$$

$$F_{\mathrm{fin}} = M_{\mathrm{fin}} \frac{\mathrm{d}^2 X_{\mathrm{fin}}}{\mathrm{d}t^2} \quad (8.15)$$

The entire model comprises of nine parameters $K_{\mathrm{P,L}}$, $K_{\mathrm{I,L}}$, $K_{\mathrm{D,L}}$, $K_{\mathrm{P,G}}$, $K_{\mathrm{I,G}}$, $K_{\mathrm{D,G}}$,$\tau_{\mathrm{F,G}}$, $\tau_{\mathrm{E,G}}$, $\tau_{\mathrm{F,L}}$. The model was trained for no-slip condition when the F_{G} is high (10 N) for optimizing the cost function C_{L}.

$$C_{\mathrm{L}} = \sigma_{\mathrm{x}} + \frac{\sqrt{(\bar{X}_{\mathrm{O,M}} - X_{\mathrm{ref}})^2}}{X_{\mathrm{ref}}} + \frac{\sqrt{(\max(\mathrm{X_{O,M}}) - X_{\mathrm{ref}})^2}}{X_{\mathrm{ref}}} \quad (8.16)$$

The above cost function for the F_{L} controller was designed to prevent the object position variability (σ_{x} = object position standard deviation), minimizing the difference between the object position ($\bar{X}_{o,M}$ = average object position between 3.5 and 4.5 s) and the target position (X_{ref}), and to minimize the overshoot.

Following determining the F_{L} parameters, F_{G} controller was trained with the obtained F_{L} parameters using cost function C_{G}. C_{G} was designed to minimize the difference between model values (subscript M) and the experimental values (subscript E) for PGF and SGF and to minimize the slip.

$$C_{\mathrm{G}} = \frac{\sqrt{(\mathrm{PGF_M} - \mathrm{PGF_E})^2}}{\mathrm{PGF_E}} + \frac{\sqrt{(\mathrm{SGF_M} - \mathrm{SGF_E})^2}}{\mathrm{SGF_E}} + \sqrt{(X_{\mathrm{o}} - X_{\mathrm{fin}})^2} \quad (8.17)$$

All the PID parameters were obtained using the genetic algorithm.

The friction coefficient (μ) from the experimental data was determined using standard slip method with the F_{G} at which the object slips given by (F_{crit}).

$$\mu = \frac{M_{\mathrm{o}}g}{2F_{\mathrm{crit}}} \quad (8.18)$$

The model was first trained on dry friction condition (when the fingers were not artificially wet) and then on wet condition (when the fingers were saturated with water after immersion in water >2 min).

The results demonstrated that the F_G output of the model is friction-sensitive and could be used to model the PD pathology.

8.2.8 *Utility-Based Decision-Making Model of Grip Force Generation in Parkinson's Patients (Gupta, Balasubramani, & Chakravarthy, 2013c)*

As discussed earlier, an optimal GF is necessary to manipulate the object. A low GF causes a poor coupling between the fingers and the object that prevents the object from lifting from initial rest, or if the object was already lifted, it slips from the fingers. Contrary to this, excessive GF might damage the fragile object and prolonged enhanced force production could result in fatigue that may in turn compromise the task.

Healthy subjects generate grip force greater than the minimum GF required to prevent slip (F_{crit}), at the same time, they abstain from generating excessive amount to prevent damage to the object, hence a narrow GF range of operation. The difference between the actual F_G generated and F_{crit} is the safety margin (SM). SM is essential to prevent the object slips due to perturbations.

Interestingly, SM in PD patients is much higher compared to the healthy subjects (Ingvarsson et al., 1997; Fellows & Noth, 2004). Recent evidences attribute cognitive and motor performance decline on impaired decision making in PD patients (Moustafa, Chakravarthy, & Phillips, et al., 2016; Moustafa, Chakravarthy, Phillips, & Gupta, et al., 2016). Therefore, in Gupta, Balasubramani, et al. (2013c), higher SM generation was considered as a DM problem. It is also worth noting that PD is a BG disorder with BG also a site for DM.

In an experimental condition with a constant object constant mass, the GF variability is a resultant of the friction between the finger–object interface and due to the output (in this case GF) selection criteria (decision making) by the brain.

In their model, Gupta, Balasubramani, et al. (2013c) modified the F_G controller to a second-order system from PID controller (Gupta et al., 2013a, 2013b). The controller output for step input is given by Eq. (8.18).

$$F_G = \frac{\omega_n^2}{s^2 + 2\omega_n \zeta s + \omega_n^2} \tag{8.19}$$

where the ω_n is the natural frequency and ζ is the damping factor. The response peak is given by M_p, and time to peak is given by T_p.

$$M_p = e^{\left(-\frac{\zeta \omega_n}{\omega_d}\right)\pi} \tag{8.20}$$

$$T_p = \frac{\pi}{\omega_d} \tag{8.21}$$

where the damped frequency is given as $\omega_d = \omega_n \sqrt{1 - \zeta^2}$.

The F_L controller and the finger–object interaction module remained unchanged (Fig. 8.5) from the previous study (Gupta et al., 2013a, 2013b).

The F_L controller was optimized similar to Gupta et al. (2013a, 2013b) by keeping the F_G = 10 N to prevent the object from slipping while lifting the object. The performance of the trial was evaluated using cost function CE that considers the total slip and overshoot encountered in the trial based on which the F_L controller parameters were updated.

$$CE(F_{Gref}) = 00.5\left(\frac{\bar{X}_{fin} - \bar{X}_o}{\bar{X}_{fin}}\right)^2 + 00.5\left(\frac{X_{ref} - \bar{X}_o}{X_{ref}}\right)^2 \tag{8.22}$$

Interestingly, if both the F_L and F_G controllers are used together in a trial, the object experiences small amount of slip during the F_G development. Therefore, the performance of the lift was obtained by repeatedly simulating the trial for F_{Gref} and μ-dependent uniform noise as the input to the controller given as

$$F_{Gref,noisy} = F_{Gref} + 3(10.44 - \mu) \tag{8.23}$$

The CE obtained for the noisy F_{Gref} was converted to a performance variable, Per, as

$$\mathrm{Per}(F_{Gref,noisy}) = e^{-CE(F_{Gref,noisy})} \tag{8.24}$$

Since humans and primates show neural correlates for computing value, risk, and risk sensitivity (d'Acremont, Lu, Li, Van der Linden, & Bechara, 2009; Wu, Delgado, & Maloney, 2009; Schultz, 2010; Zhang, Maddula, & Maloney, 2010;

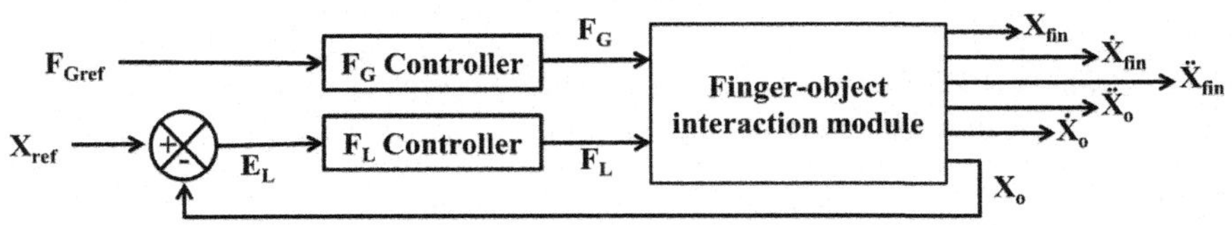

Fig. 8.5 Overall model architecture showing the grip force controller (F_G controller), F_L controller (F_L controller), and the finger–object interaction module. The inputs to the model are the reference grip force (F_{Gref}) and target position (X_{ref}), E_L is the input error to the F_L controller, the X_{fin} is finger position, and X_o is the object position. Modified from (Gupta, Balasubramani et al., 2013c)

Lakshminarayanan, Chen, & Santos, 2011; Leathers & Olson, 2012; Wolpert & Landy, 2012), a utility-based decision-making paradigm was applied to the model.

The mean (*V*) and variance (*h*) in performance were evaluated.

$$\mathrm{V}(F_{\mathrm{Gref}}) = \overline{\mathrm{Per}}\left(F_{\mathrm{Gref,noisy}}\right) \tag{8.25}$$

$$h(F_{\mathrm{Gref}}) = \mathrm{var}\left(\mathrm{Per}\left(F_{\mathrm{Gref,noisy}}\right)\right) \tag{8.26}$$

A radial basis function neural network (RBFNN) with learning capabilities ($n = 60$, range = [0.1 12] in 0.2 steps, spread = 0.7) was designed to approximate the mean and variance of the performance for the given F_{Gref}. The output of the RBFNN for the given trial, tr, is given as

$$\rho(F_{\mathrm{Gref}}(\mathrm{tr})) = \sum_{\mathrm{i}=1}^{\mathrm{n}} w_{\mathrm{V}} \phi_{\mathrm{m}}(F_{\mathrm{Gref}}(\mathrm{tr})) \tag{8.27}$$

$$\phi_{\mathrm{m}}(F_{\mathrm{Gref}}(\mathrm{tr})) = \mathrm{e}^{-\frac{\left(F_{\mathrm{Gref}}(\mathrm{tr}) - \mathrm{RBF}_{\mathrm{cent,m}}\right)^2}{\mathrm{RBF}_{\mathrm{spr,m}}}} \tag{8.28}$$

where $\mathrm{RBF}_{\mathrm{cent,m}}$ is the center and $\mathrm{RBF}_{\mathrm{spr,m}}$ is the spread of the basis function for *m*th basis function.

The weights of the RBF are updated for $\eta_{\mathrm{V}} = 0.1$.

$$\Delta w_{\mathrm{V}} = \eta_V \Delta V_{\mathrm{CE}}(F_{\mathrm{Gref}}) \phi(F_{\mathrm{Gref}}) \tag{8.29}$$

where

$$\Delta V_{\mathrm{CE}}(F_{\mathrm{Gref}}) = \mathrm{e}^{-\mathrm{CE}\left(\hat{\mathrm{F}}_{\mathrm{Gref,noisy}}\right)} - \mathrm{e}^{-\mathrm{CE}(\mathrm{F}_{\mathrm{Gref}})} \tag{8.30}$$

Risk prediction error is determined as the variance in ΔV_{CE} for the risk, h,

$$\xi(F_{\mathrm{Gref}}) = (\Delta V_{\mathrm{CE}}(F_{\mathrm{Gref}}))^2 - \mathrm{h}(F_{\mathrm{Gref}}) \tag{8.31}$$

With the weights for the risk, w_{h} (for learning rate $\eta_{\mathrm{h}} = 0.1$) is updated as

$$\Delta w_{\mathrm{h}} = \eta_{\mathrm{h}} \xi(F_{\mathrm{Gref}}) \phi(F_{\mathrm{Gref}}) \tag{8.32}$$

The trained RBFNN outputs the approximated estimates of value, $V(F_{\mathrm{Gref}})$, and risk, $h(F_{\mathrm{Gref}})$, using the following equations.

$$V(F_{\mathrm{Gref}}(\mathrm{tr})) = w_{\mathrm{V}} \phi(F_{\mathrm{Gref}}(\mathrm{tr})) \tag{8.33}$$

$$h(F_{\mathrm{Gref}}(\mathrm{tr})) = w_{\mathrm{h}} \phi(F_{\mathrm{Gref}}(\mathrm{tr})) \tag{8.34}$$

A modified utility, *U*, equation was used to determine the utility of the trial for value, risk, and risk sensitivity (Pragathi Priyadharsini, Ravindran & Srinivasa

Chakravarthy, 2012). This formulation enables risk aversion for gains and risk seeking for losses—a behavior similar to that of human decision making (Kahneman, 1979).

$$U(\mathrm{tr}) = V(\mathrm{tr}) - \alpha\,\mathrm{sign}(V(\mathrm{tr}))\sqrt{h(\mathrm{tr})} \tag{8.35}$$

In the current context, the utility was determined as

$$U(F_{\mathrm{Gref}}(\mathrm{tr})) = V(F_{\mathrm{Gref}}(\mathrm{tr})) - \alpha\sqrt{h(F_{\mathrm{Gref}}(\mathrm{tr}))} \tag{8.36}$$

This formulation enables choosing the decisions based on maximizing the utility $U(F_{\mathrm{Gref}}(\mathrm{tr}))$ with an increase in α making the decisions more risk aversive and risk seeking otherwise. This was coupled with the stochastic hill-climbing process 'Go/Explore/NoGo' (GEN) (Chakravarthy, Joseph, & Bapi, 2010; Magdoom et al., 2011; Kalva, Rengaswamy, Chakravarthy, & Gupte, 2012), a modification of the classical Actor–Critic model for BG modeling (Joel, Niv, & Ruppin, 2002).

$$\Delta \mathrm{F}_{\mathrm{Gref}}(\mathrm{tr}) = \begin{cases} \Delta \mathrm{F}_{\mathrm{Gref}}(\mathrm{tr}-1), & \text{“}\mathbf{Go}\text{”} \quad \delta_{\mathrm{v}}(\mathrm{tr}) > DA_{\mathrm{hi}} \\ \psi, & \text{“}\mathbf{Explore}\text{”} \quad \delta_{\mathrm{v}}(\mathrm{tr}) > DA_{\mathrm{lo}}^{\wedge}\delta_{\mathrm{v}}(\mathrm{tr}) > DA_{\mathrm{hi}} \\ \Delta \mathrm{F}_{\mathrm{Gref}}(\mathrm{tr}-1), & \text{“}\mathbf{NoGo}\text{”} \quad \delta_{\mathrm{v}}(\mathrm{tr}) \leq \mathrm{DA}_{\mathrm{lo}} \end{cases} \tag{8.37}$$

The above equations were combined into a single equation in Sukumar, Rengaswamy, and Chakravarthy (2012) as follows:

$$\Delta F_{\mathrm{Gref}}(\mathrm{tr}) = \begin{array}{c} A_{\mathrm{G}}\mathrm{logsig}\ (\lambda_{\mathrm{G}}\delta_{\mathrm{u}}(\mathrm{tr}))\Delta \mathrm{F}_{\mathrm{Gref}}(\mathrm{tr}-1) \\ +\ A_{\mathrm{E}}\psi e^{\wedge}(-\delta_{\mathrm{U}}^{2}(\mathrm{tr})/\sigma_{\mathrm{E}}^{2} \\ -A_{\mathrm{N}}\mathrm{logsig}\ (\lambda_{\mathrm{N}}\delta_{u}\ (\mathrm{tr}))\Delta F_{\mathrm{Gref}}(\mathrm{tr}-1) \end{array} \tag{38}$$

where A_{G}, $\underline{A}_{\mathrm{E}}$, A_{N} are the gains for Go, Explore, and NoGo components, respectively; λ_{G} and λ_{N} are sensitivities of the Go and NoGo components, respectively; ψ is random noise between [−1 1]; and σ_{E} is standard deviation for the explorer.

$$\delta_{\mathrm{U(tr)}} = U(\mathrm{F}_{\mathrm{Gref}}(\mathrm{tr})) - U(F_{\mathrm{Gref}}(\mathrm{tr}-1)) \tag{39}$$

$$\mathrm{logsig(n)} = \frac{1}{1+\mathrm{e}^{-\mathrm{n}}} \tag{40}$$

For modeling the PD condition, force selection was considered as an outcome of decision making by healthy and PD patients. In healthy controls, the dopamine availability is unconstrained (between the range $[a, b]$ with $a < b$); PD patients show a limited dopamine production due to loss of dopaminergic cells in SNc (range between $[a, \delta_{\mathrm{Lim}}]$ with $\delta_{\mathrm{Lim}} < b$); and in PD ON condition in addition to the limited dopamine available, there is a contribution of the L-DOPA medication that increases the basal levels of dopamine (range between $[a + \delta_{\mathrm{Med}}, \delta_{\mathrm{Lim}} + \delta_{\mathrm{Med}}]$), for

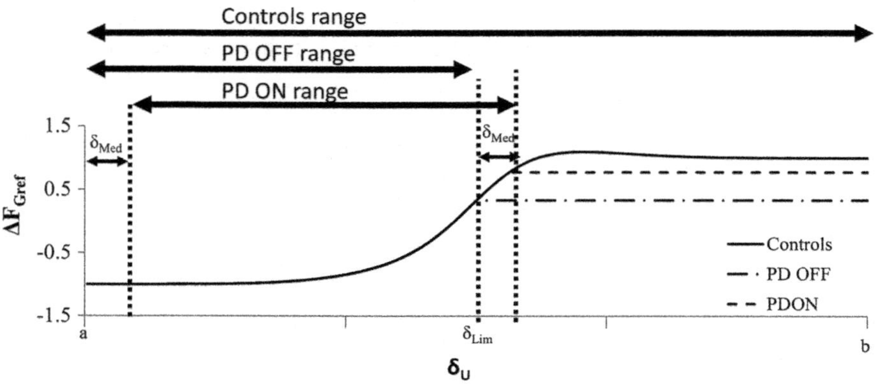

Fig. 8.6 An illustration of the limiting δ_U assumed to represent dopamine levels. Healthy controls have the entire range of dopamine available [a, b]. PD OFF condition has dopamine levels clamped [a, δ_{Lim}]. In PD ON condition due to the medication, the range gets modified to [a + δ_{Med}, δ_{Lim} + δ_{Med}]

illustration, see Fig. 8.6. In the model similar to Magdoom et al. (2011) and Sukumar et al. (2012), δ_U was interpreted as the dopamine levels; hence, δ_U is defined as follows:

$$\delta_U(tr) = \begin{cases} [a, b], & \text{for controls} \\ [a, \delta_{Lim}], & \text{for PD OFF} \\ [a + \delta_{Med}, \delta_{Lim} + \delta_{Med}], & \text{for PD ON} \end{cases} \tag{41}$$

To determine the parameter values (A_G, A_E, A_N, λ_G, λ_N, σ_E) of Eq. (8.38), CE_{GEN} was developed to minimize the mean and the standard deviation of the experimentally and model values.

$$CE_{GEN} = 2\left(\overline{SGF}_{expt} - \overline{SGF}_{sim}\right)^2 + \left(\sigma_{expt} - \sigma_{sim}\right)^2 \tag{42}$$

where SGF is defined as the average F_G generated during 4–5 s. σ is the standard deviation obtained, and subscripts *expt* and *sim* are experimental and simulated values, respectively.

The parameters were obtained using genetic algorithm using 20 as population size, 0.8 as crossover fraction, 4 as elite count, 1000 as generation time, and 10^{-6} as tolerance.

The model was subjected to multiple conditions reported in the published literature (Ingvarsson et al., 1997; Fellows et al., 1998): friction (μ = 0.44 for silk and 0.94 for sandpaper) and dopamine levels (controls, PD OFF, and PD ON), object mass [Mo = 0.33 for Fellows et al. (1998) and 0.3 for Ingvarsson et al. (1997)], and friction coefficient dependent noise ($\nu \in [-3, 3]$ for silk and $[-1.5, 1.5]$ for sandpaper). A detailed list of parameters is presented in Table 8.1.

Table 8.1 Parameters used in simulations. Modified from (Gupta, Balasubramani, et al., 2013c

Study	(Gordon et al., 1997)	(Ingvarsson et al., 1997)	(Fellows et al., 1998)
Object surface	Silk surface	Sandpaper surface	Silk surface
M_o (in kg)	0.3	0.3	0.33
μ	0.44	0.94	0.44
Noise type	Uniformly distributed	Uniformly distributed	Uniformly distributed
v	[−3, 3]	[−1.5, 1.5]	[−3, 3]
Conditions simulated	Controls, PD OFF, PD ON	Controls, PD OFF, PD ON	Controls, PD ON

The results (Fig. 8.7) from the simulation show that the model could successfully generate the average SGF and the standard deviation SGF with low errors for a plethora of different conditions.

It was also found that the PD patients have lower α compared to controls. For Fellows et al. (1998), α was obtained as 0.7 for controls and 0.312 for PD ON condition (with $\delta_{\text{Lim}} = -0.5$ and $\delta_{\text{Med}} = 0.427$). Similar results were also obtained for Ingvarsson et al. (1997), where the value of α was found to be 0.5 for controls and 0.30 for both PD ON ($\delta_{\text{Lim}} = 0.5$ and $\delta_{\text{Med}} = 0.005$) and PD OFF ($\delta_{\text{Lim}} = 0.5$ and $\delta_{\text{Med}} = 0$) conditions. Therefore, PD patients exhibit risk sensitivity different than controls.

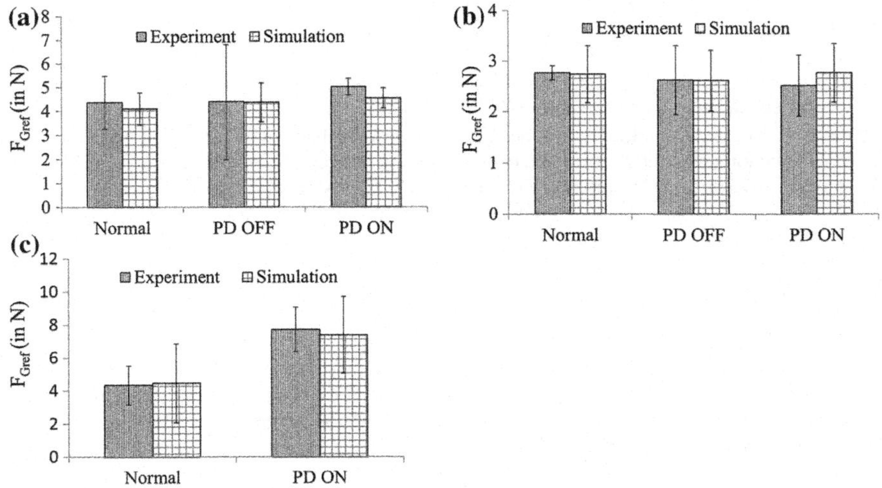

Fig. 8.7 Comparison of the experimental and simulated for **a** Ingvarsson, Gordon et al. (1997) silk surface, **b** Ingvarsson, Gordon et al. (1997) sandpaper surface, and, **c** Fellows, North et al. (1998). In Ingvarsson et al. (1997), grip force data for three dopaminergic levels (controls, PD OFF, and PD ON) were collected. In Fellows et al. (1998), only the grip force for only controls and in PD ON conditions was obtained

The model shows a robust performance for various experimental conditions, thereby suggesting the suitability of the Go/ Explore/NoGo in utility formulations for decision-making motor tasks.

One of the limitations of the current model (Gupta, Balasubramani, et al., 2013c) is the non-inclusion of cerebellum in the model. The role of cerebellum goes beyond motor control (Manto et al., 2012) and affects sensory acquisition and discrimination (Gao et al., 1996) processes as well. A recently published study suggests PD pathology-related cerebellar changes in the brain (Wu & Hallett, 2013). The authors went on to further state that the knowledge of the relationship between PD and cerebellum is very limited (Wu & Hallett, 2013). Since PD is primarily considered as a BG disorder and due to lack of evidences suggesting a strong link between the PD and cerebellum, the model assumes that the cerebellar function is unaffected in the PD. Hence, the Go/Explore/NoGo approach seems to be a resilient tool for modeling variety of motor control tasks including precision grip in healthy controls and PD patients.

References

Almecija, S., Moya-Sola, S., & Alba, D. M. (2010). Early origin for human-like precision grasping: A comparative study of pollical distal phalanges in fossil hominins. *PLoS ONE, 5*(7), e11727.

Boecker, H., Lee, A., Mühlau, M., Ceballos-Baumann, A., Ritzl, A., Spilker, M., et al. (2005). Force level independent representations of predictive grip force–load force coupling: A PET activation study. *Neuroimage, 25*(1), 243–252.

Calvin, W. H. (1982). Did throwing stones shape hominid brain evolution? *Ethology and Sociobiology, 3*(3), 115–124.

Chakravarthy, V. S., Joseph, D., & Bapi, R. S. (2010). What do the basal ganglia do? A modeling perspective. *Biological Cybernetics, 103*(3), 237–253.

d'Acremont, M., Lu, Z. L., Li, X., Van der Linden, M., & Bechara, A. (2009). Neural correlates of risk prediction error during reinforcement learning in humans. *Neuroimage, 47*(4), 1929–1939.

Davare, M., Andres, M., Cosnard, G., Thonnard, J.-L., & Olivier, E. (2006). Dissociating the role of ventral and dorsal premotor cortex in precision grasping. *The Journal of neuroscience, 26* (8), 2260–2268.

Davidson, P. R., & Wolpert, D. M. (2003). Motor learning and prediction in a variable environment. *Current Opinion in Neurobiology, 13*(2), 232–237.

de Gruijl, J. R., van der Smagt, P., & De Zeeuw, C. I. (2009). Anticipatory grip force control using a cerebellar model. *Neuroscience, 162*(3), 777–786.

Ehrsson, H. H., Fagergren, A., Johansson, R. S., & Forssberg, H. (2003). Evidence for the involvement of the posterior parietal cortex in coordination of fingertip forces for grasp stability in manipulation. *Journal of Neurophysiology, 90*(5), 2978–2986.

Fagergren, A., Ekeberg, O., & Forssberg, H. (2000). Precision grip force dynamics: A system identification approach. *Biomedical Engineering, IEEE Transactions on, 47*(10), 1366–1375.

Fagergren, A., Ekeberg, Ö., & Forssberg, H. (2003). Control strategies correcting inaccurately programmed fingertip forces: Model predictions derived from human behavior. *Journal of Neurophysiology, 89*(6), 2904–2916.

Fellows, S. J., & Noth, J. (2004). Grip force abnormalities in de novo Parkinson's disease. *Movement Disorders, 19*(5), 560–565.

Fellows, S. J., Noth, J., & Schwarz, M. (1998). Precision grip and Parkinson's disease. *Brain, 121*(9), 1771–1784.

Flanagan, J. R., Bowman, M. C., & Johansson, R. S. (2006). Control strategies in object manipulation tasks. *Current Opinion in Neurobiology, 16*(6), 650–659.

Flanagan, J. R., Vetter, P., Johansson, R. S., & Wolpert, D. M. (2003). Prediction precedes control in motor learning. *Current Biology, 13*(2), 146–150.

Flanagan, J. R., & Wing, A. M. (1997). The role of internal models in motion planning and control: Evidence from grip force adjustments during movements of hand-held loads. *The Journal of Neuroscience, 17*(4), 1519–1528.

Forssberg, H., Eliasson, A. C., Kinoshita, H., Johansson, R. S., & Westling, G. (1991). Development of human precision grip. I: Basic coordination of force. *Experimental Brain Research, 85*(2), 451–457.

Forssberg, H., Eliasson, A. C., Kinoshita, H., Westling, G., & Johansson, R. S. (1995). Development of human precision grip. IV. Tactile adaptation of isometric finger forces to the frictional condition. *Experimental Brain Research, 104*(2), 323–330.

Forssberg, H., Kinoshita, H., Eliasson, A. C., Johansson, R. S., Westling, G., & Gordon, A. M. (1992). Development of human precision grip. II. Anticipatory control of isometric forces targeted for object's weight. *Experimental Brain Research, 90*(2), 393–398.

Gao, J.-H., Parsons, L. M., Bower, J. M., Xiong, J., Li, J., & Fox, P. T. (1996). Cerebellum Implicated in Sensory Acquisition and Discrimination Rather Than Motor Control. *Science, 272*(5261), 545–547.

Gordon, A. M., Forssberg, H., Johansson, R. S., Eliasson, A. C., & Westling, G. (1992). Development of human precision grip. III. Integration of visual size cues during the programming of isometric forces. *Experimental Brain Research, 90*(2), 399–403.

Gordon, A. M., Forssberg, H., Johansson, R. S., & Westling, G. (1991). Visual size cues in the programming of manipulative forces during precision grip. *Experimental Brain Research, 83*(3), 477–482.

Gordon, A. M., Quinn, L., Reilmann, R., & Marder, K. (2000). Coordination of prehensile forces during precision grip in Huntington's disease. *Experimental Neurology, 163*(1), 136–148.

Gupta, A., Avinash, M., Kandaswamy, D., Kumar, M., Devasahayam, S., Babu, K. S. & Chakravarthy, V. S. (2013a). *Biologically inspired closed-loop model of precision grip lifting task. Advances in cognitive neurodynamics (III)* (543–550) Dordrecht: Springer.

Gupta, A., Avinash, M., Kandaswamy, D., Kumar, M., Devasahayam, S., Babu, K. S., et al. (2013b). Human precision grip performance under variable skin friction conditions: A modelling and experimental study. *International Journal of Mind, Brain and Cognition. B. Publications. New Delhi, 4,* 7–45.

Gupta, A., Balasubramani, P. P., & Chakravarthy, V. S. (2013c). Computational model of precision grip in Parkinson's disease: A utility based approach. *Frontiers in Computational Neuroscience, 7,* 172.

Ingvarsson, P. E., Gordon, A. M., & Forssberg, H. (1997). Coordination of manipulative forces in Parkinson's disease. *Experimental Neurology, 145*(2 Pt 1), 489–501.

Jenmalm, P., Goodwin, A. W., & Johansson, R. S. (1998). Control of grasp stability when humans lift objects with different surface curvatures. *Journal of Neurophysiology, 79*(4), 1643–1652.

Joel, D., Niv, Y., & Ruppin, E. (2002). Actor-critic models of the basal ganglia: New anatomical and computational perspectives. *Neural Networks, 15*(4–6), 535–547.

Johansson, R. S., & Cole, K. J. (1994). Grasp stability during manipulative actions. *Canadian Journal of Physiology and Pharmacology, 72*(5), 511–524.

Johansson, R. S., Riso, R., Hager, C., & Backstrom, L. (1992). Somatosensory control of precision grip during unpredictable pulling loads. I. Changes in load force amplitude. *Experimental Brain Research, 89*(1), 181–191.

Johansson, R. S., & Westling, G. (1984). Roles of glabrous skin receptors and sensorimotor memory in automatic control of precision grip when lifting rougher or more slippery objects. *Experimental Brain Research, 56*(3), 550–564.

Johansson, R. S., & Westling, G. (1988a). Coordinated isometric muscle commands adequately and erroneously programmed for the weight during lifting task with precision grip. *Experimental Brain Research, 71*(1), 59–71.

Johansson, R. S., & Westling, G. (1988b). Programmed and triggered actions to rapid load changes during precision grip. *Experimental Brain Research, 71*(1), 72–86.

Kahneman, D., & Tversky, A. (1979). Prospect theory: An analysis of decision under risk. *Econometrica, 47,* 263–292.

Kalva, S. K., Rengaswamy, M., Chakravarthy, V. S., & Gupte, N. (2012). On the neural substrates for exploratory dynamics in basal ganglia: A model. *Neural Networks, 32,* 65–73.

Kandel, E. R., Schwartz, J. H., & Jessell, T. M. (2000). *Principles of neural science*, New York: McGraw-Hill.

Kim, I., & Inooka, H. (1994). Determination of grasp forces for robot hands based on human capabilities. *Control Engineering Practice, 2*(3), 415–420.

Kim, I., Nakazawa, N., & Inooka, H. (2002). Control of a robot hand emulating human's hand-over motion. *Mechatronics, 12*(1), 55–69.

Kinoshita, H., Oku, N., Hashikawa, K., & Nishimura, T. (2000). Functional brain areas used for the lifting of objects using a precision grip: A PET study. *Brain Research, 857*(1), 119–130.

Lakshminarayanan, V. R., Chen, M. K., & Santos, L. R. (2011). The evolution of decision-making under risk: Framing effects in monkey risk preferences. *Journal of Experimental Social Psychology, 47*(3), 689–693.

Leathers, M. L., & Olson, C. R. (2012). In monkeys making value-based decisions, LIP neurons encode cue salience and not action value. *Science, 338*(6103), 132–135.

Lemon, R. N., Johansson, R., & Westling, G. (1995). Corticospinal control during reach, grasp, and precision lift in man. *The Journal of Neuroscience, 15*(9), 6145–6156.

Magdoom, K. N., Subramanian, D., Chakravarthy, V. S., Ravindran, B., Amari, S., & Meenakshisundaram, N. (2011). Modeling basal ganglia for understanding Parkinsonian reaching movements. *Neural Computation, 23*(2), 477–516.

Manto, M., Bower, J. M., Conforto, A. B., Delgado-García, J. M., da Guarda, S. N. F., Gerwig, M., et al. (2012). Consensus paper: Roles of the cerebellum in motor control—the diversity of ideas on cerebellar involvement in movement. *The Cerebellum, 11*(2), 457–487.

Moustafa, A. A., Chakravarthy, S., Phillips, J. R., Crouse, J. J., Gupta, A., Frank, M. J., et al. (2016a). Interrelations between cognitive dysfunction and motor symptoms of Parkinson's disease: Behavioral and neural studies. *Reviews in the Neurosciences, 27*(5), 535–548.

Moustafa, A. A., Chakravarthy, S., Phillips, J. R., Gupta, A., Keri, S., Polner, B., et al. (2016b). Motor symptoms in Parkinson's disease: A unified framework. *Neuroscience and Biobehavioral Reviews, 68,* 727–740.

Muir, R., & Lemon, R. (1983). Corticospinal neurons with a special role in precision grip. *Brain Research, 261*(2), 312–316.

Napier, J. R. (1956). The prehensile movements of the human hand. *J Bone Joint Surg Br, 38-B*(4), 902–913.

Newman-Norlund, R. D., van Schie, H. T., van Zuijlen, A. M., & Bekkering, H. (2007). The mirror neuron system is more active during complementary compared with imitative action. *Nature Neuroscience, 10*(7), 817.

Nowak, D. A., Glasauer, S., & Hermsdörfer, J. (2004). How predictive is grip force control in the complete absence of somatosensory feedback? *Brain, 127*(1), 182–192.

Nowak, D. A., & Hermsdörfer, J. (2005). Grip force behavior during object manipulation in neurological disorders: Toward an objective evaluation of manual performance deficits. *Movement Disorders, 20*(1), 11–25.

Panger, M. A., Brooks, A. S., Richmond, B. G., & Wood, B. (2002). Older than the Oldowan? Rethinking the emergence of hominin tool use. *Evolutionary Anthropology: Issues, News, and Reviews, 11*(6), 235–245.

Pope, P., Wing, A. M., Praamstra, P., & Miall, R. C. (2005). Force related activations in rhythmic sequence production. *Neuroimage, 27*(4), 909–918.

Priyadharsini, B. P., Ravindran, B., & Chakravarthy, V. S. (2012). In: A. P. Villa, W. Duch, P. Érdi, F. Masulli & G. Palm (Eds.), *Understanding the role of serotonin in basal ganglia through a unified model. Artificial Neural Networks and Machine Learning—ICANN 2012* (Vol. 7552, pp. 467–473). Springer Berlin Heidelberg.

Prodoehl, J., Yu, H., Wasson, P., Corcos, D. M., & Vaillancourt, D. E. (2008). Effects of visual and auditory feedback on sensorimotor circuits in the basal ganglia. *Journal of Neurophysiology, 99*(6), 3042–3051.

Saels, P., Thonnard, J. L., Detrembleur, C., & Smith, A. M. (1999). Impact of the surface slipperiness of grasped objects on their subsequent acceleration. *Neuropsychologia, 37*(6), 751–756.

Schmitz, C., Jenmalm, P., Ehrsson, H. H., & Forssberg, H. (2005). Brain activity during predictable and unpredictable weight changes when lifting objects. *Journal of Neurophysiology, 93*(3), 1498–1509.

Schultz, W. (2010). Dopamine signals for reward value and risk: Basic and recent data. *Behavioral and Brain Functions, 6,* 24.

Spraker, M. B., Yu, H., Corcos, D. M., & Vaillancourt, D. E. (2007). Role of individual basal ganglia nuclei in force amplitude generation. *Journal of Neurophysiology, 98*(2), 821–834.

Strick, P. L. (1976). Anatomical analysis of ventrolateral thalamic input to primate motor cortex. *Journal of Neurophysiology, 39*(5), 1020–1031.

Sukumar, D., Rengaswamy, M., & Chakravarthy, V. S. (2012). Modeling the contributions of Basal ganglia and Hippocampus to spatial navigation using reinforcement learning. *PLoS ONE, 7*(10), e47467.

Ulloa, A., Bullock, D., & Rhodes, B. J. (2003). Adaptive force generation for precision-grip lifting by a spectral timing model of the cerebellum. *Neural Networks, 16*(5–6), 521–528.

Vaillancourt, D. E., Mayka, M. A., Thulborn, K. R., & Corcos, D. M. (2004). Subthalamic nucleus and internal globus pallidus scale with the rate of change of force production in humans. *Neuroimage, 23*(1), 175–186.

Vaillancourt, D. E., Yu, H., Mayka, M. A., & Corcos, D. M. (2007). Role of the basal ganglia and frontal cortex in selecting and producing internally guided force pulses. *Neuroimage, 36*(3), 793–803.

Van Schaik, C. P., Deaner, R. O., & Merrill, M. Y. (1999). The conditions for tool use in primates: Implications for the evolution of material culture. *Journal of Human Evolution, 36*(6), 719–741.

Warren, W. H. (2006). The dynamics of perception and action. *Psychological Review, 113*(2), 358.

Wasson, P., Prodoehl, J., Yu, H., Corcos, D., & Vaillancourt, D. (2007). *Prediction and the basal ganglia*. San Diego: Society for Neuroscience.

Westling, G., & Johansson, R. S. (1984). Factors influencing the force control during precision grip. *Experimental Brain Research, 53*(2), 277–284.

Witney, A. G., Wing, A., Thonnard, J.-L., & Smith, A. M. (2004). The cutaneous contribution to adaptive precision grip. *Trends in Neurosciences, 27*(10), 637–643.

Wolpert, D. M., & Flanagan, J. R. (2001). Motor prediction. *Current Biology, 11*(18), R729–R732.

Wolpert, D. M., Ghahramani, Z., & Jordan, M. I. (1995). An internal model for sensorimotor integration. *Science, 269*(5232), 1880–1882.

Wolpert, D. M., & Landy, M. S. (2012). Motor control is decision-making. *Current Opinion in Neurobiology, 22*(6), 996–1003.

Wu, S. W., Delgado, M. R., & Maloney, L. T. (2009). Economic decision-making compared with an equivalent motor task. *Proc Natl Acad Sci U S A, 106*(15), 6088–6093.

Wu, T., & Hallett, M. (2013). The cerebellum in Parkinson's disease. *Brain, 136*(3), 696–709.

Zhang, H., Maddula, S. V., & Maloney, L. T. (2010). Planning routes across economic terrains: Maximizing utility, following heuristics. *Front Psychol, 1,* 214.

Chapter 9
Go-Explore-NoGo (GEN) Paradigm in Decision Making—A Multimodel Approach

Alekhya Mandali, S. Akila Parvathy Dharshini and V. Srinivasa Chakravarthy

Abstract In this chapter, we built a hybrid model using the combination of biophysical and Izhikevich neurons and validated our earlier hypothesis about Go-Explore-NoGo (GEN) mechanism in BG. The hybrid model consists of Hodgkin–Huxley type model for STN, GPe, and GPi and spiking model for striatum. To capture the effect of dopamine (DA) on the BG nuclei dynamics, the synaptic weights between STN–GPe and the T-type calcium current in STN known to induce bursting behavior were modulated by DA. We compared the results from hybrid model with spiking Izhikevich model and rate-coded model for binary action selection task. The results from the hybrid model further reinforced the theory of GEN showing exploration levels are dependent on the level of DA. The results from n-arm bandit task also show that by decreasing the striatum (D1) to GPi weight in the spiking model, we can increase the exploration level in the system reflected as the decreased average reward obtained by the model. The n-arm bandit results were compared with the results from rate-coded and lumped softmax model.

9.1 Introduction

We had suggested earlier that the STN–GPe loop, a coupled excitatory–inhibitory network in the indirect pathway (IP), might be the substrate for exploration (Chakravarthy, Joseph, & Bapi, 2010). It is suggested that coupled excitatory–inhibitory networks have an ability to exhibit rich dynamic behavior like oscillations and chaos (Borisyuk, Borisyuk, Khibnik, & Roose, 1995; Sinha, 1999). Simulation studies suggest that between the Go and NoGo regimes of classical models of BG, there exist a third regime, the Explore regime. This Explore regime, which is effectively a stochastic regime, actually arises out of the chaotic dynamics of STN–GPe system. This hypothesis has inspired models simulating various BG functions ranging from action selection in continuous spaces (Krishnan, Ratnadurai, Subramanian, Chakravarthy, & Rengaswamy, 2011), reaching movements (Magdoom et al., 2011), spatial navigation (Sukumar, Rengaswamy, & Chakravarthy, 2012), precision grip

V. S. Chakravarthy and A. Moustafa, *Computational Neuroscience Models of the Basal Ganglia*, Cognitive Science and Technology,
https://doi.org/10.1007/978-981-10-8494-2_9

(Gupta, Balasubramani, & Chakravarthy, 2013), and gait (Muralidharan, Balasubramani, Chakravarthy, Lewis, & Moustafa, 2013) in normal and Parkinsonian conditions.

We earlier showed using a network of rate-coding neurons that exploration emerges out of the chaotic dynamics of the STN–GPe system (Kalva, Rengaswamy, Chakravarthy, & Gupte, 2012). Most rate-coded models, by design, fail to capture dynamic phenomena like synchronization found in more realistic spiking neuron models (Bevan, Magill, Terman, Bolam, & Wilson, 2002; Choongseok Park, Worth, & Rubchinsky, 2010; Park, Worth, & Rubchinsky, 2011). The synchronization within the BG nuclei had gained attention since the discovery that STN, GPe, and GPi neurons show high levels of synchrony in Parkinsonian conditions (Bergman, Wichmann, Karmon, & DeLong, 1994; Bevan et al., 2002; Hammond, Bergman, & Brown, 2007; Tachibana, Iwamuro, Kita, Takada, & Nambu, 2011; Weinberger & Dostrovsky, 2011).

In this chapter, we show using approaches at multiple scales that STN–GPe chaotic dynamics is the source of exploration. We present two models (1) Izhikevich spiking neuron networks and (2) biophysical neural network. At each level, the models consistently reveal the emergence of an Explore regime between the classical Go and NoGo regimes. We also show the role of DA in controlling the exploration–exploitation tradeoff using a binary action selection paradigm. We further incorporate learning into the models and show how the three models can learn the n-armed bandit task, which we will explain in detail below.

9.2 Methods

In this section, we present two models of BG at multiple scales that exhibit the Go, Explore, and NoGo regimes and also preform action selection. The first model has neurons that are Izhikevich spiking neurons, calibrated to reproduce the firing rates of various BG nuclei, and the second model uses conductance-based neuron models. The models are applied to (1) the binary action selection problem and (2) the n-armed bandit problem, where the network is required to choose the most salient one of several inputs with different levels of salience. A 'tonic dopamine' signal modulates the response of the D1R and D2R expressing medium spiny neurons (MSNs) in the striatum. In classical accounts of BG function, striatal dopamine is thought to switch transmission between DP and IP, whereby higher dopamine leads to 'Go' behavior, while the 'NoGo' behavior is elicited at lower dopamine levels. However, all the three aforementioned models exhibit a novel stochastic regime, known as the Explore regime, between the Go and NoGo regimes. All the networks have a common architecture as shown in Fig. 9.1.

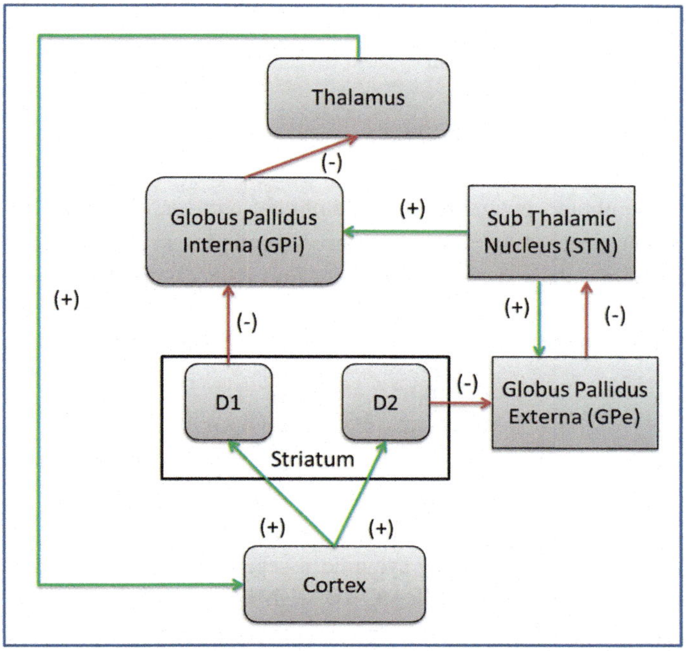

Fig. 9.1 Figure 9.1 shows the common architectures of three models of the basal ganglia. Information from the cortex is projected to the basal ganglia through the striatum (D1R and D2R-expressing MSNs) which is directed to direct (GPi) and indirect pathway (GPe → STN), respectively. The output system GPi receives input from D1 striatum and STN. The synaptic projections from striatum, GPe, and GPi were modeled as GABAergic and from STN as glutamatergic

9.2.1 Spiking Izhikevich Two-Variable Neuron Model

The first one is a network model of BG (Fig. 9.1) was built using two-variable Izhikevich spiking neurons (Izhikevich, 2003) where each nucleus was modeled as a lattice with (= 50 × 50 neurons) of neurons. Equations related to the Izhikevich spiking neuron model are described in (Mandali & Chakravarthy, 2015; Mandali, Rengaswamy, Chakravarthy, & Moustafa, 2015). The basic Izhikevich equations used in the model are described below, and the details of the model with equations are explained in Chap. 6.

$$\frac{dv_{ij}^x}{dt} = 0.04\left(v_{ij}^x\right)^2 + 5v_{ij}^x - u_{ij}^x + 140 + I_{ij}^x + I_{ij}^{syn} \tag{9.1}$$

$$\frac{du_{ij}^{x}}{dt} = a\left(bv_{ij}^{x} - u_{ij}^{x}\right) \tag{9.2}$$

$$\text{if } v_{ij}^{x} \geq v_{\text{peak}} \left\{ \begin{array}{l} v_{ij}^{x} \leftarrow c \\ u_{ij}^{x} \leftarrow u_{ij}^{x} + \mathrm{d} \end{array} \right\} \tag{9.3}$$

where v_{ij}^{x} = membrane potential, u_{ij}^{x} = membrane recovery variable, I_{ij}^{syn} = total synaptic current received, I_{ij}^{x} = external current applied to neuron x at location (i, j), v_{peak} = maximum voltage set to neuron (+30 mv) with x being STN or GPe or GPi neuron.

The synaptic connectivity between the nuclei (STN and GPe) is modeled as,

$$\tau_{\text{Recep}} * \frac{dh_{ij}^{x\rightarrow y}}{dt} = -h_{ij}^{x\rightarrow y}(t) + S_{ij}^{x}(t) \tag{9.4}$$

$$I_{ij}^{x\rightarrow y}(t) = W_{x\rightarrow y} * h_{ij}^{x\rightarrow y}(t) * \left(E_{\text{Recep}} - V_{ij}^{y}(t)\right) \tag{9.5}$$

where τ_{Recep} = decay constant for synaptic receptor, E_{Recep} = receptor associated synaptic potential (Recep = AMPA/GABA/NMDA), S_{ij}^{x} = Spiking activity of neuron 'x' at time 't,' $h_{ij}^{x\rightarrow y}$ = gating variable for the synaptic current from 'x' to 'y,' $W_{x\rightarrow y}$ = synaptic weight from neuron 'x' to 'y,' V_{ij}^{y} = membrane potential of the neuron 'y' for the neuron at the location (i, j).

9.2.2 Hybrid Biophysical Model

The second BG model is a combination of spiking and conductance-based neural network model. The striatal MSNs (both D1R and D2R-expressing) were modeled as Izhikevich neurons (Eqs. 9.9–9.11) and STN, GPe, and GPi using conductance-based neuron models (Rubin & Terman, 2004; Terman, Rubin, Yew, & Wilson, 2002). Each of the nuclei has ten neurons, and their synaptic connectivity pattern is given in Table 9.1.

Table 9.1 Table 9.1 shows the connectivity pattern between STN, GPe, and GPi neurons

Connection	Type of connection
1STN → 2GPi	Excitatory
1GPe → 2STN	Inhibitory
1GPe → 1GPi	Inhibitory
1STN → 1GPe	Excitatory
1GPe → 2GPe	Inhibitory
1STN → 2STN	Excitatory

9.2.2.1 STN Neurons

We have adapted the biophysical model by Rubin & Terman, (2004) which was capable of showing multiple firing patterns observed experimentally in STN.

$$C_{\mathrm{m}} \frac{\mathrm{d}V_{\mathrm{m}}^{\mathrm{STN}}}{\mathrm{d}t} = -I_{\mathrm{L}} - I_{\mathrm{K}} - I_{\mathrm{Na}} - I_{\mathrm{CaL}} - I_{\mathrm{CaT}} - I_{\mathrm{AHP}} - I_{\mathrm{GS}} - I_{\mathrm{Lat}} \tag{9.6}$$

where $V_{\mathrm{m}}^{\mathrm{STN}}$ is membrane potential of STN neurons, C_{m} is the membrane capacitance, I_{L} is the leakage current, I_{Na} is the leakage current, I_{CaL} is the long-lasting calcium current, I_{CaT} is the transient calcium current, I_{AHP} is the after hyperpolarization current, I_{GS} is the inhibitory GABAergic current from GPe neurons, and I_{Lat} is the excitatory glutamatergic lateral current from neighboring STN neurons.

9.2.2.2 GPe Neurons

Furthermore, the GPe, GPi membrane potential was modeled utilizing same ion channels equivalent to STN. Additionally, output from striatum was also presented as input to GPe and GPi neurons.

$$C_{\mathrm{m}} \frac{\mathrm{d}V_{\mathrm{m}}^{\mathrm{GPe}}}{\mathrm{d}t} = -I_{\mathrm{L}} - I_{\mathrm{K}} - I_{\mathrm{Na}} - I_{\mathrm{CaL}} - I_{\mathrm{CaT}} - I_{\mathrm{AHP}} - I_{\mathrm{SG}} - I_{\mathrm{Lat}} - I_{\mathrm{D2}} \tag{9.7}$$

where $V_{\mathrm{m}}^{\mathrm{GPe}}$ is membrane potential of GPe neurons, C_{m} is the membrane capacitance, I_{L} is the leakage current, I_{Na} is the sodium current, I_{CaL} is the long-lasting calcium current, I_{CaT} is the transient calcium current, I_{AHP} is the after hyperpolarization current, I_{SG} is the excitatory glutamatergic current from STN neurons, I_{Lat} is the inhibitory GABAergic lateral current from neighboring GPe neurons, and I_{D2} is the inhibitory GABAergic lateral current from D2 striatal MSNs calculated to that described similar to Eqs. (12, 13).

9.2.2.3 GPi Neurons

$$C_{\mathrm{m}} \frac{\mathrm{d}V_{\mathrm{m}}^{\mathrm{GPi}}}{\mathrm{d}t} = -I_{\mathrm{L}} - I_{\mathrm{K}} - I_{\mathrm{Na}} - I_{\mathrm{CaL}} - I_{\mathrm{CaT}} - I_{\mathrm{AHP}} - I_{\mathrm{SGi}} - I_{\mathrm{GeGi}} + I_{\mathrm{D1}} \tag{9.8}$$

where $V_{\mathrm{m}}^{\mathrm{GPi}}$ is membrane potential of GPi neurons, C_{m} is the membrane capacitance, I_{L} is the leakage current, I_{Na} is the leakage current, I_{CaL} is the long-lasting calcium current, I_{CaT} is the transient calcium current, I_{AHP} is the after hyperpolarization current, I_{SGi} is the excitatory glutamatergic current from STN neurons, and $\mathrm{I_{D1}}$ is the inhibitory GABAergic lateral current from D1 striatal MSNs calculated as described in Eqs. (12, 13). The modulation of DA on to the striatal activity was not included in this level of modeling.

DA plays an important role in maintaining the firing pattern of STN, GPe, and GPi neurons. Dopamine depletion leads to bursting behavior in STN and reduction in GPe firing rate in the Parkinson condition. To simulate the effect of dopamine, the conductance of *T*-type calcium channel and synaptic strength between STN → GPe, GPe → STN, STN → GPi and GPe → GPe was varied as a function of dopamine to exhibit the pathological oscillatory behavior.

9.2.2.4 Synaptic Currents

The synaptic currents between STN, GPe, and GPi neurons were modeled based on the equations described below

$$I_{x\to y} = g_{x\to y}\left[v_x - E_{x\to y}\right]\sum_j s_x^j \tag{9.9}$$

$$s_x' = A_x[1 - s_x]H_\infty(v_x - \theta_x) - B_x s_x \tag{9.10}$$

where $I_{x\to y}$ is the synaptic current from neuron '*x*' to '*y*,' $g_{x\to y}$ is the synaptic conductance from neuron '*x*' to '*y*,' v_x is the membrane potential of neuron *x*, $E_{x\to y}$ is the receptor potential for that synapse, s_x is the synaptic variable for the neuron '*x*.' The values for A_x, B_x, θ_x, H are adapted from Rubin & Terman, (2004).

9.2.2.5 Dopaminergic Modulation

Based on the observation that DA modulates the synaptic strength between STN, GPe, and GPi neurons and between laterals, the synaptic conductance variable 'g' (Eq. 9.17) was made a function of DA.

$$g^{\text{new}}_{\text{GPe}\to\text{STN}} = 5 * (1.1 - DA) * g^{\text{old}}_{\text{GPe}\to\text{STN}} \tag{9.11}$$

$$g^{\text{new}}_{\text{STN}\to\text{GPi}} = 20 * (1.1 - DA) * g^{\text{old}}_{\text{STN}\to\text{GPi}} \tag{9.12}$$

It is well known that the bursting activity in STN neurons is attributed to the activity of T-type calcium channels (Heida, Marani, & Usunoff, 2008; Tai, Yang, Pan, Huang, & Kuo, 2011) which is observed in PD. Recently, the influence of DA on calcium channel and its probable role in therapy (Yang, Tai, Pan, & Kuo, 2014) has been studied both by experimental and computational techniques (Dovzhenok & Rubchinsky, 2012; Heida et al., 2008; Loucif, Woodhall, Sehirli, & Stanford, 2008; Ramanathan, Tkatch, Atherton, Wilson, & Bevan, 2008; Tai et al., 2011). Bearing these observations in mind, we made the 'T'-type calcium conductance in STN neurons a function of DA.

$$g_{Ca_T}^{\text{new}} = 9 * (1.1 - DA) * g_{Ca_T}^{\text{old}} \tag{9.13}$$

where $g_{\text{GPe}\rightarrow\text{STN}}^{\text{old}}$, $g_{\text{STN}\rightarrow\text{GPi}}^{\text{old}}$, and $g_{Ca_T}^{\text{old}}$ are the conductance values taken from (Rubin & Terman, 2004; Terman et al., 2002) and DA is the dopamine level ranging from (0.1 to 1).

In all the three models (rated-coded, spiking and hybrid), the final action selection was performed using the race model (Mandali et al., 2015).

9.2.3 *Tasks*

9.2.3.1 Binary Action Selection Task

The first task we simulated was the simple binary action selection similar to Humphries, Stewart, and Gurney (2006), where two competing stimuli were presented to the model (Humphries, Stewart, & Gurney, 2006). The saliency of the input was represented in terms of their frequency. In a classical reinforcement learning paradigm, selection of higher salient stimulus among the available choices could be considered as 'exploitation' while selecting the less salient one as 'exploration' (Sutton & Barto, 1998). So the action selected is defined as 'Go' if stimulus #2 (more salient) is selected, 'Explore' if stimulus #1 (less salient) is selected, and 'NoGo' if none of them is selected. The simulation was run various values of DA, and the corresponding selected action was recorded. We also varied the lateral connection strengths in STN and observed its effect on exploration.

9.2.3.2 n-Arm Bandit Task

We then simulated the classical n-arm bandit problem, where the model is expected to learn and select the slot that delivers the maximum reward on its selection. The rewards obtained from the arm could be deterministic/probabilistic in nature, and the final goal is to learn and select the maximum reward giving arm. Experiments have showed the effect of DA on cortico-striatal weights which are similar to LTP/LTD behavior. Based on the classical experiment by Schultz, analogy between the temporal difference error term (δ) in reinforcement learning and DA has been established.

The saliency of each slot was represented by the cortico-striatal weights which was updated using the 'δ' term

$$\Delta w_{i,k+1}^{\text{D1}} = \eta \delta_k x_{i,k}^{\text{inp}} \tag{9.14}$$

$$\Delta w_{i,k+1}^{\text{D2}} = \eta \delta_k x_{i,k}^{\text{inp}} \tag{9.15}$$

The expected value and the actual reward obtained for the *k*th trial are given as

$$V_k = \sum_{i=1}^{n} w_{i,k}^{\mathrm{D1}} * x_{i,k}^{\mathrm{inp}} \tag{9.16}$$

$$R_k = \sum_{i=1}^{n} r_{i,k} * x_{i,k}^{\mathrm{inp}} \tag{9.17}$$

$$r_{i,k} = \left(\frac{i}{4} + 0.3 * \mathrm{e}\right), \mathrm{e} \sim N \tag{9.18}$$

The temporal difference term 'δ' for the given trial was calculated as

$$\delta_k = R_k + \gamma V_{k+1} - V_k \tag{9.19}$$

where $w_{i,k+1}^{\mathrm{D1}}$, $w_{i,k+1}^{\mathrm{D2}}$ were the cortico-striatal weights of D1 and D2 striatum for k + 1th trial, η is the learning rate, δ_k is the temporal difference error calculated for kth trial, V_{k+1}, V_k are the expected values for k + 1th and kth trial, and γ is the discount factor (=0).

9.3 Results

We present the results from each of the two models starting from both binary action selection and n-arm bandit task. The results from the GEN model (Chap. 5) are also included for comparison.

9.3.1 Binary Action Selection

These are results from the binary action selection problem described in Sect. 9.2.3.1. We highlight the concept of Explore regime along with Go and NoGo regime which have been explained in the classical picture of basal ganglia. The task was simulated by the two models by changing the dopamine level and observing the action that was selected (Fig. 9.2). The results obtained from the two models (Izhikevich and hybrid) along with the rate-coded one from earlier chapter are presented in Fig. 9.2a–c.

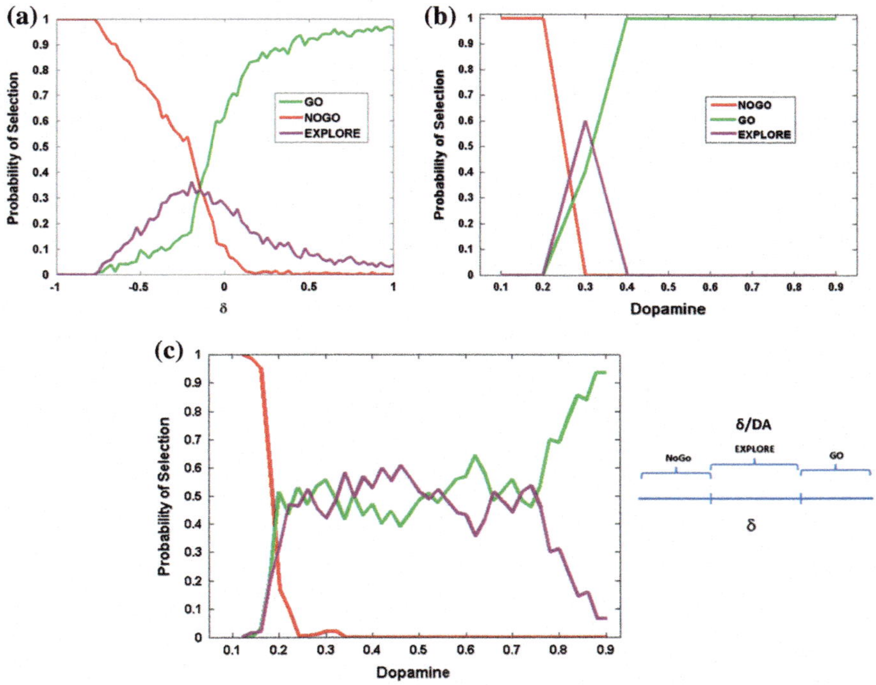

Fig. 9.2 Figure 9.2 shows the probability of selection of stimulus in the binary action selection where red indicates 'NoGo' which means no action, pink shows Explore, i.e., less salient stimulus and green indicates 'Go' the salient stimulus when the dopamine (*δ*/DA) was varied. **a** Rate-coded model (from Chap. 5) **b** Spiking model **c** hybrid model with an illustration of the three regimes with change in DA

We then increased the strength of lateral connections in STN, i.e., (w_{ss} in rate-coded model from last chapter and A_s in spiking model) and observed action selection in binary action selection task (Fig. 9.3). We observed that the amount of exploration decreased as the lateral strength was increased, and this result was consistently observed from both the results.

9.3.2 N-Arm Bandit Task

The next set of simulation involved the implementation of the simple n-arm bandit task. For this task, the total number of slots was defined to be four with the average reward as given in Eq. 9.18. The results were compared between rate-coded (earlier chapter), lumped (ε-greedy), and spiking models. The results from hybrid model are not included because of computational limitations.

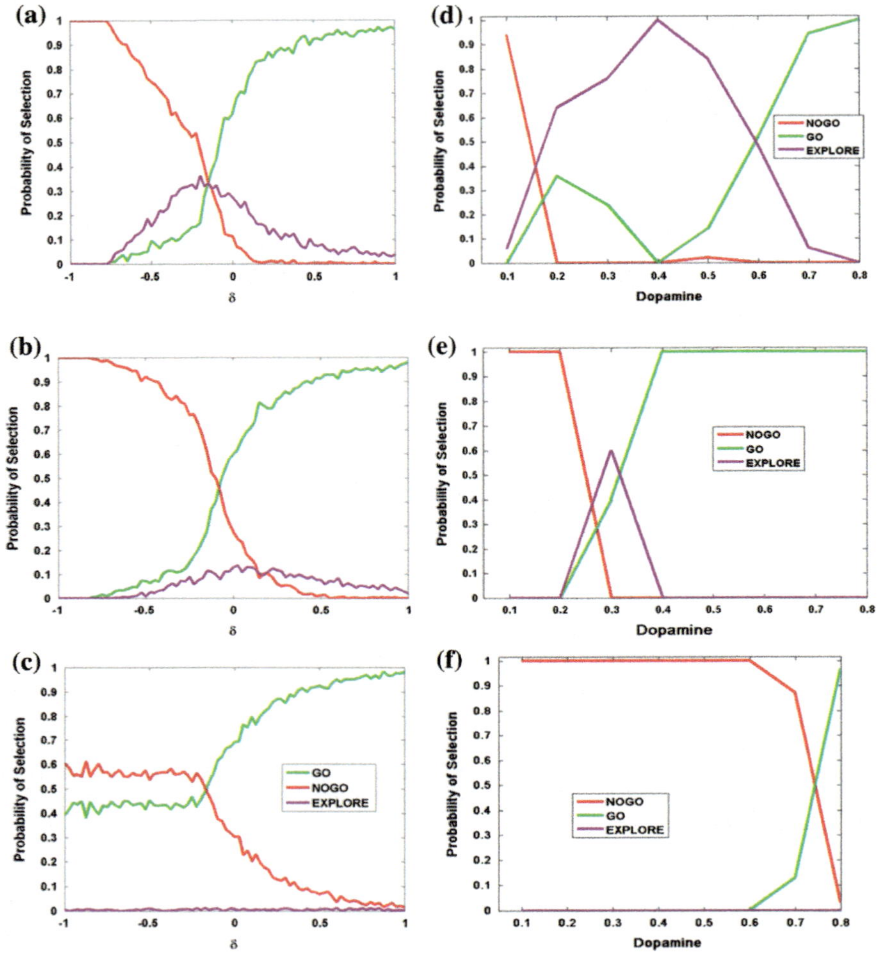

Fig. 9.3 Figure 9.3 shows the probability of action selection when the laterals strength in STN neurons was increased. The subfigures **a**, **b**, **c** are the results obtained from rate-coded model for lateral strengths ($w_{ss} = 0.001$, $w_{ss} = 0.05$, $w_{ss} = 0.95$). The figures **d**, **e**, **f** were the results from spiking model ($A_{\mathrm{STN}} = 0.15$, $A_{\mathrm{STN}} = 0.2$, $A_{\mathrm{STN}} = 0.3$).

9.4 Discussion

In this chapter, we show the consistency in our 'Go-Explore-NoGo' theory of BG using computational models at multiple levels of abstraction. The results from the two models (spiking, hybrid) for binary action selection task and n-arm bandit task were compared.

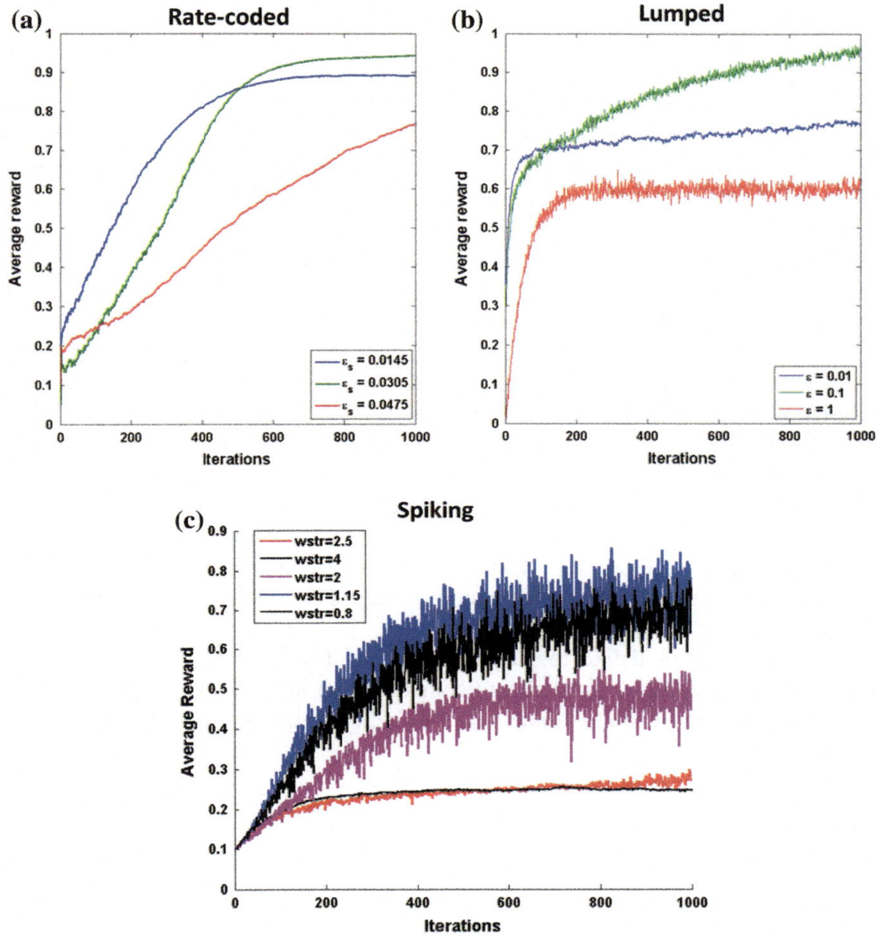

Fig. 9.4 Figure 9.4 shows the average reward obtained from the n-arm bandit task from each of the models **a** average reward obtained from rate-coded model when the lateral connections were varied which control the exploration; **b** average reward obtained from lumped model for different values of 'ε'; and **c** average reward obtained from spiking model for various values of synaptic weight (w_{str}) which controls the amount of exploration

The first task, binary action selection task was used to test the presence of Explore regime when dopamine levels were varied in the model (Fig. 9.2). All the three models showed the presence of 'NoGo' regime at lower DA levels, 'Go' at higher levels, and 'Explore' at intermediate levels. The model showed selection of no input at low DA due to very high activation of NoGo pathway and higher salient stimulus due to predominant influence of direct pathway. In between the two extremes, the models showed selection of lower salient stimulus (Explore) which is hypothesized to be due to the STN–GPe dynamics. To further understand this aspect, we studied the correlation and synchrony levels (Mandali & Chakravarthy, 2015) in STN and GPe neurons and observed an increase in synchrony at low DA levels. The desynchronous dynamics of STN–GPe neurons observed normally change to more synchronous one at low DA levels [for results please refer to (Mandali et al., 2015)]. We also observed that an increase in the synaptic strength of STN collaterals resulted in increased synchrony in the STN–GPe network (Fig. 9.3).

To further verify that the STN–GPe system influences exploration, we integrated spiking model with RL method and calculated the average reward and compared the results obtained from lumped (ε-greedy), rated-coded and Izhikevich models (Fig. 9.4) while performing the n-armed bandit for different level of synaptic strengths. The average reward obtained from rate-coded and spiking models for different synaptic strengths was compared with standard RL Explore–Exploit models (greedy and ε-greedy where ε controls the level of exploration in the model). The average reward obtained by varying the synaptic strengths (Fig. 9.4a and c), which internally controlled the STN–GPe dynamics, was comparable to the results obtained from lumped RL models with different 'ε' levels. The above results show that DA controlled the 'Explore' regime by modulating the STN–GPe dynamics. Furthermore, the synaptic strength of STN collaterals also controls the percentage of exploration.

In future, we would like to include the n-arm bandit results from the hybrid model which we could not now due to computational limitations. We plan to study the influence of individual channels of STN, GPe neurons on exploration in the n-arm bandit task.

References

Bergman, H., Wichmann, T., Karmon, B., & DeLong, M. (1994). The primate subthalamic nucleus. II. Neuronal activity in the MPTP model of parkinsonism. *Journal of Neurophysiology, 72*(2), 507–520.

Bevan, M. D., Magill, P. J., Terman, D., Bolam, J. P., & Wilson, C. J. (2002). Move to the rhythm: oscillations in the subthalamic nucleus–external globus pallidus network. *Trends in Neurosciences, 25*(10), 525–531.

Borisyuk, G. N., Borisyuk, R. M., Khibnik, A. I., & Roose, D. (1995). Dynamics and bifurcations of two coupled neural oscillators with different connection types. *Bulletin of Mathematical Biology, 57*(6), 809–840.

Chakravarthy, V., Joseph, D., & Bapi, R. S. (2010). What do the basal ganglia do? *A modeling perspective. Biological cybernetics, 103*(3), 237–253.

Dovzhenok, A., & Rubchinsky, L. L. (2012). On the Origin of Tremor in Parkinson's Disease. *PLoS ONE, 7*(7), e41598.

Gupta, A., Balasubramani, P. P., & Chakravarthy, V. S. (2013). Computational model of precision grip in Parkinson's disease: a utility based approach. *Frontiers in computational neuroscience, 7.*

Hammond, C., Bergman, H., & Brown, P. (2007). Pathological synchronization in Parkinson's disease: Networks, models and treatments. *Trends in Neurosciences, 30*(7), 357–364.

Heida, T., Marani, E., & Usunoff, K. G. (2008). *The Basal Ganglia*: Springer.

Humphries, M. D., Stewart, R. D., & Gurney, K. N. (2006). A physiologically plausible model of action selection and oscillatory activity in the basal ganglia. *The Journal of Neuroscience, 26* (50), 12921–12942.

Izhikevich, E. M. (2003). Simple model of spiking neurons. *IEEE Transactions on Neural Networks, 14*(6), 1569–1572.

Kalva, S. K., Rengaswamy, M., Chakravarthy, V. S., & Gupte, N. (2012). On the neural substrates for exploratory dynamics in basal ganglia: a model. *Neural Networks, 32,* 65–73. https://doi.org/10.1016/j.neunet.2012.02.031.

Krishnan, R., Ratnadurai, S., Subramanian, D., Chakravarthy, V. S., & Rengaswamy, M. (2011). Modeling the role of basal ganglia in saccade generation: Is the indirect pathway the explorer? *Neural Networks, 24*(8), 801–813.

Loucif, A. J., Woodhall, G. L., Sehirli, U. S., & Stanford, I. M. (2008). Depolarisation and suppression of burst firing activity in the mouse subthalamic nucleus by dopamine D1/D5 receptor activation of a cyclic-nucleotide gated non-specific cation conductance. *Neuropharmacology, 55*(1), 94–105.

Magdoom, K., Subramanian, D., Chakravarthy, V. S., Ravindran, B., Amari, S.-I., & Meenakshisundaram, N. (2011). Modeling basal ganglia for understanding parkinsonian reaching movements. *Neural Computation, 23*(2), 477–516.

Mandali, A., & Chakravarthy, V. S. (2015). *A computational basal ganglia model to assess the role of STN-DBS on Impulsivity in Parkinson's disease.* Paper presented at the Neural Networks (IJCNN), 2015 International Joint Conference on.

Mandali, A., Rengaswamy, M., Chakravarthy, V. S., & Moustafa, A. A. (2015). A spiking Basal Ganglia model of synchrony, exploration and decision making. *Frontiers in Neuroscience, 9,* 191.

Muralidharan, V., Balasubramani, P. P., Chakravarthy, V. S., Lewis, S. J., & Moustafa, A. A. (2013). A computational model of altered gait patterns in parkinson's disease patients negotiating narrow doorways. *Frontiers in computational neuroscience, 7.*

Park, C., Worth, R. M., & Rubchinsky, L. L. (2010). Fine temporal structure of beta oscillations synchronization in subthalamic nucleus in Parkinson's disease. *Journal of Neurophysiology, 103*(5), 2707–2716.

Park, C., Worth, R. M., & Rubchinsky, L. L. (2011). Neural dynamics in parkinsonian brain: the boundary between synchronized and nonsynchronized dynamics. *Physical Review E, 83*(4), 042901.

Ramanathan, S., Tkatch, T., Atherton, J. F., Wilson, C. J., & Bevan, M. D. (2008). D2-like dopamine receptors modulate SKCa channel function in subthalamic nucleus neurons through inhibition of Cav2. 2 channels. *Journal of Neurophysiology, 99*(2), 442–459.

Rubin, J. E., & Terman, D. (2004). High frequency stimulation of the subthalamic nucleus eliminates pathological thalamic rhythmicity in a computational model. *Journal of Computational Neuroscience, 16*(3), 211–235.

Sinha, S. (1999). Noise-free stochastic resonance in simple chaotic systems. *Physica A: Statistical Mechanics and its Applications, 270*(1), 204–214.

Sukumar, D., Rengaswamy, M., & Chakravarthy, V. S. (2012). Modeling the contributions of Basal ganglia and Hippocampus to spatial navigation using reinforcement learning. *PLoS ONE, 7*(10), e47467.

Sutton, R. S., & Barto, A. G. (1998). *Reinforcement learning: An introduction* (Vol. 1): Cambridge Univ Press.

Tachibana, Y., Iwamuro, H., Kita, H., Takada, M., & Nambu, A. (2011). Subthalamo-pallidal interactions underlying parkinsonian neuronal oscillations in the primate basal ganglia. *European Journal of Neuroscience, 34*(9), 1470–1484.

Tai, C.-H., Yang, Y.-C., Pan, M.-K., Huang, C.-S., & Kuo, C.-C. (2011). Modulation of subthalamic T-type Ca 2+ channels remedies locomotor deficits in a rat model of Parkinson disease. *The Journal of Clinical Investigation, 121*(8), 3289–3305.

Terman, D., Rubin, J., Yew, A., & Wilson, C. (2002). Activity patterns in a model for the subthalamopallidal network of the basal ganglia. *The Journal of neuroscience, 22*(7), 2963–2976.

Weinberger, M., & Dostrovsky, J. O. (2011). A basis for the pathological oscillations in basal ganglia: the crucial role of dopamine. *NeuroReport, 22*(4), 151.

Yang, Y.-C., Tai, C.-H., Pan, M.-K., & Kuo, C.-C. (2014). The T-type calcium channel as a new therapeutic target for Parkinson's disease. *Pflügers Archiv-European Journal of Physiology, 466*(4), 747–755.

Chapter 10
A Cortico-Basal Ganglia Model to Understand the Neural Dynamics of Targeted Reaching in Normal and Parkinson's Conditions

Vignesh Muralidharan, Alekhya Mandali, Pragathi Priyadharsini Balasubramani, Hima Mehta, V. Srinivasa Chakravarthy and Marjan Jahanshahi

Abstract We present a cortico-basal ganglia model to study the neural mechanisms behind reaching movements in normal and in Parkinson's disease conditions. The model consists of the following components: a two-joint arm model (AM), a layer of motor neurons in the spinal cord (MN), the proprioceptive cortex (PC), the motor cortex (MC), the prefrontal cortex (PFC), and the basal ganglia (BG). The model thus has an outer sensory-motor cortical loop and an inner cortico-basal ganglia loop to drive learning of reaching behavior. Sensory and motor maps are formed by the PC and MC which represent the space of arm configurations. The BG sends control signals to the MC following a stochastic gradient ascent policy applied to the value function defined over the arm configuration space. The trainable connections from PFC to MC can directly activate the motor cortex, thereby producing rapid movement avoiding the slow search conducted by the BG. The model captures the two main stages of motor learning, i.e., slow movements dominated by the BG during early stages and cortically driven fast movements with smoother trajectories at later stages. The model explains PD performance in stationary and pursuit reaching tasks. The model also shows that PD symptoms like tremor and rigidity could be attributed to synchronized oscillations in STN–GPe. The model is in line with closed-loop control and with neural representations for all the nuclei which explains Parkinsonian reaching. By virtue of its ability to capture the role of cortico-basal ganglia systems in controlling a wide range of features of reaching, the proposed model can potentially serve as a benchmark to test various motor pathologies of the BG.

10.1 Introduction

Reaching movements are for movement science, what the simple pendulum is for classical mechanics. Reaching movements reveal a lot about how the brain plans and executes movement kinematics and dynamics, in normal and pathological

V. S. Chakravarthy and A. Moustafa, *Computational Neuroscience Models of the Basal Ganglia*, Cognitive Science and Technology,
https://doi.org/10.1007/978-981-10-8494-2_10

conditions. Early experiments by Fitts and Morrasso on reaching movements showed that the hand velocity profile has a bell-shaped distribution providing a glimpse into the planning of motor trajectories (Fitts, 1954; Morasso, 1981). However, it was observed that such planning required adaptive feedback mechanisms which could relay the current state of the motor effector and the learning framework for optimal control (Todorov, 2004). As a result, internal models were introduced which minimize the error between the target and current arm position by including the factor of variability which accounts for the noise in movement (Shadmehr & Krakauer, 2008). According to the optimal feedback control framework, the current state plays a crucial role in determining the future state and eventually the trajectory and thus probed investigators to modeling reaching with Baye's approach as it allows integration of previous knowledge with current sensory information (Schaal & Schweighofer, 2005). Kording and Wolpert showed that the experimental results of visually guided reaching task in the presence of noisy feedback and explained using the Bayesian approach how subjects represented both the statistics of the sensorimotor task and the uncertainty in the task (Körding & Wolpert, 2004). A neural correlate to Bayesian processing by neurons was suggested by Knill and Pouget using a computational model by introducing Poisson noise in the neural activity (Knill & Pouget, 2004).

Though the above computational models provided insights into movement planning and execution, they do not specify the corresponding neural correlates. During the same period, experimental groups were studying the roles of various cortical and subcortical areas in motor learning and execution (Doya, 1999). Particularly, the basal ganglia (BG) are involved in the learning of new actions and sequences from cortical projections which are modulated by the midbrain dopaminergic system (Hikosaka, Nakamura, Sakai, & Nakahara, 2002). Using a computational model, Nakahara, Doya, & Hikosaka, (2002) showed that parallel learning occurs in BG-cortical systems through the visual and the motor loop. The final output is selected by the presupplementary area which acts as a coordinator for optimal acquisition and execution of well-learned sequences (Hikosaka et al., 2002; Nakahara, Doya, & Hikosaka, 2001). Chen and colleagues developed a model of closed-loop control of hand movement, in which a sensory module receives input from visual/proprioceptive areas and a Motor Module drives a mechanical two-link arm (Chen & Reggia, 1996). By randomly activating various points in motor cortex, the arm is driven to various points in the workspace; feedback from the hand is used to train the proprioceptive cortex, the motor cortex, and the motor neurons of the spinal cord module. The model is trained by unsupervised learning and was unable to describe goal-oriented reaching which requires either supervised or reinforcement learning. Izawa, Kondo, and Ito (2004) modeled a two-link arm model with detailed arm kinematics and included learning using reinforcement learning (Izawa et al., 2004).

We present a model of reaching that describes the contributions of basal ganglia (BG) and the sensory-motor cortical pathway to reaching. The model particularly highlights the role of BG in motor learning. In the model, the BG system discovers the desired motor cortical output by processing the reaching error which, we propose, is coded by nigrostriatal dopamine signals. This desired output is used by the motor cortex for training. Thus in the model, the BG leads the motor cortex in learning. The relative contributions of the cortical areas and BG evolve with learning, with the contribution of BG dwindling with learning. The model explains reaching movements in normal and Parkinsonian conditions and explores the causes of the distinct paths of evolution of PD symptoms into tremor-dominant and rigidity-dominant. The present model is a detailed network version of a simple lumped model of reaching that we proposed earlier (Magdoom et al., 2011).

10.2 Methods

The cortico-basal ganglia model consists of two major components: the outer loop which is the sensory-motor cortical loop and an inner loop which is the cortico-basal ganglia loop (Fig. 10.1). These loops are an integral part for the execution of controlled movements.

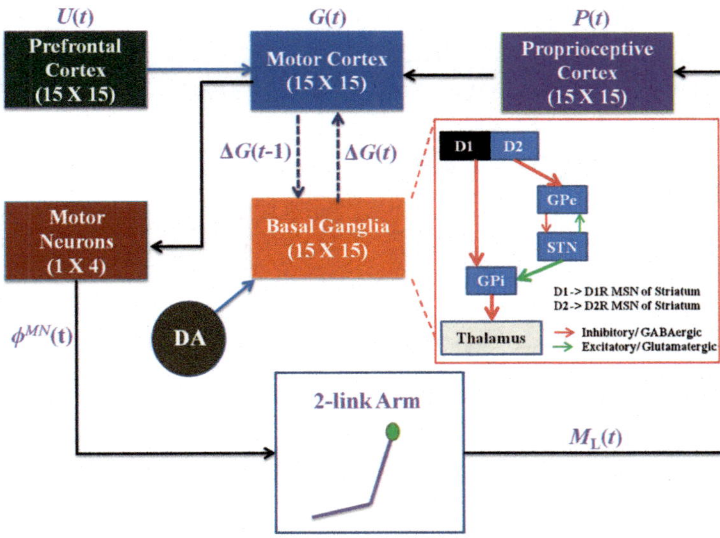

Fig. 10.1 Cortico-basal ganglia model used for simulating the reaching movements. The architecture is designed to have two loops, a sensory-motor 'outer' loop (shown by solid black arrows) and the cortico-basal ganglia 'inner' loop (shown by dashed black arrows). The basal ganglia is shown to have projections from midbrain dopaminergic (DA) neurons. The motor cortex receives projections from higher frontal areas which in the model is the prefrontal cortex. The $(m \times n)$ shows the size of the neuronal sheet used for each area in the model (in the basal ganglia all the nuclei are 15×15 in size)

10.2.1 Arm Model

A simple two-joint kinematic model of an arm is used in the model. Each joint is controlled by an agonist (Ag) and an antagonist (An) muscle pair innervated by a pair of motor neurons; the muscles in turn control the position of the arm in the 2D space. The input to the arm is a four-dimensional vector $\phi^{\mathrm{MN}}(t)$ which represents the muscle innervations for the agonist–antagonist pair for both the joints. The activation is then transformed to obtain the joint angles ($\theta^{\mathrm{JA}}_{S/E}(t)$) for shoulder and the elbow joint using Eqs. (10.1) and (10.2), respectively.

$$\theta_S^{\mathrm{JA}}(t) = \left(\phi_{\mathrm{Ag}}^{\mathrm{MN}}(t) - \phi_{\mathrm{An}}^{\mathrm{MN}}(t)\right)\frac{\pi}{2} + \frac{\pi}{2} \tag{10.1}$$

$$\theta_E^{\mathrm{JA}}(t) = \left(\phi_{\mathrm{Ag}}^{\mathrm{MN}}(t) - \phi_{\mathrm{An}}^{\mathrm{MN}}(t)\right)\frac{\pi}{2} + \frac{\pi}{2} \tag{10.2}$$

The arm covers a given set of targets in the workspace, restricted by the range of movements of the joints. The joint angle measures are subsequently used to determine the lengths (μ^E and μ^S) of each muscle [Eqs. (10.3), (10.4), (10.5), and (10.6)].

$$\mu_{\mathrm{Ag}}^S(t) = \sqrt{a_S^2 + b_S^2 + 2a_S b_S \cos\left(\theta_S^{\mathrm{JA}}\right)} \tag{10.3}$$

$$\mu_{\mathrm{An}}^S(t) = \sqrt{a_S^2 + b_S^2 - 2a_S b_S \cos\left(\theta_S^{\mathrm{JA}}\right)} \tag{10.4}$$

$$\mu_{\mathrm{Ag}}^E(t) = \sqrt{a_E^2 + b_E^2 + 2a_E b_E \cos\left(\theta_E^{\mathrm{JA}}\right)} \tag{10.5}$$

$$\mu_{\mathrm{An}}^E(t) = \sqrt{a_E^2 + b_E^2 - 2a_E b_E \cos\left(\theta_E^{\mathrm{JA}}\right)} \tag{10.6}$$

These muscle lengths form the four-dimensional vector ($M_L = [\mu_{\mathrm{Ag}}^S\ \mu_{\mathrm{An}}^S\ \mu_{\mathrm{Ag}}^E\ \mu_{\mathrm{An}}^E]$) which is used to develop a sensory (proprioceptive) map of the arm. Furthermore, the end effector position ($X^{\mathrm{arm}} = [x_1^{\mathrm{arm}}\ x_2^{\mathrm{arm}}]$) is also estimated in Eqs. (10.7) and (10.8).

$$x_1^{\mathrm{arm}} = (l_S - a_S)\cos\left(\theta_S^{\mathrm{JA}}\right) + l_E \cos\left(\theta_S^{\mathrm{JA}} + \theta_E^{\mathrm{JA}}\right) \tag{10.7}$$

$$x_2^{\mathrm{arm}} = (l_S - a_S)\sin\left(\theta_S^{\mathrm{JA}}\right) + l_E \sin\left(\theta_S^{\mathrm{JA}} + \theta_E^{\mathrm{JA}}\right) \tag{10.8}$$

10.2.2 The Sensory-Motor Cortical Loop

Sensory and Motor Maps

The sensory-motor cortical loop comprises of the arm, the proprioceptive cortex (PC), i.e., the proprioceptive area of the primary somatosensory cortex, the motor cortex, and the spinal motor neurons. PC is modeled as a self-organizing map (SOM) of size $N_{PC} \times N_{PC}$ (Kohonen, 1990). In order to develop a sensory map of the arm which we will from now on refer to as the proprioceptive map/cortex (PC), the muscle length vector ($M_L(t)$) received from the arm is used as feature vector to train the PC. The activation of a single node i in the PC is given by Eq. (10.9).

$$P_i(t) = \exp\left(\frac{-\left\|M_L(t) - W_{PC,i}\right\|^2}{\sigma_{PC}^2}\right) \tag{10.9}$$

where $W_{PC,i}$ is the weight connection between the muscle length vector of the arm and the ith node of the PC, and σ_{PC} is the width of the Gaussian response.

The motor cortex (MC) is modeled as a combination of a continuous attractor neural network (CANN) (Trappenberg, 2003) and a SOM of size $N_{MC} \times N_{MC}$. This represents two distinct characteristics of cortical areas which are known to have low-dimensional representation of the input space and dynamics based on the connectivity in these areas. The CANN architecture is characterized by short-range excitation and long-range inhibition. Its weight kernel (W_{MC}^C) is parameterized by the strength of the excitatory connections (A_{lat}^C), the radius of the excitatory connections (σ_{lat}^C), and the global inhibition constant (K^C). A dynamic model like the CANN is used to model MC, instead of a static model like SOM, so as to be able to dynamically integrate the afferent inputs coming from the PC, BG, and the prefrontal cortex (PFC). The PC activity is used as the input to generate the low-dimensional feature maps at the level of the MC. The MC uses this sensory map information to develop a motor map of the arm. This is done by giving the output of PC, a matrix of size $N_{PC} \times N_{PC}$, converted into a vector of size $N_{PC}^2 \times 1$, as input vector to the SOM part of MC. Training of this SOM is performed by the standard SOM algorithm (Kohonen, 1990). Output of the PC (G_{PC}), in addition to two other inputs, is presented as input to the CANN part of the MC (I_{MC}). The network is fully connected from the arm to the PC; similarly, every PC neuron projects to every neuron in MC. The activation of a node i in the SOM part of the MC is given by Eq. (10.10).

$$G_{PC,i}(t) = \exp\left(\frac{-\left\|P(t) - W_{MC,i}\right\|^2}{\sigma_{MC}^2}\right) \tag{10.10}$$

Here, $W_{MC,i}$ is the weight connection between the PC and the ith node of the SOM part of MC and σ_{MC} is the width of the Gaussian response. The MC activation via the attractor dynamics is driven by the PC, the BG, and the PFC. Therefore, the total input coming into the MC is $I_{MC}(t) = A_{PC}G_{PC}(t) + A_{BG}G_{BG}(t) + A_{PFC}G_{PFC}(t)$, where A_{PC}, A_{BG}, and A_{PFC} are the respective gains of the PC, BG, and PFC networks. With these inputs, the activation dynamics of the MC is given in (10.11).

$$\tau_{MC}\frac{dg_{MC}}{dt} = -g_{MC} + W_{MC}^{C} \otimes G + I_{MC} \tag{10.11}$$

where g_{MC} is the internal state of the MC neurons, W_{MC}^{C} is the weight kernel given by $W_{MC,i,j}^{C} = A_{lat}^{C}\exp\left(\frac{-\|(i_{MC}-i_h)+(j_{MC}-j_h)\|^2}{2\left(\sigma_{lat}^{C}\right)^2}\right) - K^C$ which determines the local excitation/global inhibition dynamics, $[i_{MC}, j_{MC}]$ are the locations of the nodes in MC, and $[i_h, j_h]$ corresponds to the central node. The output activity of the MC ($G(t)$) is obtained by performing a divisive normalization [Eq. (10.12)], which is done often to produce biologically realistic activity bumps (Pouget & Latham, 1999).

$$G(t) = \frac{g_{MC}^2}{1 + \left(\frac{2\pi}{N_{MC}^2}\right) b_{MC} \sum g_{MC}^2} \tag{10.12}$$

Neurons of the motor cortex project to the motor neuronal layer (MN). The motor neurons—there are just four of them—in turn project one each to the four muscles of the arm as described by Eq. (10.13).

$$\phi^{MN}(t) = A_{MN} W_{MC\to MN} G(t) \tag{10.13}$$

In order to close this loop, i.e., to train the connections between the MC and the MN ($W_{MC\to MN}$) layers [Eq. (10.14)], we initially provide the input at the MN layer as the desired activation for the arm ($\varphi_D^{MN}(t)$). This produces a sensory activity in the PC which in turn generates a motor activity in the MC ($G(t)$). The weights between the MC and the MN layers are trained in a supervised manner by comparing the network-derived MN activation $\varphi^{MN}(t)$ to the desired activation $\varphi_D^{MN}(t)$ (Eq. 10.14). This gives a loop which is consistent in mapping the external arm space to the neuronal space and vice versa.

$$\Delta W_{MC\to MN} = \eta_{MC\to MN}\left(\phi_D^{MN}(t) - \phi^{MN}(t)\right)G(t) \tag{10.14}$$

10.2.3 Training the Cortical Loop

The training schema for the entire model is shown in Fig. 10.2. The steps for training the sensory-motor loop are as follows.

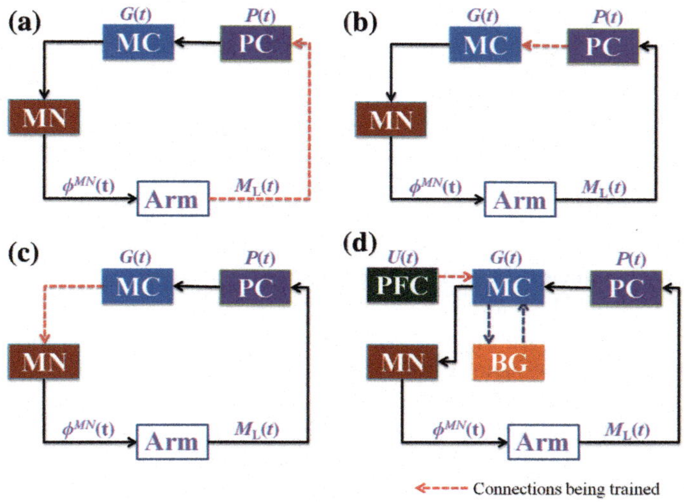

Fig. 10.2 Training schema in the cortico-basal ganglia model. It initially starts with (**a**) training the arm to PC connections, followed by (**b**) training the PC-to-MC connections, and finally closing the loop by (**c**) training the MC-to-MN weights. Then, the BG module is introduced and the PFC-to-MC connections are trained (**d**). In every figure, the dashed arrows indicate the connections that are being trained

1. Randomly generate n different muscle activations of the arm, which result in n-arm configurations. Each configuration of the arm provides a feature vector of muscle lengths, M_L.
2. The feature vector of muscle lengths, M_L, is presented as input to the PC layer, which is trained using the SOM algorithm.
3. The output state of PC layer is then presented as input to MC. Output of MC is presented as input to MN layer via a weight stage ($W_{\mathrm{MC}\rightarrow\mathrm{MN}}$). $W_{\mathrm{MC}\rightarrow\mathrm{MN}}$ are trained by the following procedure. A random activation vector (φ^{MN}) is given to the MN layer. The output of the MN layer then activates the arm and puts it in an equilibrium configuration. Starting from the muscle lengths from the arm, we track the signal flow via the PC, MC, and back to MN layers. Output of the MN layer, ideally, must be equal to the random activation vector (φ^{MN}) given to the arm. The matrix of the sensory activations is then passed on to the MC layer (SOM) to evolve the motor map of the arm.
4. Finally, the loop is closed by training the weight connection between the MC and the MN layers by generating a desired MN activity ($\varphi_D^{\mathrm{MN}}(t)$) and approximating the network-derived MN activity $\varphi^{\mathrm{MN}}(t)$ as in Eq. (10.14).

10.2.4 The Basal Ganglia

The BG module has the following components: the striatum, the Globus Pallidus internal and external segments (GPi and GPe), the subthalamic nucleus (STN), and the thalamus (Fig. 10.3a). The output of the BG modulates the MC activity to provide the appropriate control signal for the arm to reach the target. We begin with an outline of learning and operation of the BG module. In line with our earlier models of BG, in the proposed model, the BG module is trained by reinforcement learning to choose the optimal actions (Balasubramani, Chakravarthy, Ravindran, & Moustafa, 2014; Chakravarthy & Balasubramani, 2015; Chakravarthy, Joseph, & Bapi, 2010; Gupta, Balasubramani, & Chakravarthy, 2013; Magdoom et al., 2011; Muralidharan, Balasubramani, Chakravarthy, Lewis, & Moustafa, 2013). The BG, acting via the cortical loop, drives the arm so that the hand reaches the desired target (Fig. 10.3b). Prefrontal inputs which represent the target or the goal position and the current hand position information from the sensory cortex are thought to be combined in the BG to compute a value function that codes for the error between the desired and the actual hand position (Fig. 10.3c). The output of the BG performs a form of stochastic hill-climbing over the value function (Magdoom et al., 2011). Thus, by way of searching for the maxima in the value function, in the early stages of learning, the BG module drives the motor cortex to make reaching movements to the target. In the model however this value computation ($V^{\mathrm{arm}}(t)$) is not a result of training; value is presented as an explicit function of distance between the end effector (X^{arm}) and the goal position (X^{targ}) in Eq. (10.15). The σ_V term defines the spatial range over which the value function is sensitive for that particular target.

$$V^{\mathrm{arm}}(t) = \exp\left(\frac{-\|X^{\mathrm{targ}} - X^{\mathrm{arm}}\|^2}{\sigma_V^2}\right) \qquad (10.15)$$

Stochastic Hill-Climbing

The input to the striatum comes from the motor cortex in the form of a difference vector. We hypothesize that this difference vector which is the change in the motor cortical activity ($\Delta G(t)$) is the drive for the sustenance of motor activity. This information is then modulated via the direct (projections from D1-expressing neurons in the striatum) and the indirect (projections from D2-expressing neurons in the striatum) pathways as a function of the dopamine signal [Eqs. (10.17) and (10.18)]. From previous studies (Chakravarthy & Balasubramani, 2015; Magdoom et al., 2011), we have shown that this switching between direct and indirect pathways can be carried out using a form of the temporal difference signal called the *value difference* [Eq. (10.16)].

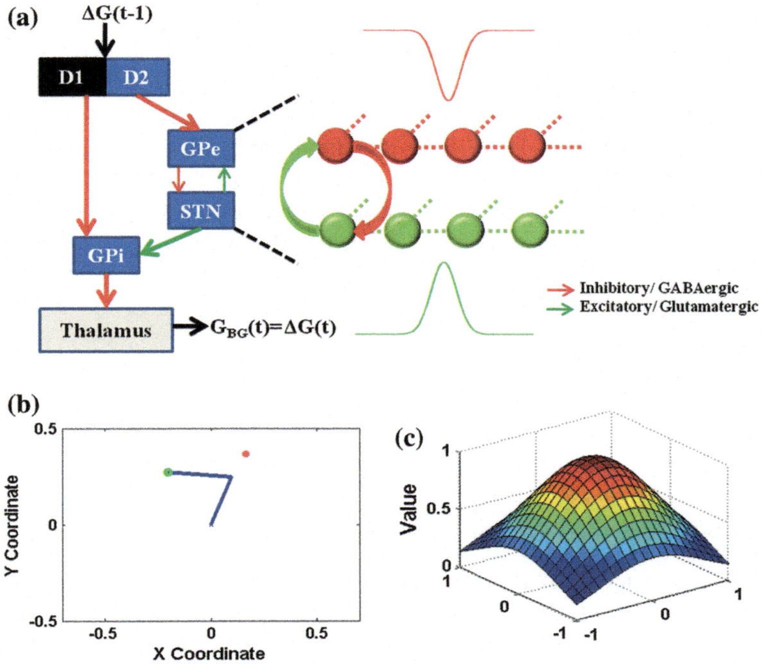

Fig. 10.3 Basal ganglia and value function. The network of the basal ganglia (**a**) which receives cortical input, $\Delta G(t)$, and via the direct and indirect pathways computes $\Delta G(t + 1)$. The indirect pathway has a 2D sheet of reciprocally connected STN–GPe neurons whose dynamics is governed by the lateral (neighborhood Gaussian connectivity, green STN, red GPe) and interconnectivity in these layers. The two-link arm (**b**) represented by blue lines for the links and green circle for the end effector, approaching a goal position (red circle) and computing the value function which peaks at the target location (**c**) in this case is a target at [0 0]. The BG dynamics essentially constitutes a stochastic hill-climbing mechanism that seeks to maximize the value function in order to reach the target

$$\delta_V = V^{\text{arm}}(t) - V^{\text{arm}}(t-1) \tag{10.16}$$

The quantity δ_v in Eq. (10.16) is called the *value difference*, which is subtly different from the temporal difference error. We proposed earlier that value difference also correlates with dopamine signals just as TD error has been suggested to be represented by dopamine signals (Chakravarthy & Balasubramani, 2015; Magdoom et al., 2011; Muralidharan et al., 2013). Value difference signals are thought to be carried by nigrostriatal connections to the striatum, where they modulate the responses of striatal projection neurons to cortical inputs as follows:

$$y_{\text{D1}} = \frac{1}{1 + \exp(-\lambda_{\text{D1}}(\delta_V - t_{\text{D1}}))} \Delta G(t) \tag{10.17}$$

$$y_{\mathrm{D2}} = \frac{1}{1 + \exp(-\lambda_{\mathrm{D2}}(\delta_V - t_{\mathrm{D2}}))}\Delta G(t) \tag{10.18}$$

where y_{D1} and y_{D2} represent the outputs of D1R- and D2R-expressing medium spiny neurons (MSNs), respectively. In the nonlinearity, λ_{D1} and t_{D1} and λ_{D2} and t_{D2} are the gains and the thresholds of the direct and indirect pathways, respectively. Also $\lambda_{\mathrm{D1}} = -\lambda_{\mathrm{D2}}$ which suggests that when δ_v is positive (negative), the direct (indirect) pathway is selected. Since D2R-expressing MSNs of the striatum project to the GPe, y_{D2} influences GPe neural dynamics, which in turn influences STN neural dynamics, as shown below.

$$\tau_{\mathrm{GPe}}\frac{dx_{\mathrm{GPe}}}{dt} = -x_{\mathrm{GPe}} + \varepsilon_g \sum\sum W^{\mathrm{glat}} x_{\mathrm{GPe}} + w_{sg} y_{\mathrm{STN}} + y_{\mathrm{D2}} \tag{10.19}$$

$$\tau_{\mathrm{STN}}\frac{dx_{\mathrm{STN}}}{dt} = -x_{\mathrm{STN}} + \varepsilon_s \sum\sum W^{\mathrm{slat}} y_{\mathrm{STN}} - w_{gs} x_{\mathrm{GPe}} \tag{10.20}$$

$$y_{\mathrm{STN}} = \tan h(\lambda_{\mathrm{STN}} x_{\mathrm{STN}}) \tag{10.21}$$

λ_{STN} controls the slope of the sigmoid, thus the STN output. τ_{STN} and τ_{GPe} are the respective timescales of STN and GPe. The weight parameters that control the connection strengths *between* the STN and GPe are w_{sg} and w_{gs}, and the weights that control lateral connections *within* both the STN and the GPe layer are W^{slat} and W^{glat} with connection strengths ϵ_s and ϵ_g, respectively, which have a Gaussian neighborhood as defined in (10.22).

$$W_{i,j,k,l}^{\mathrm{glat/slat}} = \exp\left(-\frac{\left(i_{g/s} - k_{g/s}\right)^2 + \left(j_{g/s} - l_{g/s}\right)^2}{\left(\sigma_{\mathrm{lat}}^{g/s}\right)^2} \right) \tag{10.22}$$

The indirect pathway consisting of the STN and GPe forms a coupled excitatory–inhibitory pair of neuronal pools [Eqs. (10.19) and (10.20)]. Such excitatory–inhibitory pairs of neuron pools are known to exhibit complex oscillations (Kalva, Rengaswamy, Chakravarthy, & Gupte, 2012). The dynamics of these oscillators is highly dependent on the input, which constitutes the projections from the D2-expressing neurons of the striatum. The STN layer in the model exhibits correlated activity for high striatal input, and uncorrelated oscillatory activity for low striatal inputs (see Appendix in Chap. 5). The uncorrelated oscillations of the STN are a key source of exploratory drive that randomly pushes the arm around in the workspace.

Here, $\sigma_{\mathrm{lat}}^{g/s}$ is the spread of the lateral connections, respectively, for the STN–GPe network. So for a given neuron *i*, *j* the weights represent a 2D Gaussian whose maximum is centered on (*i*, *j*). The output of the STN is combined in the GPi with the signal arriving via the direct pathway from the D1R-expressing MSNs in the striatum as follows:

$$y_{\text{GPi}} = A_{\text{D1}} y_{\text{D1}} - A_{\text{D2}} y_{\text{STN}} \tag{10.23}$$

At the level of the GPi, the DP output, i.e., y_{D1} and the STN output (y_{STN}) are combined (Eq. 10.23) and then passed on to the thalamus. The thalamus is modeled as a continuous attractor network which is necessary to integrate as well as filter information from the GPi output.

10.2.5 Prefrontal Cortex—Information of Goal Position

The motor command is thought to arise from the PFC, in the sense that the goal of the movement is represented in the PFC (Asplund, Todd, Snyder, & Marois, 2010; Matsumoto, Suzuki, & Tanaka, 2003). The PFC specifies information regarding the position of the goal to be reached. Similar to the PC and the MC layers, the PFC layer is trained like a SOM with weights W_{PFC}, but the input features are the spatial locations that the arm could reach in the model. Training in the model occurs in weights linking the PFC and the MC. The weights between the PFC and the MC ($W_{\text{PFC}\rightarrow\text{MC}}$) are trained as follows. A target, $X^{\text{targ}} = [x_1^{\text{targ}} \; x_2^{\text{targ}}]$ activates corresponding neurons in the PFC with activation $U(t)$ using Eq. (10.24).

$$U_i(t) = \exp\left(\frac{-\left\|X^{\text{targ}}(t) - W_{\text{PFC},i}\right\|^2}{\sigma_{\text{PFC}}^2}\right) \tag{10.24}$$

The arm initially makes reaching movements which are driven by dynamics aided by a stochastic hill-climbing procedure called 'Go-Explore-NoGo (GEN)' applied to the value function (Chakravarthy & Balasubramani, 2015). Furthermore, whenever the arm reaches the target position, the connections from PFC and MC are also trained, so that the motor command can directly activate the motor cortex, thereby producing rapid movement, without the slow search conducted by the BG. In this case, the training is initiated only when the arm reaches the target, i.e., the end effector is within a small radius, ξ (=0.1 units) of the target location. Similar to the MC and MN layer weights, the $W_{\text{PFC}\rightarrow\text{MC}}$ are trained in a supervised fashion (Eq. 10.25). Let G_{PFC} be the activity that PFC activation induces in MC; let G_{targ} be the activity in MC that drives the arm to the target location. Therefore, G_{targ} serves as a target vector for G_{PFC}. The weights from PFC to MC are therefore trained as follows:

$$\Delta W_{\text{PFC}\rightarrow\text{MC}} = \eta_{\text{PFC}\rightarrow\text{MC}}\left(G^{\text{targ}}(t) - G^{\text{PFC}}(t)\right)U(t) \tag{10.25}$$

Here, $G^{\text{PFC}}(t)$ is the PFC-driven MC activity and as this learning progresses, the arm reaches the goal position faster and faster. Therefore, the model exhibits two stages of motor learning: Slow movements dominated by the BG are seen in the early stages, while the cortically driven fast movements dominate the later stages. The PFC contribution increases as a function of the reaching error as in Eq. (10.26).

$$A_{\mathrm{PFC}} = A_{\mathrm{PFC}} + \frac{f_t}{r_e} \tag{10.26}$$

Here, f_t is a factor for controlling the speed of growth of A_{PFC} and r_e is the reaching error estimated as average distance to the target.

10.2.6 Timescales of Motor Movement in the Cortex and the BG

Reaching movements, like several other behavioral events, involve dynamics at multiple timescales: the neuronal activity which is generally in milliseconds, and the actual movement which unfolds over the order of seconds. In the model, the cortical loop is assumed to run slightly slower than the BG module. The integration time step used is 1 ms. As the dynamics of the STN–GPe loop in the indirect pathway needs some time to settle, we run this loop for 50 iterations, before sending the output to the MC. Thus, a single update of the MC activity happens after every 50 ms during which the BG dynamics run. All the results presented are at the timescale of the MC.

10.2.7 Simulating Pathology—Parkinsonian Condition

The *value difference* term 'δ_v', as we have mentioned earlier, is a correlate of the dopamine signal. To simulate the dopamine-deficient state of PD in the model, the 'δ_v' term is clamped to a lower value. Thus if $[d_{\mathrm{low}}\ d_{\mathrm{high}}]$ represents the normal range of the dopamine signal exhibited by the control subjects, then PD OFF conditions are simulated having a smaller range $[d_{\mathrm{low}}\ \delta_v^*]$, where δ_v^* denotes the clamped limit that is lesser than d_{high} [Eq. (10.27)]. In addition to this, PD ON conditions could also be simulated [Eq. (10.28)] in the model by adding a constant additive term which we call the medication factor to the value difference (δ_v^{med}).

$$\text{PD OFF}: \begin{array}{l} \text{If } \delta_V > \delta_V^* \\ \quad \delta_V = \delta_V^* \end{array} \tag{10.27}$$

$$\text{PD ON}: \begin{array}{l} \text{If } \delta_V > \delta_V^* \\ \quad \delta_V = \delta_V^* + \delta_V^{\mathrm{med}} \\ \text{else} \\ \quad \delta_V = \delta_V + \delta_V^{\mathrm{med}} \end{array} \tag{10.28}$$

Furthermore, the degeneration of the SNc neurons is not the only pathology linked with PD. Others areas of the BG such as the STN and GPe are shown to have pathological synchronized oscillations in the PD patients (Weinberger, Hutchison, & Dostrovsky, 2009). Pathological β-band oscillations in these loops have also been linked to PD tremor and rigidity (Mallet et al., 2008; Weinberger et al., 2009). Therefore, in addition to clamping δ_v, we also investigated other parameters such as the lateral connection strengths in STN and GPe neurons (W_{slat} & W_{glat}), the interconnection strengths w_{sg} and w_{gs}, and the relative contributions of the direct and the indirect pathways on the final motor action. Finally, there is definitely an influence of dopamine on the excitability of the cortical neurons. As a result to study these effects, we also introduced a variable for tonic dopamine levels (δ_{ton}), which is updated using the value difference using Eq. (10.29), to understand the effect of dopamine depletion in the higher cortical areas.

$$\tau_{\text{ton}} \frac{\mathrm{d}\delta_{\text{ton}}}{\mathrm{d}t} = -\delta_{\text{ton}} + A_{\text{ton}}\delta_V \tag{10.29}$$

This gives an estimate of the averaged gradient information or the value function, which controls the dynamics of the MC. We made the tonic dopamine variable δ_{ton} control the strength of connectivity in the MC, i.e., A_{lat}^C which controls the strength of lateral connectivity within the attractor network of MC using Eq. (10.30).

$$\tau_{\text{MC}} \frac{\mathrm{d}A_{\text{lat}}^C}{\mathrm{d}t} = -A_{\text{lat}}^C + f(\lambda_{\text{ton}}(\delta_{\text{ton}} - \theta_{\text{ton}})) + k \tag{10.30}$$

where f is a sigmoid function $\left(f(\cdot) = \frac{1}{1+\exp(-\cdot)}\right)$, with slope λ_{ton} and threshold θ_{ton} and k is a constant to maintain baseline values of A_{lat}^C.

10.3 Results

10.3.1 Mapping of the Joint Configurations in the PC and MC

The sensory-motor cortical loop is initially tested (Fig. 10.4a) as a stand-alone network, and the MC is activated to investigate the range of movements of the arm in the workspace. Activation to the MC is given as $I_{\text{MC}} = I_{\text{PFC}} + I_{\text{BG}} + I_{\text{PC}} + I_{\text{app}}$ where $I_{\text{PFC}} = I_{\text{BG}} = I_{\text{PC}} = 0$ and I_{app} is a Gaussian current and a matrix of size $N_{\text{MC}} \times N_{\text{MC}}$ in which is centered on random nodes to activate different regions of the MC. In Eq. (10.31), i_{MC} and j_{MC} represent the nodes in the MC and i_r and j_r are random nodes over which the Gaussian current is centered.

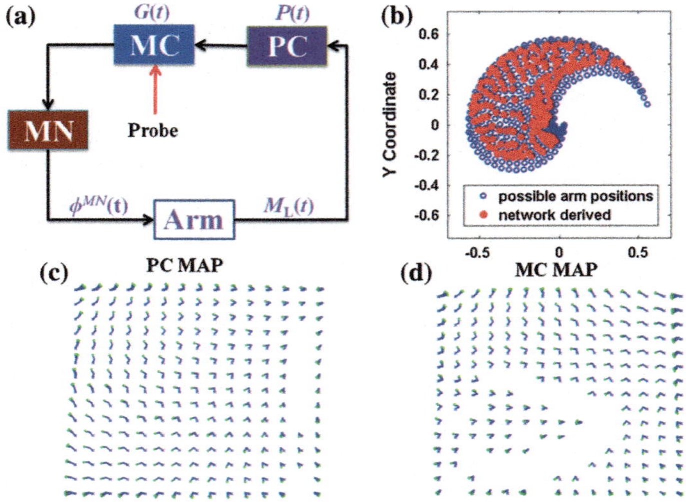

Fig. 10.4 Sensory and motor maps. The sensory-motor loop is probed at the level of the MC (**a**) and the mapping of the end effector positions approximated by the network is compared to all possible positions in the arm workspace (**b**). The joint configuration maps formed for both the PC (**c**) and the MC (**d**) layers, where blue lines indicate the two links and the green dot denotes the end effector position

$$I_{\text{app}} = \exp\left(-\frac{(i_{\text{MC}} - i_r)^2 + (j_{\text{MC}} - j_r)^2}{\left(\sigma_{\text{lat}}^{g/s}\right)^2} \right) \tag{10.31}$$

We observe that the arm is capable of reaching most of the positions in the output space (Fig. 10.4b), suggesting a consistent mapping of arm configurations in the feature space. Furthermore, to understand the loop's ability to represent arm positions uniquely, the activity generated upon probing the MC and the activity generated via the loop, i.e., arm → PC → MC are compared and are found to be the same. In addition, the activity of the PC and MC is mapped back onto the joint configuration space which produces map structures shown in Fig. 10.4c, d. In case of the PC map, the joint configuration space is fairly uniform and topography is well maintained, whereas just one level above in the MC the map starts to become more complex with both regions of continuous change in the configuration space and areas of fractures or discontinuities. The regions of overlapping representations seem to have increased from the PC map to the MC map.

10.3.2 Reaching Movements of the Arm

We initially tested the model by providing multiple targets to reach and to test if the arm reaches these areas. Figure 10.5a (1–6) shows a snapshot of the network in action as the arm reaches the target. The MC activity corresponds to the arm configuration that has successfully reached the target. The PFC activity codes for the goal position (represented by the red star). Initial movements of the arm are solely driven by the gradient information present in the value function (Eq. 10.15). The indirect pathway of the BG provides activity with low correlation under certain parametric conditions of STN–GPe connections (see BG column, STN–GPe in Table 10.1) to enable sufficient exploration of the arm in the workspace (Chakravarthy & Balasubramani, 2015). This in turn leads to training the connections between the PFC and the MC (Eq. 10.25). The PFC input to the MC specifies the activity that the motor cortex should evolve in order to reach the target. The GPi activity, which forms the input to the thalamus from the BG, is integrated in the thalamus. It is important to note that, when the MC activity and the PFC input into MC are the same, it means that the network has learned to approximate the activity needed to reach the target (Eq. 10.25). There are 50 trials in total in the simulation, where the initial 20 trials are used for learning the target location and the trajectory to follow for a successful reach, during which the amplitude of PFC input is increased as per Eq. (10.26) and the PFC-to-MC connections ($W_{\text{PFC}\rightarrow\text{MC}}$) are trained. In the next 30 trials, the arm is tested for its performance. For each trial, the arm is initialized to a starting position and provided with a specific target (the target position is kept constant for all trials) to reach. A successful reach is signified by the arm coming within at least ξ units of distance from the target. The trial is then terminated in two cases: (a) when the arm reaches the target successfully or (b) when the target is not reached and the simulation crosses the maximum time limit.

The end effector trajectories become smoother as learning progressed in controls and, furthermore, there is decrease in hand path variability as learning progresses (Fig. 10.5b). Here, the spatial variance is represented as ellipses (Georgopoulos, Kalaska, & Massey, 1981) and we see that the variance decreases with trials. We investigated the velocity profiles of the arm while performing the reach, and the characteristic of bell-shaped curve is observed in the profile (Fig. 10.5c). Additionally, it is known from previous work that these velocity profiles fit well to a delta-lognormal distribution (Plamondon, 1998), with the two lognormal components corresponding to the agonist and the antagonist bursts, respectively. We found that the reaching profiles obtained in the model fit well to this distribution (Fig. 10.5c). The performance of the arm also improves with trials seen as a decrease in the time taken to reach the target (Fig. 10.5d). There is a reduction in the variance of the time to reach across blocks of trials suggesting lower motor variability upon learning (Fig. 10.5d).

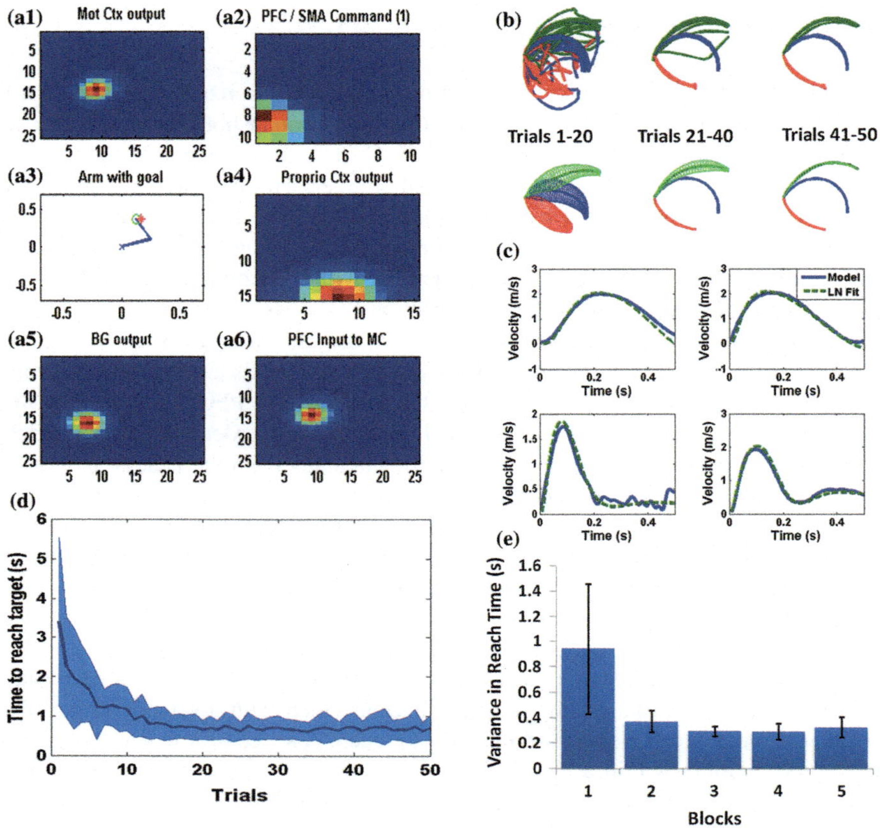

Fig. 10.5 Reaching behavior in controls. The simulation snapshot of the model while performing the reaching task (**a**) and the activities of multiples areas in the model (**a**.1–6). The end effector trajectories (**b**) obtained in the case of controls for reaching three different targets (represented by three different colors—blue, green, and red) across trials as the learning of the PFC-to-MC connections ($W_{PFC \to MC}$) takes place. The ellipses show the spatial ± SD as the model performs reaching across 50 trials. The velocity profiles during a reach (blue line) compared to the lognormal distribution (LN Fit, green dotted line) (**c**). Performance of the model in control conditions as a function of the time to reach target location (**d**) and the variability in reach times (**e**) through trials (here each block refers to 10 trials)

10.3.3 Velocity Profiles of Controls and PD Patients

Majsak and colleagues (Majsak, Kaminski, Gentile, & Flanagan, 1998) performed reaching experiments in both healthy controls and PD subjects at self-determined speeds to estimate changes in kinematics of the subjects under conditions where the objects are stationary and moving. The PD subjects were on their dopamine medication (Sinemet®) and performed six trials in each case (a) when object was

Table 10.1 Parameter values used for simulating the cortico-basal ganglia model

MC		BG		PFC		PC		MN		Arm	
SOM		Value function/ DA		SOM		SOM		MC → MN net		Kinematics	
N_{MC}	15			N_{PFC}	15	N_{PC}	15				
σ_{MC}	1	σ_V	2	σ_{PFC}	0.1	σ_{PC}	0.02	A_{MN}	0.01	a_S	0.04
CANN		A_{ton}	3	PFC → MC net				$\eta_{MC \to MN}$	0.1	b_S	0.07
A_{PC}	0.1	λ_{ton}	−50	$\eta_{PFC \to MC}$	0.1					a_E	0.03
A_{BG}	1	θ_{ton}	0.5							b_E	0.08
A_{PFC}	$0.1 - A^{T*}_{PFC}$	k	9							l_S	0.3
A^C_{lat}	10	Striatum								l_E	0.3
σ^C_{lat}	2	λ_{D1}	50								
K^C	0.5	λ_{D2}	−50								
b_{MC}	0.5	t_{D1}	0.05								
τ_{MC}	0.005	t_{D2}	0								
		STN–GPe									
		ϵ_g	1								
		ϵ_s	1								
		w_{sg}	1								
		w_{gs}	1								
		$\sigma^{g/s}_{lat}$	1								
		$\tau_{STN/GPe}$	0.005								
		GPi									
		A_{D1}	15								
		A_{D2}	1								

stationary, (b) moving, and (c) stationary again. In the model however we compared the performance of the arm in the stationary case to compare the basal-level activities. The PD ON condition in the model is simulated using Eqs. (10.27) and (10.28) where the clamped dopamine limit, δ^*_V, is set to 0.1 and the medication factor (δ^{med}_V) to 0.5. In order to account for the slowness and reduced velocity in PD movement, the tonic dopamine variable (δ_{ton}) is introduced via Eq. (10.29) and affects the MC dynamics using Eq. (10.30).

The δ_{ton} values in controls and PD condition reveal that the controls show increase in the levels of tonic dopamine as the task progresses, whereas it is much smaller in case of PD (Fig. 10.6a). This affects the attractor dynamics of the MC, where the values of A^C_{lat} are always high for the PD case compared to the control case where after some time the value falls due to the increase in δ_{ton} (Fig. 10.6b).

Lower values of A_{lat}^{C} lead to easy translation of the neural activity bump over the neural space, and high values make it difficult for the BG to trigger movements as a result of the local excitation and global inhibition dynamics. The velocity profiles of the subject groups for a self-determined speed are shown in Fig. 10.6c. The simulated controls reach the target faster than in the PD case, and their peak velocities are also higher. The kinematic variables of the reach task are shown in Fig. 10.6d–f. A significant difference is seen in the movement time between the controls and the PD subjects, suggesting slow or bradykinetic movements in the PD case.

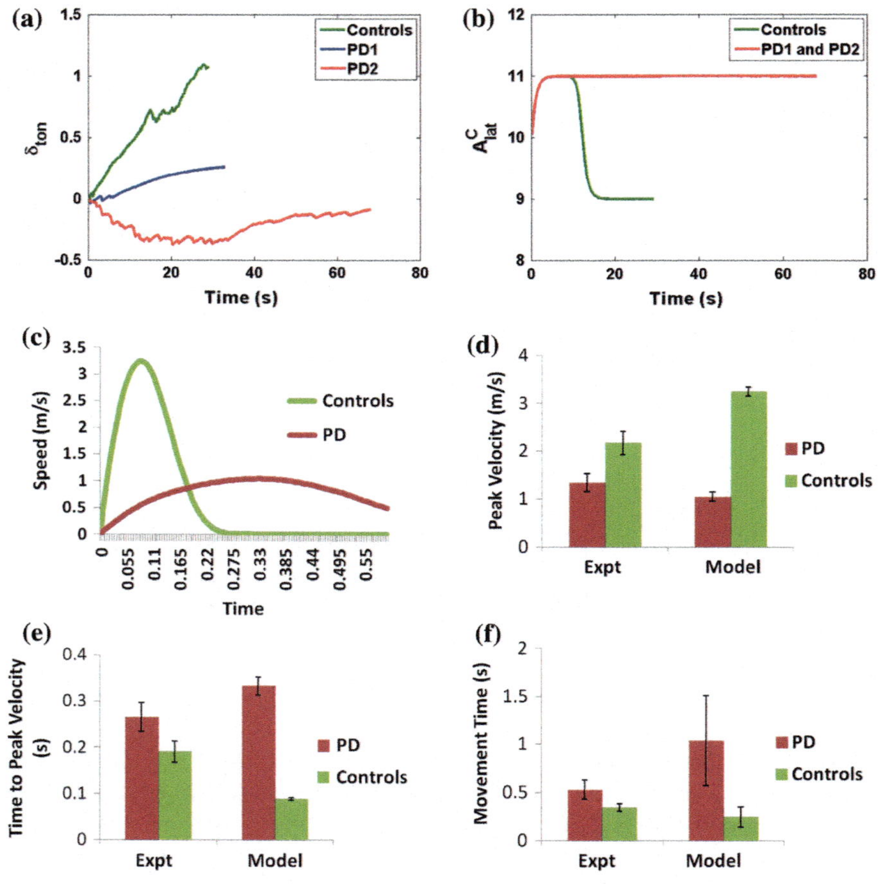

Fig. 10.6 Stationary target reaching task. The evolution of the tonic dopamine variable (δ_{ton}) (**a**) and the MC dynamics variable (A_{lat}^{C}) (**b**) for the entire task duration (50 trials) for controls, PD1 and PD2. Here, PD1 and PD2 refer to different clamped dopamine levels (δ_V^*), which is 0.1 (blue line) and 0.01 (red line), respectively. Kinematics of reaching movements with velocity profiles of controls and PD (**c**), time to peak velocity (**d**), peak velocity (**e**), and movement time (**f**) from experiment (adapted from Majsak et al., 1998) and the corresponding model performance

10.3.4 Model Performance on the Pursuit Task

Another set of experiments captured by the model included the pursuit task conducted by Soliveri and group where they tracked the ability of PD subjects to pursue a moving target using a manipulandum (Soliveri, Brown, Jahanshahi, Caraceni, & Marsden, 1997). In the model, this task is abstracted to the arm trying to reach for a series of continuously changing target positions. The target moves back and forth in a straight line in a sinusoidal fashion with a frequency of 0.25 Hz, thereby making it predictable. There were three blocks in the experiment each including 10 trials. The behavioral variable measured both in the experiment and model is percentage of time on target which is defined as the time spent on the target using the manipulandum (experiment) or arm (model). As previously described, in the model the arm is thought to be on the target if the distance to the target is within ξ (=0.1) units. The PD subjects included in the study were on their DA medication. In the model, we designed the target to move similarly as in the experiment, in a sinusoidal fashion. As the target shifted, the activation in the PFC also changed at every instant of time as it codes for the target location. This meant that the peak of the value function (Eq. 10.15) also changes continuously with time giving the arm the necessary information to track the moving target. The δ_V^* is set to -0.2 and the δ_V^{med} to 0.01. From Fig. 10.7a, it is evident that the controls are capable of pursuing the moving target more efficiently than PD ON subjects which the model captures (Fig. 10.7b), even though both subject groups showed learning as the trials progressed. This phenomenon is also observed in the experiment where the PD subjects also learn the task across blocks (Fig. 10.7c, d). The last block in the experiment suggests a performance drop in both the control and PD subjects which the authors attribute to fatigue. This is not captured in the model as we did not take into account the factor of fatigue in the muscle model. See Table 10.1 for parameter values used for simulating the experimental conditions.

10.3.5 Motor Initiation with the Cortico-BG Loop

The relative strengths of the BG along with the PFC inputs into the MC are analyzed to understand the dynamics of the cortico-basal ganglia loop on movement initiation. The arm was initialized to a starting configuration, and the PFC input is provided as pulses of duration (50 ms) with varying amplitudes. The displacement of the arm from its starting position is tracked. In the presence of only PFC (i.e., $A_{BG} = 0$), the amplitude of PFC input to initiate sufficient movement has to be high ($A_{PFC} > 0.9$). The introduction of the BG with varying degree of strengths ($A_{BG} = 0.01$, 0.05, 0.1, and 0.2) leads to movement at lower amplitude of PFC input, thus making motor initiation easier, though with higher contribution of BG the movement variability increases (Fig. 10.8a). With the introduction of PD condition ($\delta_V^* = 0$, $\delta_V^* = -1$ and $w_{sg} = w_{gs} = 3$, which causes synchronized

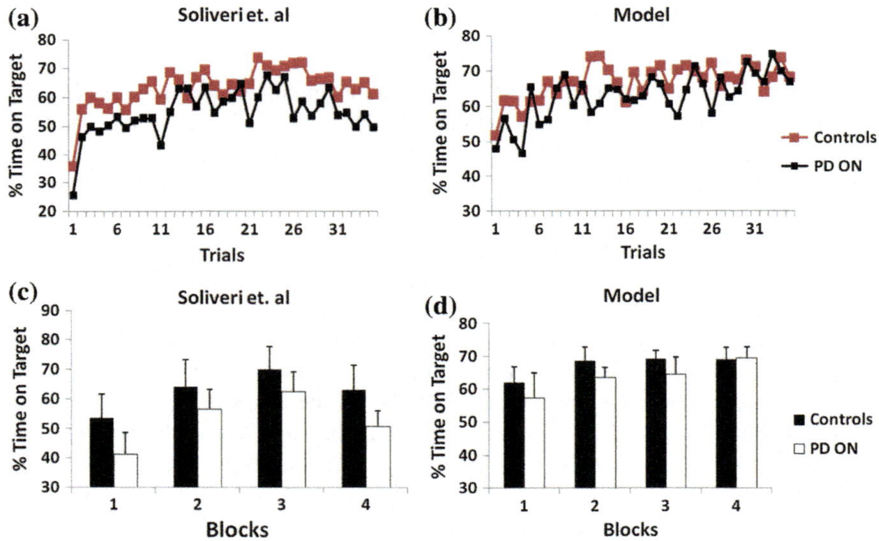

Fig. 10.7 Pursuit reaching task. The performance of subjects on the pursuit task (adapted from Soliveri et al., 1997) and the differences observed in control and PD behavior in experiment (**a** and **c**) and the model (**b** and **d**)

oscillations of the STN–GPe loop), keeping A_{BG} constant, we again see a tendency toward higher input strengths of PFC required to initiate movement (Fig. 10.8b). We observe that there needs to be compensation from higher cortical areas in disease conditions for reaching movement and could be interpreted as a deficiency in voluntary movement initiation due to the impaired BG.

10.3.6 PD Symptoms

The model is further extended to understand several motor symptoms commonly seen in PD. In the model, three cardinal symptoms of PD movement are simulated: tremor, rigidity, and bradykinesia. Initially to simulate PD condition, the value difference is clamped, but that alone does not reproduce all the above symptoms in the model. Other parameters also must be varied as shown below.

Tremor, Rigidity, and Bradykinesia

PD symptoms start appearing in the model when the dopamine signal ($\delta_V^* = -1$) is clamped and the connection strengths between the STN and GPe (w_{sg} and w_{gs}) and the lateral connection strengths in the STN (ϵ_s) are manipulated. In all the symptom cases, the interconnections, i.e., w_{sg} and w_{gs} are increased from the control levels (w_{sg} and $w_{gs} = 1$ in controls, w_{sg} and $w_{gs} = 3$ in PD). Controls have a smooth

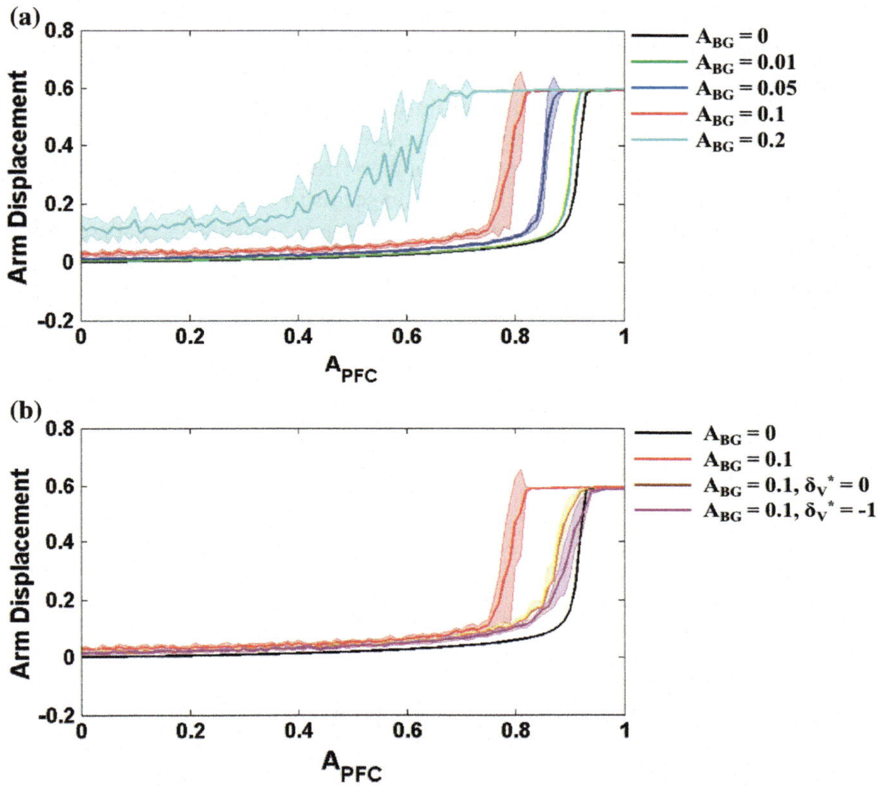

Fig. 10.8 Motor initiation in the cortico-BG loop. The displacement of the arm from a starting position with varying degree of the PFC and BG input strengths (A_{PFC}, A_{BG}) (**a**) and the effect of PD condition (**b**) on motor initiation

trajectory to the target (Fig. 10.9a), and tremor starts to appear initially with just the increase in the connection strength between STN and GPe, i.e., w_{sg} and w_{gs} (Fig. 10.9b). Rigidity and bradykinesia seem to coexist in the model and start to emerge upon decreasing the contribution of the lateral connections within the STN (ϵ_s) compared to the tremor case (Fig. 10.9c) (for rigidity $\epsilon_s = 0.7$). We estimated the frequency spectrum of the velocity of the arm during the reaching task and in the tremor case (Fig. 10.9e). There is increased power in the 4–10 Hz range which is seen clinically in PD patients as well (Jankovic, 2008). This also brings about differences in the spectrogram of the average STN activity in the tremor and rigidity scenarios compared to control scenario. In the control case, the STN activity remains sufficiently decorrelated (Fig. 10.9g). However in both the symptomatic cases, power of the spectrum seems to be concentrated within a narrow frequency range: tremor (frequency range = 10–40 Hz) and rigidity (frequency range = 20–55 Hz) (Fig. 10.9h, i). Various studies have observed pathological oscillations in the indirect pathway especially in the STN–GPe loop and are generally related to the symptoms observed in the PD patients (Hammond, Bergman, & Brown, 2007; Mallet et al., 2008; Weinberger et al., 2009). These pathological oscillations are in the β-band (13–30 Hz), and this is seen in the model where the frequency spectrogram of the average STN activity shows increased power in the β-range. In the rigidity case, the spectrum shifts to higher frequency band compared to Parkinsonian tremor. This shift in the spectrum seems to result in the arm restricted to a very small part of the state space, thereby reflecting movements that are slower and more rigid.

The three symptomatic conditions are also presented as a function of distance to the target position (Fig. 10.10a). On further exploration of the entire range of values

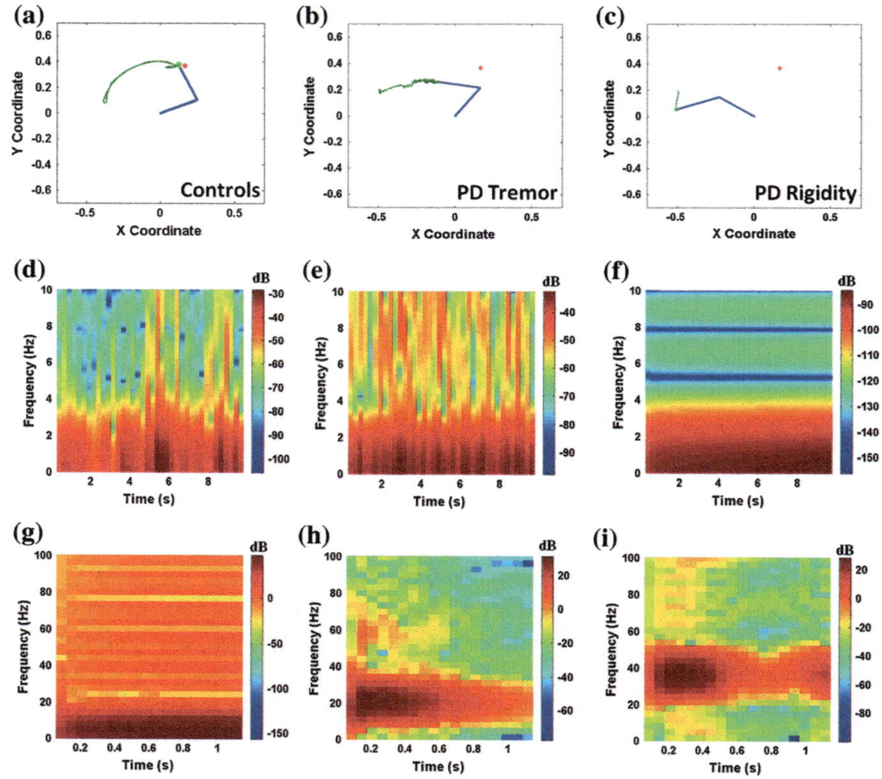

Fig. 10.9 Emergence of PD symptoms in the model. Movement trajectories (dark green trace) of the arm during reaching a target (red dot) in controls (no δ_V^*, w_{sg} and $w_{gs} = 1$, $\epsilon_s = 1$) (**a**), PD tremor ($\delta_V^* = -1$, w_{sg} and $w_{gs} = 3$, $\epsilon_s = 1$) (**b**), and PD rigidity ($\delta_V^* = -1$, w_{sg} and $w_{gs} = 3$, $\epsilon_s = 0.7$) (**c**) conditions. The frequency spectrograms of movement in controls (**d**), PD tremor (**e**), and PD rigidity (**f**), where PD tremor shows increased power in the 4–10 Hz regions. The spectrogram of the averaged STN activity in controls (**g**), during PD tremor (**h**), and rigidity (**i**). The spectrum shows a shift into higher frequencies in case of the PD symptoms (rigidity > tremor)

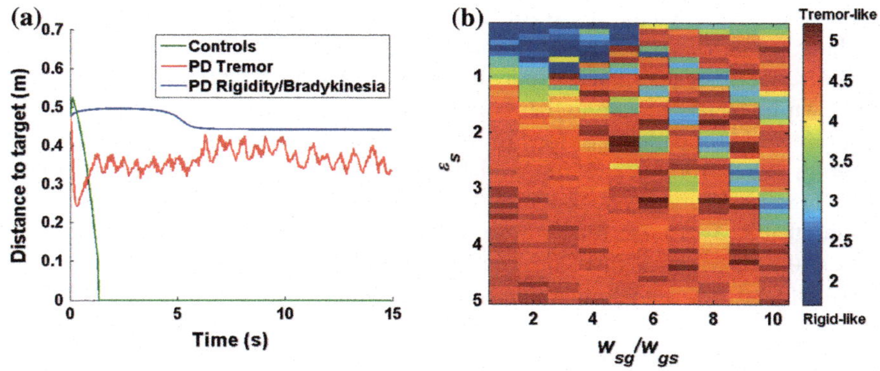

Fig. 10.10 Influence on network parameters on PD symptoms. PD symptoms viewed as a function of the distance to the target position (**a**). The analysis of the strength of interconnections within STN–GPe (w_{sg} and w_{gs}) and the lateral connection strength within the STN (ϵ_s) and their effect on the type of symptoms manifested in PD is represented as the expected value of the Fourier spectrum of velocity (**b**). The blue and red range represents rigidity/bradykinetic, and tremor movements, respectively. The control regime lies in the green range

for the STN–GPe interconnections (w_{sg} and w_{gs}) and the intraconnections (ϵ_s), the regimes of disease states seem to appear (Fig. 10.10b). In Fig. 10.10b, the values are obtained as mean of the Fourier spectrum of the arm velocity for each condition. Therefore, higher mean values of the Fourier spectrum suggest more tremor-like behavior and the lower values suggest more rigid and bradykinetic movements. Intermediate values could be control-like behavior with more a balanced frequency spectrum as shown in Fig. 10.9d. It suggests that at a given w_{sg} the range of ϵ_s at which tremor or rigidity appears is different. This could be a reason for high symptom variability among PD patients and the bias toward development of certain symptoms earlier in the disease compared to later. However, the general trend suggests that decrease in the lower lateral strengths in the STN may be a major causative reason for rigidity and higher interconnection strength for tremor.

10.4 Discussion

We present a cortico-basal ganglia model that performed reaching tasks. By inducing PD conditions in the model, we are able to simulate the impairments seen in PD reaching movements. There are two different loops in the model: the sensory-motor cortical loop and the cortico-basal ganglia loop. So the model is on the lines of optimal control theory and closed-loop control where the sensory-motor integration provides the necessary feedback mechanisms for control; the cortico-basal ganglia loop sets the optimality criterion to maximize performance (Todorov, 2004). In this case, the criterion becomes the value function: Stochastic gradient descent dynamics executed by the BG model essentially drives the reaching movements of the arm toward a target. One of the assumptions in the model is that this value function may be readily available to the BG module by the top-down information from higher cortical areas. A plausible mechanism could be the prefrontal cortical connections to the ventral striatum (Alexander, Crutcher, & DeLong, 1991; Botvinick, 2008), which code the goal information in the form of value function at the level of ventral striatum. Alternatively, it is not essential to assume that such a value function can be constructed readily from the goal information; the value can also be constructed in the ventral striatum by the plasticity of cortico-striatal connections that combine the goal information from the prefrontal cortex with the sensory-motor information from the sensory-motor cortical projections to the ventral striatum.

The acquisition of motor skill requires learning at several levels (Hikosaka et al., 2002). One of the key points of the model is during the early phase of learning, when the movements are slower and more dynamic, the BG is dominant. As the cortical learning progresses, specifically PFC-to-MC training in the model, the movements become quicker and more directed toward the target location. This phenomenon is actually seen in monkeys performing associative learning tasks, where initially the neuronal activity is higher in the striatal areas, suggesting responses to rewards, and a slow increase in the activity of the prefrontal cortex (Pasupathy & Miller, 2005). These studies also suggest that the output of BG trains the higher cortical areas and this is precisely captured in the model. There is a significant contribution of striatum during initial stages of learning, and as the cortical systems start to take over, actions become habitual and automatic (Ashby, Turner, & Horvitz, 2010). The initial slower movements in the model are driven by climbing the value function (i.e., dopamine-dependent), and as the PFC contribution increases, it can activate other cortical areas like MC which can directly influence spinal motor neurons for faster movements.

10.4.1 Cortico-Basal Ganglia Loop as an Attractor Network

The proposed model highlights the idea that the final hand position due to reaching is an attracting state of the cortico-basal ganglia dynamics. Therefore, the attractor dynamics of the cortico-basal ganglia loop must be well understood in order to understand reaching dynamics in normal and Parkinsonian conditions. The attractor dynamics that drives the reaching movements in the model arises from three sources: (1) the lateral connections in the CANN model of MC, (2) the sensory-motor cortical loop dynamics, and (3) the cortico-basal ganglia loop dynamics. The attractor dynamics of the CANN model has been explored extensively in other studies (Gupta et al., 2013; Jankovic, 2008). The attractor dynamics of the sensory-motor cortical loop, even in the absence of the cortico-basal ganglia loop, is demonstrated in Fig. 10.8a, where it is shown that the PFC input to MC must exceed a threshold to move the hand from its current state. Addition of the cortico-basal ganglia loop seems to lower the threshold; it is easier initiate the hand movement for a given PFC activation, if there is assistance from the cortico-basal ganglia loop. This result explains the relative difficulty observed in PD patients in initiating hand movements (Chen & Reggia, 1996). In Fig. 10.8b, the PD pathology is investigated slightly differently. Instead of removing the BG input to MC, the δ_V term, which represents dopamine projections to the striatum, is clamped at two levels (clamp value, $\delta_V^* = 0$, and $\delta_V^* = -1$; $w_{sg} = w_{gs} = 3$). These parameter settings suppress the dopamine signal (δ_V) and also change the dynamics of STN–GPe loop to synchronized dynamics, emulating PD conditions (Chakravarthy & Balasubramani, 2015). Under these conditions also, it can be seen that it is harder to initiate movement, even though the strength of BG input to MC is unaffected. These results show that the dynamics of the cortico-basal ganglia loop amplifies the output of MC, thereby facilitating movement in normal condition. This amplification is suppressed in PD conditions, due to reduced dopamine levels and increased synchronization of the STN–GPe loop. These results resonate well with the model of the role of BG in willed action proposed in (Chakravarthy, 2013).

10.4.2 Indirect Pathway for Exploration and Emergence of PD Symptoms

The STN–GPe loop modeled as a network of coupled oscillators induces exploratory dynamics in the model [Eqs. (10.19), (10.20), and (10.21)]. We have in previous studies substantiated the role of the indirect pathway as an Explorer that performs random search over the action space which is necessary when viewing basal ganglia as a reinforcement learning engine (Chakravarthy & Balasubramani, 2015). The exploration in the model comes from the fact that there is a stochastic drift in the activity of MC, influenced by the complex dynamics of the STN–GPe, thereby driving the arm to visit all possible arm configurations. As a result, the indirect pathway becomes very important in the initial trials to drive arm movements where the movement variability is also high (Fig. 10.5d, e). In the initial part of a reaching trial, value difference (δ_V) is small which makes the striatal output to GPe and GPi low. Therefore, in the initial part of a trial, the output of BG is dominated by the output of the STN–GPe loop, which facilitates movement initiation. However, once movement begins, δ_V changes significantly strongly reflecting the gradient of the value function. In the PD case, the clamping of the value difference (δv) in the model enhances the outputs of D2R neurons in the striatum and amplifies the contributions of the indirect pathway to BG output. Thus, in PD conditions, the BG output depends on the dynamics of the STN–GPe loop, with altered dynamics of STN–GPe manifesting as impaired movement (Table 10.2).

Table 10.2, which summarizes the results from Fig. 10.9, shows the parameters of STN–GPe loop under control and PD conditions. Particularly, it shows that the internucleus connections (w_{gs} and w_{sg}) are high compared to the control for both the

Table 10.2 STN–GPe parameters and relation to PD symptoms (rigidity and tremor)

	w_{gs} (=w_{sg})	ϵ_s	δ_v^*
Controls	1	1	No bound on δ_v^*
Tremor	3	1	−1
Rigidity	3	0.7	−1

symptom categories (rigidity and tremor). Furthermore, in case of rigidity, the STN lateral connectivity strength (ϵ_s) is lower than in case of tremor. It is known that symptoms in PD could be correlated with synchronized oscillations in the STN, which is often seen in the β-range (Mallet et al., 2008; Weinberger et al., 2009). Studies show that the oscillatory activity in the STN ranges from the low frequency 3–7 Hz to beta (13–30 Hz) in the more dorsal regions to even gamma (30–100 Hz) in the ventral areas (Zaidel, Spivak, Grieb, Bergman, & Israel, 2010). The model concurs with this where we see the STN activity ranging from desynchronized in control case (Fig. 10.9g) to synchronized beta in tremor (Fig. 10.9h) condition and high beta (20.5–28 Hz), bordering on gamma (25–100 Hz) during rigidity (Fig. 10.9i). There is not much evidence on how such pathological oscillations give rise to both tremor and rigidity. An interesting observation from the model was that both tremor and rigidity were associated with different frequency bands of the STN activity (as shown in Fig. 10.9h, i), with rigidity associated closer to the gamma range compared to the tremor. It remains to be verified whether such firing patterns exist in the basal ganglia under conditions of rigidity and tremor.

10.4.3 *Effect of Dopamine on Motor Performance*

In the model, the dopamine signal (δ_v) aids in switching between the direct and the indirect pathways of the BG. In order to model the effect of dopamine on the motor cortex, we define the tonic dopamine variable, δ_{ton}, [Eq. (10.29)], which controls the lateral inhibition in the CANN component of the MC. This tonic dopamine variable is a local-time-averaged version of phasic dopamine (Eq. 10.29). There seems to be a higher degree of intracortical inhibition with the application of dopamine agonists to the motor cortex and a significant decrease in this inhibition upon the administration of dopamine antagonist (Ziemann, Tergau, Bruns, Baudewig, & Paulus, 1997). In PD ON subjects, the peak velocity of reaching and the acceleration of movement are higher and the time spent in deceleration is lower compared to the OFF case suggesting its benefit in reducing bradykinesia (Castiello, Bennett, Bonfiglioli, & Peppard, 2000). In the model, these effects are reproduced by making the A_{lat}^{C} parameter in MC a function of the tonic dopamine variable δ_{ton}. The A_{lat}^{C} parameter modulated by the tonic dopamine affects the intrinsic excitability of the CANN component of MC (Fig. 10.6b), where larger values of A_{lat}^{C} make the CANN dynamics more stable and resistant to any changes in the input that comes from areas, viz. BG, PC and PFC.

10.4.4 Limitations and Future Directions

The immediate limitation of the model is the lack of a distinct striatal module; instead we have used striatal activation functions to modulate the cortical input entering the BG. Since the arm used is a 2D kinematic model (it could be extended to 3D naturally), the introduction of nonlinear muscle model with force dynamics would aid in understanding the agonist–antagonist interaction during movement. In future, we would like to extend the model as a test bench to analyze reaching movement impairments in other basal ganglia pathologies like Huntington's chorea, ballismus, dystonia, and even drug-induced dyskinesias (Jankovic, 2008). Since the model is generalized in its approach, by involving different end effectors, like locomotor apparatus or articulators, we can understand motor behavior such as gait and speech, respectively (Canter, 1963; Hausdorff, Cudkowicz, Firtion, Wei, & Goldberger, 1998). One of the interesting results from the model is that with BG impairment, movement initiation becomes difficult and would require more voluntary effort to do so. This could be tested by stimulating motor areas using techniques like TMS (transcranial magnetic stimulation) in PD patients and see whether movement initiation requires more amplitude of stimulation than controls. This would also enhance the theory of the BG as an active player in regulating willed action (Chakravarthy, 2013). We show that shift in PD symptoms from tremor to rigidity could be caused by an increase in the correlated activity of the STN neurons. Experiments could target the changing activity in the STN and look for similar changes in movement behavior. Finally, the attractor dynamics, local excitation and global inhibition, of the MC in the model could be manipulated by using DA agonist and antagonists to see which aspects of these dynamics does dopamine have an influence on.

Conflict of Interest The authors declare that the research was conducted in the absence of any commercial or financial relationships that could be construed as a potential conflict of interest.

References

Alexander, G. E., Crutcher, M. D., & DeLong, M. R. (1991). Basal ganglia-thalamocortical circuits: parallel substrates for motor, oculomotor, "prefrontal" and "limbic" functions. *Progress in Brain Research, 85,* 119–146.

Ashby, F. G., Turner, B. O., & Horvitz, J. C. (2010). Cortical and basal ganglia contributions to habit learning and automaticity. *Trends in Cognitive Sciences, 14*(5), 208–215.

Asplund, C. L., Todd, J. J., Snyder, A. P., & Marois, R. (2010). A central role for the lateral prefrontal cortex in goal-directed and stimulus-driven attention. *Nature Neuroscience, 13*(4), 507–512.

Balasubramani, P. P., Chakravarthy, V. S., Ravindran, B., & Moustafa, A. A. (2014). An extended reinforcement learning model of basal ganglia to understand the contributions of serotonin and dopamine in risk-based decision making, reward prediction, and punishment learning. *Frontiers in Computational Neuroscience, 8,* 47.

Botvinick, M. M. (2008). Hierarchical models of behavior and prefrontal function. *Trends in Cognitive Sciences, 12*(5), 201–208.

Canter, G. J. (1963). Speech characteristics of patients with Parkinson's disease: I. Intensity, pitch, and duration. *Journal of Speech & Hearing Disorders*.

Castiello, U., Bennett, K., Bonfiglioli, C., & Peppard, R. (2000). The reach-to-grasp movement in Parkinson's disease before and after dopaminergic medication. *Neuropsychologia, 38*(1), 46–59.

Chakravarthy, V. S. (2013). Do basal Ganglia amplify willed action by stochastic resonance? A model. *PloS one, 8*(11), e75657.

Chakravarthy, V. S., & Balasubramani, P. P. (2015). Basal ganglia system as an engine for exploration. *Encyclopedia of Computational Neuroscience*, 315–327.

Chakravarthy, V. S., Joseph, D., & Bapi, R. S. (2010). What do the basal ganglia do? *A modeling perspective. Biological cybernetics, 103*(3), 237–253.

Chen, Y., & Reggia, J. A. (1996). Alignment of coexisting cortical maps in a motor control model. *Neural Computation, 8*(4), 731–755.

Doya, K. (1999). What are the computations of the cerebellum, the basal ganglia and the cerebral cortex? *Neural Networks, 12*(7), 961–974.

Fitts, P. M. (1954). The information capacity of the human motor system in controlling the amplitude of movement. *Journal of Experimental Psychology, 47*(6), 381.

Georgopoulos, A. P., Kalaska, J. F., & Massey, J. T. (1981). Spatial trajectories and reaction times of aimed movements: Effects of practice, uncertainty, and change in target location. *Journal of Neurophysiology, 46*(4), 725–743.

Gupta, A., Balasubramani, P. P., & Chakravarthy, S. (2013). Computational model of precision grip in Parkinson's disease: A utility based approach. *Frontiers in computational neuroscience, 7,* 172.

Hammond, C., Bergman, H., & Brown, P. (2007). Pathological synchronization in Parkinson's disease: Networks, models and treatments. *Trends in Neurosciences, 30*(7), 357–364.

Hausdorff, J. M., Cudkowicz, M. E., Firtion, R., Wei, J. Y., & Goldberger, A. L. (1998). Gait variability and basal ganglia disorders: Stride-to-stride variations of gait cycle timing in Parkinson's disease and Huntington's disease. *Movement Disorders, 13*(3), 428–437.

Hikosaka, O., Nakamura, K., Sakai, K., & Nakahara, H. (2002). Central mechanisms of motor skill learning. *Current Opinion in Neurobiology, 12*(2), 217–222.

Izawa, J., Kondo, T., & Ito, K. (2004). Biological arm motion through reinforcement learning. *Biological Cybernetics, 91*(1), 10–22.

Jankovic, J. (2008). Parkinson's disease: Clinical features and diagnosis. *Journal of Neurology, Neurosurgery and Psychiatry, 79*(4), 368–376.

Kalva, S. K., Rengaswamy, M., Chakravarthy, V. S., & Gupte, N. (2012). On the neural substrates for exploratory dynamics in basal ganglia: A model. *Neural Networks, 32,* 65–73.

Knill, D. C., & Pouget, A. (2004). The Bayesian brain: The role of uncertainty in neural coding and computation. *Trends in Neurosciences, 27*(12), 712–719.

Kohonen, T. (1990). The self-organizing map. *Proceedings of the IEEE, 78*(9), 1464–1480.

Körding, K. P., & Wolpert, D. M. (2004). Bayesian integration in sensorimotor learning. *Nature, 427*(6971), 244–247.

Magdoom, K., Subramanian, D., Chakravarthy, V. S., Ravindran, B., Amari, S.-I., & Meenakshisundaram, N. (2011). Modeling basal ganglia for understanding Parkinsonian reaching movements. *Neural Computation, 23*(2), 477–516.

Majsak, M. J., Kaminski, T., Gentile, A. M., & Flanagan, J. R. (1998). The reaching movements of patients with Parkinson's disease under self-determined maximal speed and visually cued conditions. *Brain, 121*(4), 755–766.

Mallet, N., Pogosyan, A., Márton, L. F., Bolam, J. P., Brown, P., & Magill, P. J. (2008). Parkinsonian beta oscillations in the external globus pallidus and their relationship with subthalamic nucleus activity. *The Journal of Neuroscience, 28*(52), 14245–14258.

Matsumoto, K., Suzuki, W., & Tanaka, K. (2003). Neuronal correlates of goal-based motor selection in the prefrontal cortex. *Science, 301*(5630), 229–232.

Morasso, P. (1981). Spatial control of arm movements. *Experimental Brain Research, 42*(2), 223–227.

Muralidharan, V., Balasubramani, P. P., Chakravarthy, V. S., Lewis, S. J., & Moustafa, A. A. (2013). A computational model of altered gait patterns in Parkinson's disease patients negotiating narrow doorways. *Frontiers in Computational Neuroscience, 7*.

Nakahara, H., Doya, K., & Hikosaka, O. (2001). Parallel cortico-basal ganglia mechanisms for acquisition and execution of visuomotor sequences—A computational approach. *Journal of Cognitive Neuroscience, 13*(5), 626–647.

Pasupathy, A., & Miller, E. K. (2005). Different time courses of learning-related activity in the prefrontal cortex and striatum. *Nature, 433*(7028), 873–876.

Plamondon, R. (1998). A kinematic theory of rapid human movements: Part III. Kinetic outcomes. *Biological Cybernetics, 78*(2), 133–145.

Pouget, S. D. A., & Latham, P. (1999). *Divisive normalization, line attractor networks and ideal observers.* Paper presented at the Advances in Neural Information Processing Systems 11: Proceedings of the 1998 Conference.

Schaal, S., & Schweighofer, N. (2005). Computational motor control in humans and robots. *Current Opinion in Neurobiology, 15*(6), 675–682.

Shadmehr, R., & Krakauer, J. W. (2008). A computational neuroanatomy for motor control. *Experimental Brain Research, 185*(3), 359–381.

Soliveri, P., Brown, R., Jahanshahi, M., Caraceni, T., & Marsden, C. (1997). Learning manual pursuit tracking skills in patients with Parkinson's disease. *Brain, 120*(8), 1325–1337.

Todorov, E. (2004). Optimality principles in sensorimotor control. *Nature Neuroscience, 7*(9), 907–915.

Trappenberg, T. (2003). Continuous attractor neural networks. In *Recent developments in biologically inspired computing* (pp. 398–425).

Weinberger, M., Hutchison, W. D., & Dostrovsky, J. O. (2009). Pathological subthalamic nucleus oscillations in PD: Can they be the cause of bradykinesia and akinesia? *Experimental Neurology, 219*(1), 58–61.

Zaidel, A., Spivak, A., Grieb, B., Bergman, H., & Israel, Z. (2010). Subthalamic span of β oscillations predicts deep brain stimulation efficacy for patients with Parkinson's disease. *Brain*, awq144.

Ziemann, U., Tergau, F., Bruns, D., Baudewig, J., & Paulus, W. (1997). Changes in human motor cortex excitability induced by dopaminergic and anti-dopaminergic drugs. *Electroencephalography and Clinical Neurophysiology/Electromyography and Motor Control, 105*(6), 430–437.

Chapter 11
Studying the Effect of Dopaminergic Medication and STN–DBS on Cognitive Function Using a Spiking Basal Ganglia Model

Alekhya Mandali and V. Srinivasa Chakravarthy

Abstract Using the spiking Izhikevich model used in the earlier chapters, we studied the effect of medication [L-Dopa and dopamine agonists (DAA)] and subthalamic nucleus (STN) deep brain stimulation on decision making using two cognitive tasks, i.e., Iowa gambling task (IGT) and the probabilistic learning task (PLT) and were validated using the experimental results. Based on the experimental observations that dopaminergic activity is analogous to temporal difference (TD) and induces cortico-striatal plasticity, we introduced learning in the cortico-striatal weights using the reinforcement learning framework. For PLT and IGT, the model in PD condition under medication (L-Dopa) was unable to learn from punishments which is attributed to excess dopamine levels in striatum even during punishment. The model under DAA was impulsive reflected in the lower RT in PLT and negative score in IGT. We varied two parameters during DBS (1) the electrode position within STN and (2) antidromic activation of GPe neurons. The performance in both IGT (Score) and PLT (reaction time) was dependent on the position of the electrode and amplitude of the current for a specific electrode position. We also observed that a higher antidromic activation of GPe neurons does not impact the learning ability but decreases reaction time as reported in DBS patients for PLT. These results suggest a probable role of electrode and antidromic activation in modulating the STN activity and eventually affecting the patient's performance.

11.1 Introduction

The surgical technique, deep brain stimulation (DBS) of the subthalamic nucleus (STN), is now widely used in the treatment of Parkinson's disease (PD) when either dopamine replacement therapy does not provide continuous relief from motor symptoms or leads to drug-induced dyskinesias (Benabid, 2003). Irrespective of its wide clinical usage (Garcia, D'Alessandro, Bioulac, & Hammond, 2005), the exact

V. S. Chakravarthy and A. Moustafa, *Computational Neuroscience Models of the Basal Ganglia*, Cognitive Science and Technology,
https://doi.org/10.1007/978-981-10-8494-2_11

mechanism of DBS action is still under debate. Furthermore, various experimental studies show a controversial effect of DBS on cognition (Jahanshahi et al., 2000) particularly on impulsivity (Brittain et al., 2012; Frank, Samanta, Moustafa, & Sherman, 2007; Smeding, Speelman, Huizenga, Schuurman, & Schmand, 2009). Several reports have highlighted the development of new onset, often transient, impulse control disorders (ICDs) following STN stimulation (Combs et al., 2015; Hershey et al., 2004; Smeding et al., 2007). This was thought to be due to stimulation parameters such as current spread and electrode position which affected the outcome in cognitive tasks (Sudhyadhom et al., 2007; Witt et al., 2013; York, Wilde, Simpson, & Jankovic, 2009). STN stimulation can also increase risk-taking behavior in Iowa gambling task (IGT) (Evens et al., 2015). Patients with STN–DBS tended to overestimate their performance with a preference toward competitive environments (Florin et al., 2013). On the other hand, pre-existing ICDs were reported to resolve following STN–DBS, as a result of reduction in dopaminergic medication (Castrioto et al., 2015). Thus, STN–DBS may lead to varying net effects on impulsivity in PD through different mechanisms.

Probabilistic learning task (PLT) (Frank et al., 2007; Frank, Seeberger, & O'reilly, 2004) also captures the decision-making ability and impulsivity features. PLT tests the learning capability of the performer not only in choosing rewarding choices but also in avoiding punishing ones. Experimental results show that the performance of normals and PD OFF subjects during PLT is similar in terms of choosing rewarding and avoiding punishing choice (Frank et al., 2007). The performance of PD ON subjects was opposite to that of PD OFF with a preference toward the rewarding choice, which was accounted to the presence of excess DA levels in the striatum due to medication. This excess DA (due to medication) prevents the PD subjects to learn from punishments. Another critical feature captured by PLT is the reaction time (RT). It has been observed that normal subjects take more time when presented with multiple equally rewarding stimuli (high conflict) and are expected to choose one among them (Frank et al., 2007). Frank et al. (2007) hypothesized that STN increases its activity and buys the extra time needed ('holding the horses') during such situations. This was further shown by Zaghloul et al. (2012), where an increase in STN activity in PD patients during high conflict conditions was observed (Zaghloul et al., 2012). Experiments conducted by Frank et al. (2007) showed that the performance of DBS subjects on PLT was not significantly different in terms of learning ability but showed impulsive behavior in terms of RT.

The aim of this study is twofold, firstly to show that the spiking BG model is able to replicate the performance of normal, PD OFF, PD ON (L-DOPA) conditions as in experimental studies (Frank et al., 2007) and secondly to hypothesize the effect of DAA, DBS electrode on learning, impulsivity, and behavior. The details of the model are described in the previous section and below.

11.2 Materials and Methods

11.2.1 Spiking Neuron Model of the Basal Ganglia

The network model of BG (Mandali, Rengaswamy, Chakravarthy, & Moustafa, 2015) described earlier was used to simulate the cognitive tasks. For details of the model and its related equations please refer to earlier sections. The details of the cognitive tasks (IGT and PLT) and the related measures are explained below.

11.2.2 Behavioral Tasks

11.2.2.1 Iowa Gambling Task (IGT)

IGT involved the presentation of four decks of cards wherein each of the decks A/B/C/D is associated with a combination of reward and penalty. IGT was conducted for a total of 100 trials (5 bins of 20 trials each).The net outcome of a certain card selection (reward + penalty) in each trial was calculated. Over a few trials, the cards from the decks A and B (C and D) are disadvantageous (advantageous) as the corresponding expected value is negative (positive).

11.2.2.2 Probabilistic Learning Task (PLT)

The experiment consists of two stages, training and testing. During the training stage, the model was presented only with three pairs of stimuli (AB/CD/EF) one at a time in a random fashion. Each of the six choices (A/B/C/D/E/F) was associated with a reward with a priori probability. For example, selection of choice 'A' leads to reward (= +1) 80% of the time, whereas choice 'B' leads to a reward only 20% of the time. Similarly, choice 'C' ('E') gives reward with a probability of 70% (60%) and choice 'D' ('F') leads to reward only 30% (40%) of the time and punishment (= −1) for rest of the trials. The model was expected to learn these reward probabilities by the end of training.

During the testing stage, the model was tested with 15 novel combinations (e.g., AC, CE, DE) which were not presented during the training stage. No feedback was provided for the response made after each stimulus. The model was tested for its learning ability based on whether it chose (avoided) a rewarding (punishing) choice from the presented combination pair. For example, if a novel combination of choice 'A' with another choice was presented; the model was expected to choose 'A' as the probability of obtaining a reward was the highest for 'A'. Similarly, when the stimuli with combination of 'B' with other choices were presented, the model is expected to avoid selecting 'B' as its reward probability was the lowest. Apart from testing for the learning ability, the model was also tested for performance during high conflict (HC) and low conflict (LC) situations. For example, the stimulus combination 'AC' falls under the category of HC as both choice 'A' (80%) and 'C'(70%) have high reward probabilities but stimulus combination 'BC' comes under the category LC as reward probabilities ('B' = 20% and 'C' = 70%) are significantly different. The reaction time was measured for each of the conditions (HC/LC).

During this stage, the model was tested for the following conditions:

- Testing accuracy where the model was presented with 15 novel combinations not used during the training phase.
- Choice/Avoidance Accuracy of the model to select choice 'A' and avoid choice 'B' when presented with all possible novel combinations containing either 'A' or 'B'.
- Decision-making efficiency in term of reaction times during HC and LC situations.

11.2.3 Simulating Tasks Using Spiking Neuron Network Model

Based on the task, each of the nuclei is divided into either two or four quadrants. For example, since IGT consists of four decks, each nucleus [STN/GPe/GPi/Striatum (both D1 and D2)] in the network was divided equally into four quadrants, where each quadrant received input from one of the decks (Fig. 11.1b). The expected value of each card was represented by the cortico-striatal weight which was modulated by DA term 'δ'. The input to GPe and GPi (i.e., the output of D2 and D1 striatum) was modeled as Poisson spike train (Reti, 2015), whose frequency was proportional to the cortico-striatal weight $\left(w_{i,k}^{\mathrm{D1}}, w_{i,k}^{\mathrm{D2}}\right)$ of the corresponding card(i) and trial(k). The striatal neuronal firing rate was restricted to 2–40 Hz as per the experimental literature (Kravitz et al., 2010). Since DA is known to modulate plasticity in cortico-striatal conditions, the error term 'δ' (in the model) was used to update the cortico-striatal synapses (Surmeier, Ding, Day, Wang, & Shen, 2007). DA also modulated the synaptic strengths within various BG nuclei such as STN (Cragg, Baufreton, Xue, Bolam, & Bevan, 2004), GPe (Smith & Kieval, 2000).

11.2.3.1 Cortico-Striatal Weight Update and Temporal Difference Error

Each deck was associated with two cortico-striatal weights $\left(w_{i,0}^{\mathrm{D1}}, w_{i,0}^{\mathrm{D2}}\right)$ which were initialized with random values from a uniform distribution over (0, 1). The two cortico-striatal weights were trained as,

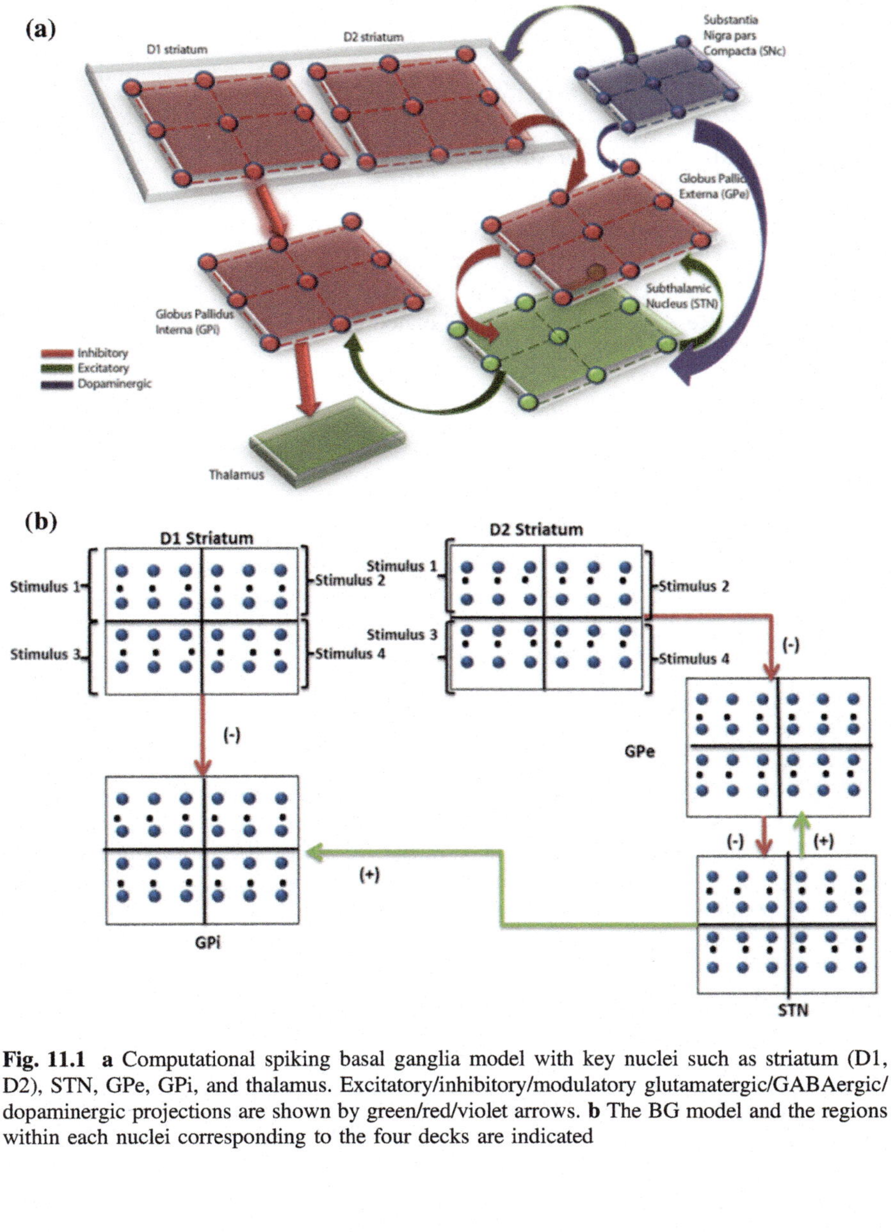

Fig. 11.1 **a** Computational spiking basal ganglia model with key nuclei such as striatum (D1, D2), STN, GPe, GPi, and thalamus. Excitatory/inhibitory/modulatory glutamatergic/GABAergic/dopaminergic projections are shown by green/red/violet arrows. **b** The BG model and the regions within each nuclei corresponding to the four decks are indicated

$$\Delta w_{i,k+1}^{\mathrm{D1}} = \eta \delta_k x_{i,k}^{\mathrm{inp}} \tag{11.1}$$

$$\Delta w_{i,k+1}^{\mathrm{D2}} = -\eta \delta_k x_{i,k}^{\mathrm{inp}} \tag{11.2}$$

The expected value (V_k) for kth trial was calculated as

$$V_k = \sum_{i=1}^{4} w_{i,k}^{\mathrm{D1}} * x_{i,k}^{\mathrm{sel}} \tag{11.3}$$

The reward (Re_k) for kth trial was calculated as

$$\mathrm{Re}_k = \sum_{i=1}^{4} r_{i,k} * x_{i,k}^{\mathrm{sel}} \tag{11.4}$$

The loss (L_k) for the kth trial was calculated as,

$$L_k = \sum_{i=1}^{4} l_{i,k} * x_{i,k}^{\mathrm{sel}} \tag{11.5}$$

The error (δ) for kth trial was defined as

$$\delta_k = \mathrm{Re}_k + L_k - V_k \tag{11.6}$$

where $w_{i,k+1}^{\mathrm{D1}}\left(w_{i,k+1}^{\mathrm{D2}}\right), w_{i,k}^{\mathrm{D1}}\left(w_{i,k}^{\mathrm{D2}}\right)$ were the cortico-striatal weights of D1 (D2) striatum for ith card in k + 1th and kth trial, $r_{i,k}$ and $l_{i,k}$ were the reward and loss obtained for the selected ith card in kth trial, x^{inp} was the input binary vector representing the four decks, x^{sel} was the binary vector representing the selected card, e.g., if the card 'A' in IGT is selected x^{sel} = [1 0 0 0].

11.2.3.2 Simulating Untreated PD and Medically Treated PD Conditions

Bearing in mind that 'δ' is similar to DA activity (Niv, 2009; Schultz, 1998), and there is loss of DA neurons in PD, we simulated PD condition by clamping the 'δ' value (Eq. 11.6) to a low limit (δ_{lim}) which resembles the untreated PD condition (Eq. 11.7).

$$\delta_{\mathrm{lim}} = \min(\delta, DA_{\mathrm{ceil}}) \tag{11.7}$$

where $\min(y, a)$ is defined as $\begin{matrix} z = y & if & y < a \\ a & if & y > a \end{matrix}$ and DA_{ceil} is the upper limit of 'δ'.

Medically treated PD condition clinically involves external intake of dopamine precursors such as L-DOPA which was simulated by adding a positive 'δ_{med}' term to the δ_{lim} (Eq. 11.7).

$$\delta_{\text{new}} = \delta_{\text{lim}} + \delta_{\text{med}} \tag{11.8}$$

Another class of medication prescribed to PD patients is DAA, which has differential affinity for dopamine receptors. We simulated DAA with preferential affinity for D2 receptors, also known to be linked to impulsivity (MacMahon & Macphee, 2008). The δ_{new} in the Eq. (11.8) was used to update only D2 cortico-striatal weight (w^{D2}) unlike for L-DOPA where both w^{D1} and w^{D2} were updated.

11.2.3.3 DBS Current

An external current which mimics the clinically delivered DBS current was applied to the STN neurons in the model. The parameters (frequency, pulse duration, and amplitude) of the stimulation current were chosen to be similar to the typical values used in a clinical setting (Garcia et al., 2005). The stimulation current was applied to the entire/part of STN module in the form of Gaussian distribution (Foutz & McIntyre, 2010). The mean of the Gaussian coincides with the lattice position (i_c, j_c) which was assumed to be the center of the electrode and extent of current spread was controlled by the variance parameter (σ).

$$I_{ij}^{\text{DBS}} = A_{\text{DBS}} * \mathrm{e}^{\frac{-((i-i_c)^2 + (j-j_c)^2)}{\sigma^2}} \tag{11.9}$$

where I_{ij}^{DBS} is the current received by the neuron at position (i, j), A_{DBS} is the amplitude of the current (pA), σ controls the current spread, and (i_c, j_c) is the mean/center point of the electrode. The effect of electrode position (i_c, j_c) and stimulation parameters A_{DBS} and σ on STN activity and on behavior was explored.

Electrode Position

Experimental results show that change in the electrode position alters behavior (Hershey et al., 2004, 2010), and this can be attributed to the difference in pattern and volume of STN activation due to the electrode position (Miocinovic et al., 2006). Also, the final action or choice selection depends on the activity of GPi neurons which receive weighted input from STN and D1R-expressing MSNs. Bearing these points in mind, we chose three electrode positions where the lattice point indicates the center of the electrode, i.e., Pos 1 in the upper half of the STN nucleus at lattice point (13, 13), Pos 2 with electrode contact center at the lattice point (25, 25), and Pos 3 in the lower half of the STN nucleus at lattice point (38, 38). Each module (StrD1, StrD2, GPe, and STN) in the model is divided into four quadrants corresponding, respectively, to the four panels in the probabilistic learning task. This is a modeling assumption that has to be made in the absence of experimental data about how the four action choices might be represented in the basal ganglia nuclei. The electrode position that we study in the model is also described with reference to such representations. Thus, the four quadrants in the modules do not correspond to the well-known basal ganglia loops like sensorimotor, associative, limbic.

11.2.4 Performance Measures

In this section, we explain all the performance measures used in this study to quantify and validate the results obtained from the model for all the conditions.

11.2.4.1 IGT Score

The performance was measured as IGT total score (number of selections from 'C', 'D'—number of selections from 'A', 'B').

11.2.4.2 PLT-Learning

The model was trained for 120 trials (= 40 per combination (AB/CD/EF)), and the learning ability of the model was checked during the training stage in terms of training accuracy where the probability of selecting the correct choice was plotted as the training progressed (trials were divided into five equal bins). The performance of the model was compared with the results (Fig. 11.2a) from (Zaghloul et al., 2012).

PLT-Testing Accuracy and Difference in Reward Expectation (DRE)

Difference in Reward Expectation (DRE): After training, the a priori choice selection probability was calculated based on the number of times the corresponding choice was presented and selected. We then calculated the difference in reward expectation (DRE), which is the difference between the two a priori choice probabilities for that particular presented stimulus. DRE captures the amount of conflict between the presented choices, the higher (lower) the DRE for that stimulus the lower (higher) is the conflict. For example, if stimulus 'BC' was presented then DRE_{BC}, which is the difference between P(B) and P(C), would be low, thereby reducing the probability of choice 'B' getting selected.

Testing Accuracy: Once the training phase is completed, the model was tested by presenting 15 novel combinations. The objective was to calculate the probability with which the first choice in the presented stimulus was selected. For example,

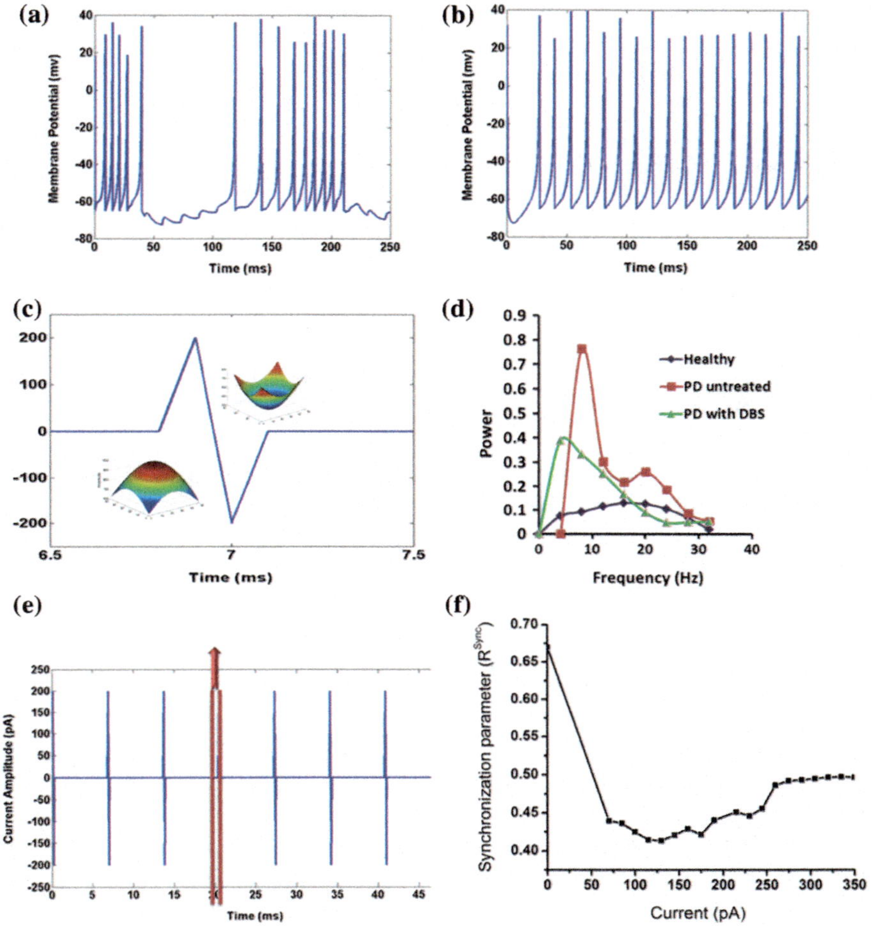

Fig. 11.2 Activity of STN neuron healthy, with and without DBS in PD. **a** The activity of STN neurons in healthy condition. **b** The bursting activity of STN neurons in PD condition. **c** STN neurons resume to tonic firing after DBS. **d** The reduction in the frequency content at tremor frequency (4 Hz) in STN neurons in mentioned conditions. **e** The DBS current in biphasic mode (frequency = 130 Hz with amplitude of 200 pA). **f** The synchronization levels in the standalone STN–GPe network with increase in DBS current

if stimulus 'AC' was presented for 20 times and choice 'A' was selected for 16 times, then the testing accuracy for choice 'A' would be 0.8 (= 16/20).

The learning ability of a system to select the most rewarding choices while avoiding the punitive ones can be obtained by just evaluating the relationship between DRE and testing accuracy. For example, the testing accuracy (of choice 'A') for the stimulus 'AF' (whose DRE > 0) would be expected to be high because the reward probability associated with choice A is also high. So for an optimally trained system, one can expect a linear relationship between testing accuracy and DRE.

PLT-Choice/Avoidance Accuracy

This quantity measures the ability of the model to select the most rewarding option 'A' and avoid the punitive choice 'B' when presented with novel combinations not used during training.

11.3 Results

We first present the results relating to the changes in synchronization levels of STN neurons due to DBS. Then, we report the performance measures obtained from model in healthy, PD untreated, treated (medication and stimulation conditions in both IGT and PLT.

11.3.1 De-synchronization by DBS Current

The membrane potential of STN neurons (Fig. 11.2a, d) in PD untreated condition showed bursting activity and frequency content showed a peak at around 4 Hz with high synchrony level (= 0.67) (Mandali et al., 2015).

On stimulating the STN neurons in PD condition, the peak around tremor frequency (= 4 Hz) was significantly reduced (Fig. 11.2d) ($P < 0.00001$). Similarly, the bursting activity in Fig. 11.2a was overridden and suppressed by the stimulating current (Fig. 11.2b). The synchrony level R^{sync} in the presence of DBS current decreased from 0.67 (in PD condition and stimulation-OFF) to 0.42 (stimulation-ON) (Fig. 11.2e) but increased at higher current amplitudes.

We then studied how PD affected decision-making ability in normals, PD OFF, PD ON (L-Dopa and DAA), and STN–DBS conditions for IGT and PLT.

In the IGT experiment, the subject is presented with four decks of cards wherein each of the decks A/B/C/D is associated with a combination of reward and penalty and is conducted for a total of 100 trials (5 bins of 20 trials each). The net outcome of a certain card selection (reward + penalty) in each trial was calculated. Over a few trials, cards from the decks A and B were disadvantageous as the expected value is negative and decks C and D were advantageous with a positive expected value. The performance was measured as IGT total score (# of selections from 'C' and 'D'—# of selections from decks 'A' and 'B').

The model's performance was compared with that of an experimental result (Fig. 11.3). In healthy control (HC) condition, the score was negative in the first

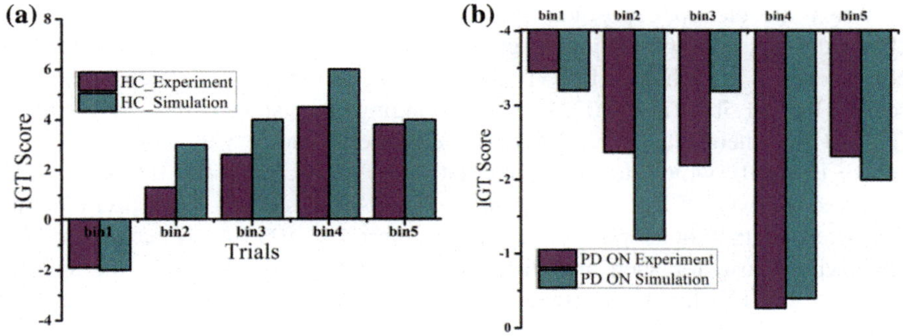

Fig. 11.3 IGT performance results were redrawn from Gescheidt et al. (2012). **a** HC from experiment and simulation. **b** Medically treated PD patients from experiment and simulation (L-DOPA)

bin, but began to increase positively from the second bin (Fig. 11.3) which was absent in both PD OFF and ON conditions. The lowest scores were obtained for PD ON (L-Dopa and DAA) (result not shown) compared to others. The mean IGT score values (Fig. 11.3) for HC, obtained from experiment (Gescheidt et al., 2012) and simulation were not significantly different ($P = 0.19$). Similarly, the mean IGT score values obtained from medically treated PD subjects and simulation were not significantly different ($P = 0.74$).

Now, the IGT task was simulated for STN–DBS condition The mean IGT score values obtained from PD 'ON' subjects from experiment (Oyama et al., 2011) and simulation were statistically similar ($P = 0.42$) (Fig. 11.4). Similarly, the mean IGT score values for PD with STN–DBS experiment and simulation were similar ($P = 0.55$).

Based on experimental observation (Hershey et al., 2010), the effect of DBS electrode parameters on IGT score was also checked in the model.

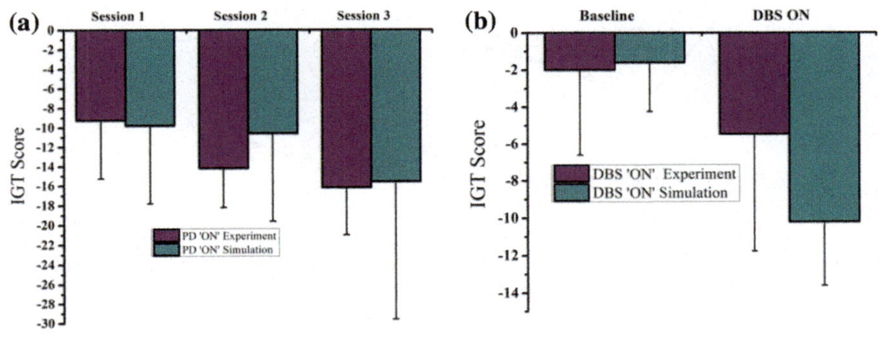

Fig. 11.4 IGT scores calculated for PD 'ON' and PD ON+DBS condition. The experimental results are redrawn from Oyama et al. (2011). **a** PD 'ON' controls from experiment and simulation for three sessions. **b** DBS subjects for baseline and DBS ON

When the electrode position (positions explained in Fig. 11.5 legend) was changed such that stimulation is given selectively to part of the STN module corresponding to each deck in IGT, we observed a significant variation in the IGT score (Fig. 11.5b) ($P < 0.0001$). On changing the spread of DBS current (Fig. 11.5c), there was a trend toward better performance with lower radius of spread ($\sigma = 10$), which did not reach statistical significance ($P = 0.67$). We also observed a lower IGT score at higher (= 300 pA) and lower (= 70 pA) currents compared to that obtained from optimal current (= 100 pA) ($P < 0.001$). The underlying cause for such an effect was investigated by observing the spiking activity of STN for both optimal (= 100 pA) and high (= 300 pA) current scenarios.

The second task in the impulsivity study is the PLT. The experiment consists of two stages, training and testing. During the training stage, the model was presented with three pairs of stimuli (AB/CD/EF) one at a time in a random fashion. Each of

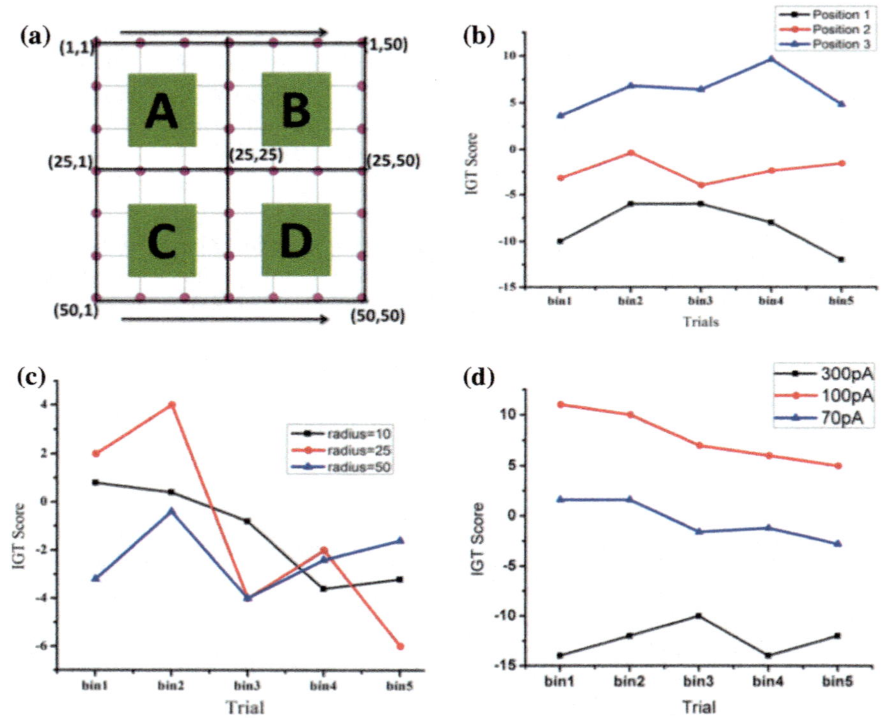

Fig. 11.5 IGT score. **a** The STN network (= 50 × 50) with quadrants which receive input from each of the corresponding decks (A, B, C, D). **b** The IGT score for three electrode positions (Position 1—in first quadrant with electrode center at lattice point (13, 13), Position 2—center of the electrode at the lattice point (25, 25), and Position 3—center of the electrode at the lattice point (38, 38) in the fourth quadrant. **c** For the electrode at position 2, the spread of the current was changed. **d** The effect of DBS current amplitude (70, 100, and 300 pA) on IGT scores when the electrode is placed in position 3

the six choices (A/B/C/D/E/F) was associated with a reward with a priori probability. For example, selection of choice 'A' leads to reward (= +1) 80% of the time, whereas choice 'B' leads to a reward only 20% of the time. Similarly, choice 'C' ('E') gives reward with a probability of 70% (60%) and choice 'D' ('F') leads to reward only 30% (40%) of the time and punishment (−1) for rest of the trials. During the testing stage, the model was tested with novel combinations (e.g., AC, BD) which were not presented during the training stage. The model was tested for its learning ability based on whether it chose (avoided) a rewarding (punishing) choice from the presented stimulus pair. Apart from testing the learning ability, the model was also tested for the performance during high conflict (HC) and low conflict (LC) situations. For example, the stimulus combination 'AC' falls under the category of HC as both choice 'A' (80%) and 'C'(70%) have high reward probabilities but stimulus combination 'BC' comes under the category LC as reward probabilities ('B' = 20% and 'C' = 70%) are significantly different. The ability of the model to select the most rewarding option 'A' and to avoid the punitive choice 'B' was then evaluated.

To this end, the model was presented with novel combinations of choices 'A' and 'B' was implemented on normals, PD OFF, PD ON (L-DOPA and DAA) conditions (Fig. 11.6). The modeling results were compared with experimental results from Frank et al. (2007). The results obtained from simulation (Fig. 11.6b) were found to be similar to that obtained from the experiment (Fig. 11.6a).

Various experimental and clinical studies reported impulsivity in PD patients after stimulation of STN which was soon contradicted. Keeping this in mind, we studied the effect of electrode position on reward and punishment learning. We choose three electrode positions where lattice point indicates the center of the electrode, i.e., Pos 1 in the upper half of the STN nucleus at lattice point (13, 13), Pos 2 with electrode contact center at the lattice point (25, 25), and Pos 3 in the

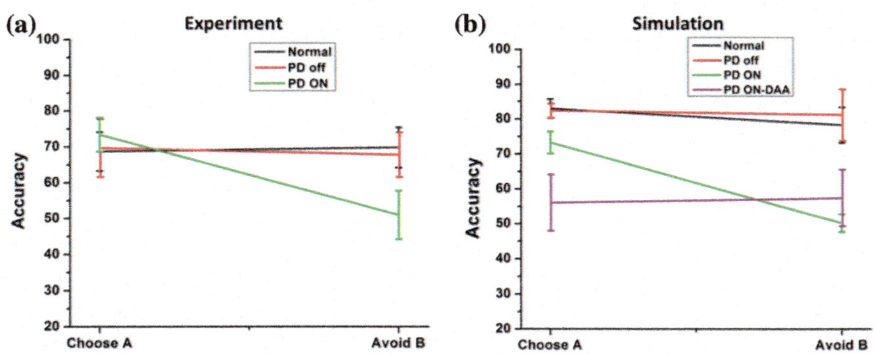

Fig. 11.6 Testing accuracy of the model in for HC, PD OFF, PD ON (L-DOPA and DAA). **a** The experiment results redrawn from Frank et al. (2007). **b** Simulation results obtained from the spiking BG model

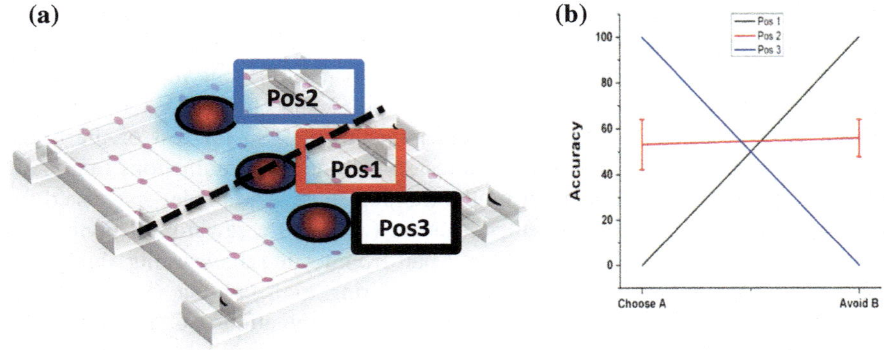

Fig. 11.7 Effect of electrode position on performance. **a** Graphical representation of the electrode position in STN lattice where Pos 1 has the electrode center at the lattice point (25, 25), Pos 2 at lattice point (13, 13), and Pos 3 at (38, 38). **b** The performance of the model during stimulation of STN for each of the three positions

lower half of the STN nucleus at lattice point (38, 38). The performance of the model shifted from reward based to punishment based on changing the DBS electrode position (Fig. 11.7).

We then studied the role of STN activity in HC condition (Brittain et al., 2012; Cavanagh et al., 2011; Frank et al., 2007; Zaghloul et al., 2012) in each of the four conditions [Normals/PD OFF/PD ON (L-DOPA and DAA)] for correct and error trials (Fig. 11.8). The reaction time for PD-DAA condition was the lowest for HC among all the other cases, suggesting impulsive behavior known to be present in dopamine agonist-treated subjects (Voon et al., 2007).

We then checked for the effect of electrode position on RT, by considering the three positions as described earlier (Fig. 11.7). The electrode position was shifted between the three positions keeping all other stimulation parameters constant, and the reaction time was measured in LC and HC trials. We observed that the reaction time decreased for HC condition, decreased for a specific electrode position (= Pos 3) (for both correct and error trials as plotted in Fig. 11.9).

11.4 Discussion

The model was used to study impulsivity in PD OFF and ON conditions. The IGT performance was poor in PD OFF as well as in ON conditions compared to HC, with worse being in ON. The model in treated PD ON condition does not learn from its action outcomes (rewards/punishments) and wanders among the decks, which is reflected in the negative IGT score (Fig. 11.3). Physiologically, this negative behavior is attributed to excess DA levels in striatum (Frank et al., 2007). In the model, striatal weights were positively updated even in punishment situation due to

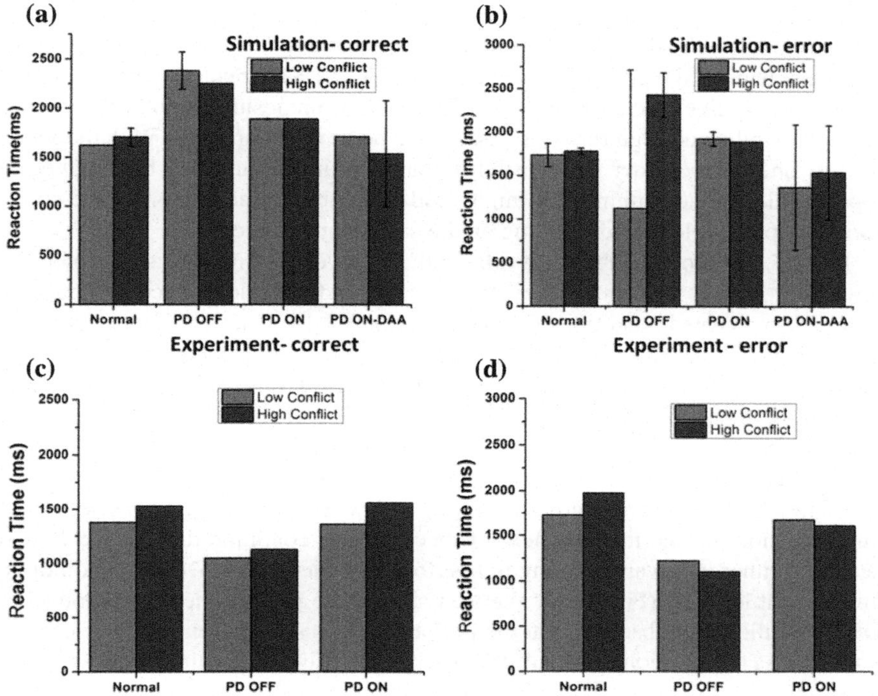

Fig. 11.8 Reaction time in milliseconds (ms) for various conditions applied on the spiking BG model. **a** Reaction time (ms) obtained from the model for all the four conditions normals, PD OFF, PD ON (L-Dopa, DAA) for LC and HC condition in correct trials. **b** Reaction time (ms) measured from the model for all the four conditions (Normals, PD OFF, PD ON (L-Dopa, DAA) for LC and HC condition in error trials. **c** Experimental reaction time (ms) obtained from normal, PD OFF, and PD ON condition for correct. **d** Error trials from (Frank et al., 2007)

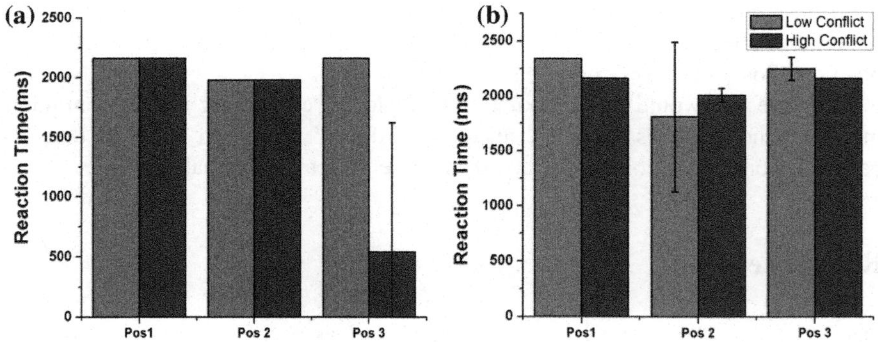

Fig. 11.9 Reaction time for LC and HC conditions when the position of the electrode was changed. Pos 1, Pos 2, Pos 3 are also described in Fig. 8. **a** Correct, **b** error trials

dopaminergic medication (δ_{med}), leading to the selection of wrong choice. Stimulation decreased the IGT performance compared to ON condition (Fig. 11.4). We then changed the position of electrode within the STN nucleus (Fig. 11.5) and observed a significant change in IGT score. Apart from position, another parameter thought to influence cognition was observed to be current amplitude. With the above results, one can consider the possibility that stimulation current when applied to topographical areas within STN might lead to inhibition/facilitation of the corresponding panel selection depending on the current amplitude.

In PLT, the model's ability to differentiate between a high rewarding and low rewarding choice in each of the physiological and pathological conditions [PD OFF, PD-medicated conditions (L-DOPA and DAA)] was tested. The accuracy of normals and the accuracy of PD OFF conditions of the model were not significantly different for both choosing A and avoiding B cases. But this behavior was absent in both of the medicated conditions (L-DOPA and DAA) (Fig. 11.6). This behavior was also experimentally observed where PD patients under medication tend to learn more from rewards than punishments (Frank et al., 2004, 2007). This can be accounted for by the medication term ($\delta_{med} = 3$) which prevents the dip selection of punitive choices. The model's performance in DAA condition did not yield good accuracy either in reward learning but performed better than L-DOPA condition in punishment learning (Fig. 11.6). We then studied the performance of the model in DBS condition and observed that the performance was dependent on the electrode position. As the position of electrode was changed (Fig. 11.7.), the model switched from reward-based to punishment-based learning. Based on the experimental data that an increase in STN activity was observed during HC conditions, we analyzed the reaction time for each of the conditions in LC and HC cases in each of the five conditions. We observed that the model in normal conditions took more time to make a choice during HC case compared to that in LC in both correct and error trials (Fig. 11.8). The impulsivity behavior observed clinically due to DAA medication (Ondo & Lai, 2008; Voon et al., 2007) was captured by the model wherein it was observed that a lower RT for HC case (Fig. 11.8a). We observed that the RTs were different for different electrode positions (Fig. 11.9), and a lower RT was obtained for HC case during both correct and error trials for a specific electrode position (Pos 3) in DBS condition.

In future, we would like to extend our model by combining it with continuum models which utilizes patient's anatomical data. This could help in deriving patient-specific protocols which could increase the patient's quality of life.

References

Benabid, A. L. (2003). Deep brain stimulation for Parkinson's disease. *Current Opinion in Neurobiology, 13*(6), 696–706.

Brittain, J.-S., Watkins, K. E., Joundi, R. A., Ray, N. J., Holland, P., Green, A. L., et al. (2012). A role for the subthalamic nucleus in response inhibition during conflict. *The Journal of Neuroscience, 32*(39), 13396–13401.

Castrioto, A., Funkiewiez, A., Debû, B., Cools, R., Lhommée, E., Ardouin, C., ... Pollak, P. (2015). Iowa gambling task impairment in Parkinson's disease can be normalised by reduction of dopaminergic medication after subthalamic stimulation. *Journal of Neurology, Neurosurgery & Psychiatry, 86*(2), 186–190.

Cavanagh, J. F., Wiecki, T. V., Cohen, M. X., Figueroa, C. M., Samanta, J., Sherman, S. J., & Frank, M. J. (2011). Subthalamic nucleus stimulation reverses mediofrontal influence over decision threshold. *Nature Neuroscience, 14*(11), 1462–1467.

Combs, H. L., Folley, B. S., Berry, D. T., Segerstrom, S. C., Han, D. Y., Anderson-Mooney, A. J., ... van Horne, C. (2015). Cognition and depression following deep brain stimulation of the subthalamic nucleus and globus pallidus pars internus in Parkinson's disease: A meta-analysis. *Neuropsychology Review*, 1–16.

Cragg, S. J., Baufreton, J., Xue, Y., Bolam, J. P., & Bevan, M. D. (2004). Synaptic release of dopamine in the subthalamic nucleus. *European Journal of Neuroscience, 20*(7), 1788–1802.

Evens, R., Stankevich, Y., Dshemuchadse, M., Storch, A., Wolz, M., Reichmann, H., ... Lueken, U. (2015). The impact of Parkinson's disease and subthalamic deep brain stimulation on reward processing. *Neuropsychologia.*

Florin, E., Müller, D., Pfeifer, J., Barbe, M. T., Fink, G. R., & Timmermann, L. (2013). Subthalamic stimulation modulates self-estimation of patients with Parkinson's disease and induces risk-seeking behaviour. *Brain*, awt241.

Foutz, T. J., & McIntyre, C. C. (2010). Evaluation of novel stimulus waveforms for deep brain stimulation. *Journal of Neural Engineering, 7*(6), 066008.

Frank, M. J., Samanta, J., Moustafa, A. A., & Sherman, S. J. (2007). Hold your horses: Impulsivity, deep brain stimulation, and medication in Parkinsonism. *Science, 318*(5854), 1309–1312.

Frank, M. J., Seeberger, L. C., & O'reilly, R. C. (2004). By carrot or by stick: Cognitive reinforcement learning in Parkinsonism. *Science, 306*(5703), 1940–1943.

Garcia, L., D'Alessandro, G., Bioulac, B., & Hammond, C. (2005). High-frequency stimulation in Parkinson's disease: More or less? *Trends in Neurosciences, 28*(4), 209–216.

Gescheidt, T., Czekóová, K., Urbánek, T., Mareček, R., Mikl, M., Kubíková, R., ... Bareš, M. (2012). Iowa Gambling Task in patients with early-onset Parkinson's disease: Strategy analysis. *Neurological Sciences, 33*(6), 1329–1335.

Hershey, T., Campbell, M. C., Videen, T. O., Lugar, H. M., Weaver, P. M., Hartlein, J., ... Perlmutter, J. S. (2010). Mapping Go–No–Go performance within the subthalamic nucleus region. *Brain, 133*(12), 3625–3634.

Hershey, T., Revilla, F., Wernle, A., Gibson, P. S., Dowling, J., & Perlmutter, J. (2004). Stimulation of STN impairs aspects of cognitive control in PD. *Neurology, 62*(7), 1110–1114.

Jahanshahi, M., Ardouin, C., Brown, R., Rothwell, J., Obeso, J., Albanese, A., ... Pollak, P. (2000). The impact of deep brain stimulation on executive function in Parkinson's disease. *Brain, 123* (6), 1142–1154.

Kravitz, A. V., Freeze, B. S., Parker, P. R., Kay, K., Thwin, M. T., Deisseroth, K., & Kreitzer, A. C. (2010). Regulation of Parkinsonian motor behaviours by optogenetic control of basal ganglia circuitry. *Nature, 466*(7306), 622–626.

MacMahon, D. G., & Macphee, G. J. (2008). Dopamine agonists and impulse control disorders in Parkinson's disease. *Progress in Neurology and Psychiatry, 12*(9), 5–9.

Mandali, A., Rengaswamy, M., Chakravarthy, S., & Moustafa, A. A. (2015). A spiking basal ganglia model of synchrony, exploration and decision making. *Frontiers in Neuroscience, 9,* 191.

Miocinovic, S., Parent, M., Butson, C. R., Hahn, P. J., Russo, G. S., Vitek, J. L., et al. (2006). Computational analysis of subthalamic nucleus and lenticular fasciculus activation during therapeutic deep brain stimulation. *Journal of Neurophysiology, 96*(3), 1569–1580.

Niv, Y. (2009). Reinforcement learning in the brain. *Journal of Mathematical Psychology, 53*(3), 139–154.

Ondo, W. G., & Lai, D. (2008). Predictors of impulsivity and reward seeking behavior with dopamine agonists. *Parkinsonism & Related Disorders, 14*(1), 28–32.

Oyama, G., Shimo, Y., Natori, S., Nakajima, M., Ishii, H., Arai, H., et al. (2011). Acute effects of bilateral subthalamic stimulation on decision-making in Parkinson's disease. *Parkinsonism & Related Disorders, 17*(3), 189–193.

Reti, I. (2015). *Brain stimulation: Methodologies and interventions*. New York: Wiley.

Schultz, W. (1998). Predictive reward signal of dopamine neurons. *Journal of Neurophysiology, 80*(1), 1–27.

Smeding, H., Goudriaan, A., Foncke, E., Schuurman, P., Speelman, J., & Schmand, B. (2007). Pathological gambling after bilateral subthalamic nucleus stimulation in Parkinson disease. *Journal of Neurology, Neurosurgery and Psychiatry, 78*(5), 517–519.

Smeding, H. M., Speelman, J. D., Huizenga, H. M., Schuurman, P. R., & Schmand, B. (2009). Predictors of cognitive and psychosocial outcome after STN DBS in Parkinson disease. *Journal of Neurology, Neurosurgery & Psychiatry*.

Smith, Y., & Kieval, J. Z. (2000). Anatomy of the dopamine system in the basal ganglia. *Trends in Neurosciences, 23,* S28–S33.

Sudhyadhom, A., Bova, F. J., Foote, K. D., Rosado, C. A., Kirsch-Darrow, L., & Okun, M. S. (2007). Limbic, associative, and motor territories within the targets for deep brain stimulation: Potential clinical implications. *Current Neurology and Neuroscience Reports, 7*(4), 278–289.

Surmeier, D. J., Ding, J., Day, M., Wang, Z., & Shen, W. (2007). D1 and D2 dopamine-receptor modulation of striatal glutamatergic signaling in striatal medium spiny neurons. *Trends in Neurosciences, 30*(5), 228–235.

Voon, V., Thomsen, T., Miyasaki, J. M., de Souza, M., Shafro, A., Fox, S. H., … Zurowski, M. (2007). Factors associated with dopaminergic drug–related pathological gambling in Parkinson disease. *Archives of Neurology, 64*(2), 212–216.

Witt, K., Granert, O., Daniels, C., Volkmann, J., Falk, D., van Eimeren, T., et al. (2013). Relation of lead trajectory and electrode position to neuropsychological outcomes of subthalamic neurostimulation in Parkinson's disease: Results from a randomized trial. *Brain, 136*(7), 2109–2119.

York, M. K., Wilde, E. A., Simpson, R., & Jankovic, J. (2009). Relationship between neuropsychological outcome and DBS surgical trajectory and electrode location. *Journal of the Neurological Sciences, 287*(1), 159–171.

Zaghloul, K. A., Weidemann, C. T., Lega, B. C., Jaggi, J. L., Baltuch, G. H., & Kahana, M. J. (2012). Neuronal activity in the human subthalamic nucleus encodes decision conflict during action selection. *The Journal of Neuroscience, 32*(7), 2453–2460.

Chapter 12
Modeling Serotonin's Contributions to Basal Ganglia Dynamics

Pragathi Priyadharsini Balasubramani, V. Srinivasa Chakravarthy, Balaraman Ravindran and Ahmed A. Moustafa

Abstract In addition to dopaminergic input, serotonergic (5-HT) fibers also widely arborize through the basal ganglia circuits and strongly control their dynamics. Although empirical studies show that 5-HT plays many functional roles in risk-based decision making, reward, and punishment learning, prior computational models mostly focus on its role in behavioral inhibition or timescale of prediction. This chapter presents an extended reinforcement learning (RL)-based model of DA and 5-HT function in the BG, which reconciles some of the diverse roles of 5-HT. The model uses the concept of utility function—a weighted sum of the traditional value function expressing the expected sum of the rewards, and a risk function expressing the variance observed in reward outcomes. Serotonin is represented by a weight parameter, used in this combination of value and risk functions, while the neuromodulator dopamine (DA) is represented as reward prediction error as in the classical models. Consistent with this abstract model, a network model is also presented in which medium spiny neurons (MSN) co-expressing both D1 and D2 receptors (D1R–D2R) is suggested to compute risk, while those expressing only D1 receptors are suggested to compute value. This BG model includes nuclei such as striatum, Globus Pallidus externa, Globus Pallidus interna, and subthalamic nuclei. DA and 5-HT are modeled to affect both the direct pathway (DP) and the indirect pathway (IP) composing of D1R, D2R, D1R–D2R projections differentially. Both abstract and network models are applied to data from different experimental paradigms used to study the role of 5-HT: (1) risk-sensitive decision making, where 5-HT controls the risk sensitivity; (2) temporal reward prediction, where 5-HT controls timescale of reward prediction, and (3) reward–punishment sensitivity, where punishment prediction error depends on 5-HT levels. Both the extended RL model (Balasubramani, Chakravarthy, Ravindran, & Moustafa, in Front Comput Neurosci 8:47, 2014; Balasubramani, Ravindran, & Chakravarthy, in Understanding the role of serotonin in basal ganglia through a unified model, 2012) along with their network correlates (Balasubramani, Chakravarthy, Ravindran, & Moustafa, in Front Comput Neurosci 9:76, 2015; Balasubramani, Chakravarthy, Ali, Ravindran, & Moustafa, in PLoS ONE 10(6):e0127542, 2015) successfully explain the three diverse roles of 5-HT in a single framework.

V. S. Chakravarthy and A. Moustafa, *Computational Neuroscience Models of the Basal Ganglia*, Cognitive Science and Technology,
https://doi.org/10.1007/978-981-10-8494-2_12

12.1 Introduction

In addition to dopamine (DA), serotonergic projections to the BG are also known to have an important role in decision making (Rogers, 2011). Serotonin or 5-hydroxytryptamine (5-HT) is a monoamine neurotransmitter with at least seven main classes of receptors (Buhot, 1997). The major synthesis site of serotonin is dorsal raphe nucleus (DRN) in the brainstem. Understanding 5-HT functioning can be done at genetic, molecular, synaptic, cellular, network, or systems level. Abnormality of serotonergic system is found in many disorders such as depression, obsessive compulsion, bipolar disorder, Parkinson's disease, Schizophrenia, to name a few (Aghajanian & Marek, 2000; Dalley, Everitt, & Robbins, 2011; Fox, Chuang, & Brotchie, 2009; Lopez-Ibor, 1992). Reconciliation of their inferences at multiple scales is essential for successful designing of therapeutic measures to disorders in cognition, perception, motor, and social domains.

The neuromodulator 5-HT is an ancient molecule that existed even in plants (Angiolillo & Vanderkooi, 1996). Through its precursor tryptophan, 5-HT is linked to some of the fundamental processes of life itself. Tryptophan-based molecules in plants are crucial for capturing the light energy necessary for glucose metabolism and oxygen production (Angiolillo & Vanderkooi, 1996). Thus, by virtue of its fundamental role in energy conversion, 5-HT is integral to mitosis, maturation, and apoptosis. In lower organisms, it modulates feeding behavior and other social behaviors such as dominance posture and escape responses (Azmitia, 2001; Chao, Komatsu, Fukuto, Dionne, & Hart, 2004; Kravitz, 2000). Due to its role as a homeostatic regulator in higher animals and in mammals, 5-HT is also associated with appetite suppression (Azmitia, 1999; Gillette, 2006; Halford, Harrold, Lawton, & Blundell, 2005). Furthermore, 5-HT has been implicated in regulating moods and emotions such as anxiety, depression, and also processes such as hallucination (Cools, Nakamura, & Daw, 2011; Tops, Russo, Boksem, & Tucker, 2009).

It is important to understand and reconcile the roles of DA and 5-HT in the BG. There are at least three ways to control 5-HT function in humans, including behavioral neurogenetics (relationship between genes coding for 5-HT system and behavior), tryptophan depletion (a drink which reduces 5-HT levels in the brain), and psychopharmacological (the administration of 5-HT agonists and antagonists to healthy human subjects) studies. Increasing 5-HT level leads to decreasing punishment prediction, though recent evidence pointing to the role of DA in processing aversive stimuli makes the picture more complicated (Boureau & Dayan, 2011; So et al., 2009). The tendency to pay more attention to negative than positive experiences or other kinds of information (negative cognitive biases) is observed at lower levels of 5-HT (Cools, Robinson, & Sahakian, 2008; Robinson, Cools, & Sahakian, 2012). Serotonin is also known to control the timescale of reward prediction (Tanaka et al., 2007) and to play a role in risk-sensitive behavior. Studies found that under conditions of tryptophan depletion, which is known to reduce brain 5-HT level, risky choices are preferred to safer ones in decision-making tasks (Long, Kuhn, & Platt, 2009; Murphy et al., 2009; Rogers, 2011). Reports about

5-HT transporter gene influencing risk-based decision making also exist (He et al., 2010; Kuhnen, Samanez-Larkin, & Knutson, 2013). 5-HT mediates differentiation to decisions on gains and losses, and is known to influence nonlinearity in risk-based decision making—risk adversity in the case of gains and risk seeking during losses, while presented with choices of equal means (Murphy et al., 2009; Zhong, Israel, Xue, Ebstein, & Chew, 2009; Zhong, Israel, Xue, Sham, et al., 2009). In summary, 5-HT is not only important for behavioral inhibition, but is also related to timescales of reward prediction, risk, anxiety, attention, etc., and to non-cognitive functions like energy conversion, apoptosis, feeding, and fatigue.

Some abstract models suggested that whereas the DA responds to appetitive stimuli, 5-HT responds to aversive or punitive stimuli (Daw, Kakade, & Dayan, 2002). Unlike computational models that argue for complementary roles of DA and 5-HT, empirical studies show that both neuromodulators play cardinal roles in coding the signals associated with the reward (Cools et al., 2011; Rogers, 2011; Tops et al., 2009). Genes that control neurotransmission of both molecules are known to affect processing of both rewarding and aversive stimuli (Cools et al., 2011). Complex interactions between DA and 5-HT make it difficult to precisely tease apart the relative roles of the two molecules in reward evaluation. Some subtypes of 5-HT receptors facilitate DA release from the midbrain DA-releasing sites, while others inhibit them (Alex & Pehek, 2007). In summary, it is clear that the relationship between DA and 5-HT is not one of simple complementarity. Both synergistic and opposing interactions exist between these two molecules in the brain (Boureau & Dayan, 2011).

There have been prior models to elucidate the function of 5-HT. Daw et al. (2002) developed a line of modeling that explores an opponent relationship (Daw et al., 2002; Dayan & Huys, 2008) between DA and 5-HT at tonic and phasic levels by representing reward and punishment prediction errors, respectively. Doya (2002) associated 5-HT with discount factor, which is a measure of the timescale of reward integration (Doya, 2002; Tanaka et al., 2007) in reinforcement learning (RL) framework. There was no single computational theory that integrates and reconciles the existing abstract computational perspectives of 5-HT function under a single umbrella.

12.2 Methods

12.2.1 Modeling the Joint Functions of DA and 5-HT in the BG: An Abstract Model (Model I)

The major circuit performing the above-described framework is the cortico-basal ganglia–thalamic loops. We focus on them in this chapter. Serotonin's function can be majorly thought to control behavior (sensory and motor levels) through

1. punishment and reward prediction (Boureau & Dayan, 2011; Cools et al., 2011; Rogers, 2011; Seymour, Daw, Roiser, Dayan, & Dolan, 2012)
2. response inhibition (Cools et al., 2008)
3. timescale of reward prediction (Tanaka et al., 2007)
4. risk sensitivity and utility construction (Long et al., 2009).

In our model, DA represents TD error as in the most extant literature of DA signaling and RL (Schultz, 1998; Sutton & Barto, 1998), and 5-HT controls risk prediction error. Action selection is controlled by the utility function, which is a weighted combination of both the value and risk functions (Bell, 2001; d'Acremont, Lu, Li, Van der Linden, & Bechara, 2009; Preuschoff, Bossaerts, & Quartz, 2006). In this modified utility function (Balasubramani et al., 2014), the weight of the risk function depends on the sign of the value function and a trade-off parameter, which is associated with 5-HT functioning. The utility function is proposed to be computed in the striatum. Three representative experiments linking 5-HT to (1) risk sensitivity (Long et al., 2009), (2) timescale of reward prediction (Tanaka et al., 2007), and (3) punishment sensitivity (Cools et al., 2008) were tested with the model (Balasubramani et al., 2014) and are briefly outlined in this chapter too.

The model hypothesizes 5-HT to control risk sensitivity and prediction error as follows: On the lines of the utility models described by (Bell, 1995) and (d'Acremont et al., 2009), the model of the utility function 'U_t' is presented as a trade-off between the expected payoff and the variance of the payoff (the subscript 't' refers to time). The original utility formulation used in (Bell, 1995; d'Acremont et al., 2009) is given by Eq. (12.1)

$$U_t(s,a) = Q_t(s,a) - \kappa\sqrt{h_t(s,a)} \tag{12.1}$$

where Q_t is the expected cumulative reward and h_t is the risk function or reward variance, for state, 's', action, 'a'; and 'κ' is the risk preference. Note that in Eq. (12.1), we represent the state and action explicitly as opposed to that presented in (Bell, 1995; d'Acremont et al., 2009). Following action execution policy 'π', the action value function 'Q' at time 't' of a state, 's', and action, 'a' may be expressed as

$$Q_{t+1}(s_t,a_t) = Q_t(s_t,a_t) + \eta_Q\delta_t \tag{12.2}$$

where 's_t' is the state at time 't', 'a_t' is the action performed at time 't', and 'η_Q' is the learning rate of the action value function ($0 < \eta_Q < 1$). The temporal difference (TD) error measure of DA is defined by δ_t in the following equation for the case of immediate reward problems.

$$\delta_t = r_t - Q_t(s_t,a_t) \tag{12.3}$$

In the case of delayed reward problems, the TD error is represented as

$$\delta_t = r_{t+1} + \gamma Q_t(s_{t+1}, a_{t+1}) - Q(s_t, a_t) \tag{12.4}$$

where 's_{t+1}' is the state at time '$t + 1$', 'a_{t+1}' is the action performed at time '$t + 1$', Similar to the value function, the risk function 'h_t' has an incremental update as defined by Eq. (12.5).

$$h_{t+1}(s_t, a_t) = h_t(s_t, a_t) + \eta_h \xi_t \tag{12.5}$$

where 'η_h' is the learning rate of the risk function ($0 < \eta_h < 1$), and 'ξ_t' is the risk prediction error expressed by Eq. (12.6),

$$\xi_t = \delta_t^2 - h_t(s_t, a_t) \tag{12.6}$$

The extended form of utility function is obtained by substituting $\kappa = \alpha$ sign ($Q_t(s_t,a_t)$) in Eq. (12.1), whose reasoning is given below.

$$U_t(s_t, a_t) = Q_t(s_t, a_t) - \alpha \, \mathrm{sign}(Q_t(s_t, a_t))\sqrt{h_t(s_t, a_t)} \tag{12.7}$$

In the above equation, the risk preference includes three components—the 'α' term, the 'sign(Q_t)' term, and the risk term '$\sqrt{H_t}$'. The sign(Q_t) term achieves a familiar feature of human decision making, viz. risk aversion for gains and risk seeking for losses (Kahneman & Tversky, 1979; Markowitz, 1952). In other words, when sign(Q_t) is positive (negative), U_t is maximized (minimized) by minimizing (maximizing) risk. Note that the expected action value Q_t would be positive for gains that earn rewards greater than a reward base (here = 0) and would be negative otherwise during losses. The model (model A) associates 5-HT level with α, a constant that controls the relative weightage between action value and risk (Eq. 12.7) to reconcile and unify various functions of serotonin in a single framework. Hence, the 5-HT activity in the striatum of the BG is related to controlling the risk sensitivity for the construction of utility.

12.2.2 *Modeling the Joint Functions of DA and 5-HT in the BG: A Network Model (Model II)*

A network model of the BG consistent with the earlier lumped model (Balasubramani, Chakravarthy, Ravindran, et al., 2015) is presented in this section. The model builds on a novel proposal that the medium spiny neurons (MSNs) of the striatum can compute either value or risk depending on the type of DA receptors they express. Whereas the MSNs that express D1-receptor (D1R) of DA compute value as being earlier reported in modeling studies (Krishnan, Ratnadurai, Subramanian, Chakravarthy, & Rengaswamy, 2011), those that co-express D1R and D2R contributing anatomically to the direct and the IPs (Bertran-Gonzalez et al., 2008; Calabresi, Maj, Pisani, Mercuri, & Bernardi, 1992; Hasbi, O'Dowd, & George, 2010, 2011; Nadjar et al., 2006; Perreault et al., 2010; Perreault, Hasbi, O'Dowd, & George, 2011) are shown to be capable of computing risk. Constructing a network model (model B) of the BG that works on extended utility (Eq. 12.7) can have the following motivation for value computation in striatum:

(1) Occurrence of TD error information in the form of DA signal at the striatum (Schultz et al., 1997),
(2) Availability of information related to the cortical sensory state in the striatum (Divac, Fonnum, & Storm-Mathisen, 1977; McGeorge & Faull, 1989), and
(3) DA-dependent plasticity in cortico-striatal connections (Reynolds & Wickens, 2002).

A typical formulation of DA-dependent learning (Reynolds & Wickens, 2002) may be expressed as the change in cortico-striatal connection strength, w (δw),

$$\Delta w = \eta \delta s \tag{12.8}$$

where 's' in Eq. (12.8) represents the cortical sensory input and is used in this section as a logical variable for neural encoding of the underlying state 's', $s = 1$ (if $s = s_{t)}$ else $s = 0$; 'δ' is the TD error (refer Eqs. (12.3 and 12.4) representing DA activity); and 'η' is the learning rate. Similar formulations have been proposed from purely RL theory considerations [see Chap. 9 of (Abbott, 2001)]. A slight variation of the above equation would be as follows:

$$\Delta w = \eta \lambda^{\text{Str}}(\delta) x \tag{12.9}$$

where 'λ^{Str}' is a function of δ, that represents the effect of DA on the neural firing rate (Reynolds, Hyland, & Wickens, 2001). Thus, the learning rule of Eq. (12.9) has a Hebb-like form, where neuromodulation is modeled in terms of the effect of the neuromodulator on the firing rate of the post-synaptic neuron. The form of the function λ^{Str} varies depending on the type of DA family receptors (R) expressed in MSNs as we explain below. In neurons with D1R expression, higher DA level increases the probability of MSN excitation by a given cortical input (Moyer, Wolf, & Finkel, 2007; Surmeier, Ding, Day, Wang, & Shen, 2007). Hence, in models that represent MSNs, λ^{Str} is described as an increasing sigmoid function of DA for neurons that express D1R. In cells with D2R, the activation is higher under conditions of low DA levels (Hernandez-Echeagaray, Starling, Cepeda, & Levine, 2004), and therefore, the λ^{Str} function is modeled as a decreasing function of DA (Frank, 2005; Frank, Samanta, Moustafa, & Sherman, 2007; Frank, Seeberger, & O'Reilly, 2004). These sigmoid λ^{Str} functions are expressed as,

$$\begin{aligned}
\lambda_{\mathrm{D1}}^{\mathrm{Str}}(\delta) &= \frac{2c_1}{1+\exp(c_2(\delta+c_3))} - c_1 \\
\lambda_{\mathrm{D2}}^{\mathrm{Str}}(\delta) &= \frac{2c_1}{1+\exp(c_2(\delta+c_3))} - c_1 \\
\lambda_{h-D1}^{\mathrm{Str}}(\delta) &= \frac{c_1}{1+\exp(c_2(\delta+c_3))} \\
\lambda_{h-D2}^{\mathrm{Str}}(\delta) &= \frac{c_1}{1+\exp(c_2(\delta+c_3))}
\end{aligned} \tag{12.10}$$

where c_1, c_2, c_3 are constants subjective to the receptor type. The gain functions of D1R MSNs and D2R MSNs are given by $\lambda_{\mathrm{D1}}^{\mathrm{Str}}, \lambda_{\mathrm{D2}}^{\mathrm{Str}}$ and that of the D1R and the D2R components of co-expressing MSNs are given by $\lambda_{h\text{-D1}}$, $\lambda_{h\text{-D2}}$, respectively. The gain function expression for risk coding MSNs $\left(\lambda_{h-D1}^{\mathrm{Str}}, \lambda_{h-D2}^{\mathrm{Str}}\right)$ is logarithmic sigmoid that lie within the limits of nonnegative real number space while that of the other MSNs (λ_{D1}, λ_{D2}) is coded by tangential sigmoid. Examples for such sigmoid λ functions with parameters for the D1R, D2R, and the D1R–D2R MSNs are shown in Fig. 12.1a. MSNs with D1R expression are appropriately suited for value computation (Kalva, Rengaswamy, Chakravarthy, & Gupte, 2012; Krishnan et al., 2011). They express $\lambda_{\mathrm{D1}}(\delta)$ as an increasing function of δ.

Such cellular models of MSNs along with the BG model explained in Balasubramani et al. (2015) are shown to consistently explain the results of extended RL model of serotonin (Eq. 12.7). Serotonin's input to D1-, D2-, D1–D2-co-expressing receptors containing MSNs are represented as α_{D1}, α_{D2}, α_{D1D2}, respectively. Whereas D1 MSNs following direct pathway, D2 and co-expressing D1–D2 receptors containing MSNs are modeled to follow IP of the BG dynamics (Fig. 12.2).

The D1R MSNs receive cortico-striatal connections whose weight is denoted by 'w_{D1}'. The value 'Q' computed by such an MSN is given by (Eq. 12.11).

$$Q = w_{\mathrm{D1}}\, s \tag{12.11}$$

Change in weight for such a neuron is given by (Eq. 12.12).

$$\Delta w_{\mathrm{D1}} = \eta_{\mathrm{D1}}\, \lambda_{\mathrm{D1}}^{\mathrm{Str}}(\delta)\, s \tag{12.12}$$

where η_{D1} is the learning rate. We will now show that a similar neuron model in which D1R and D2R are co-expressed can simulate risk computations. In case of a neuron that would compute risk, the λ^{Str} function is represented as '$\lambda_{\mathrm{D1D2}}^{\mathrm{Str}}$'. We assume that a neuron with D1R–D2R co-expression combines the characteristics of purely D1R- and D2R-expressing MSNs. Therefore, in D1R–D2R-co-expressed MSNs, the function '$\lambda_{\mathrm{D1D2}}^{\mathrm{Str}}$' is an even function of 'δ', with $\lambda_{\mathrm{D1D2}}^{\mathrm{Str}}(\delta)$ increasing with increasing magnitude of δ. In a MSN with co-expression of D1R and D2R, $\lambda_{\mathrm{D1D2}}^{\mathrm{Str}}$ (Eq. 12.13) can be expressed as the summation of functions corresponding to a D1R component $\lambda_{h-\mathrm{D1}}^{\mathrm{Str}}$ and a D2R component $\lambda_{h-\mathrm{D2}}^{\mathrm{Str}}$ as follows.

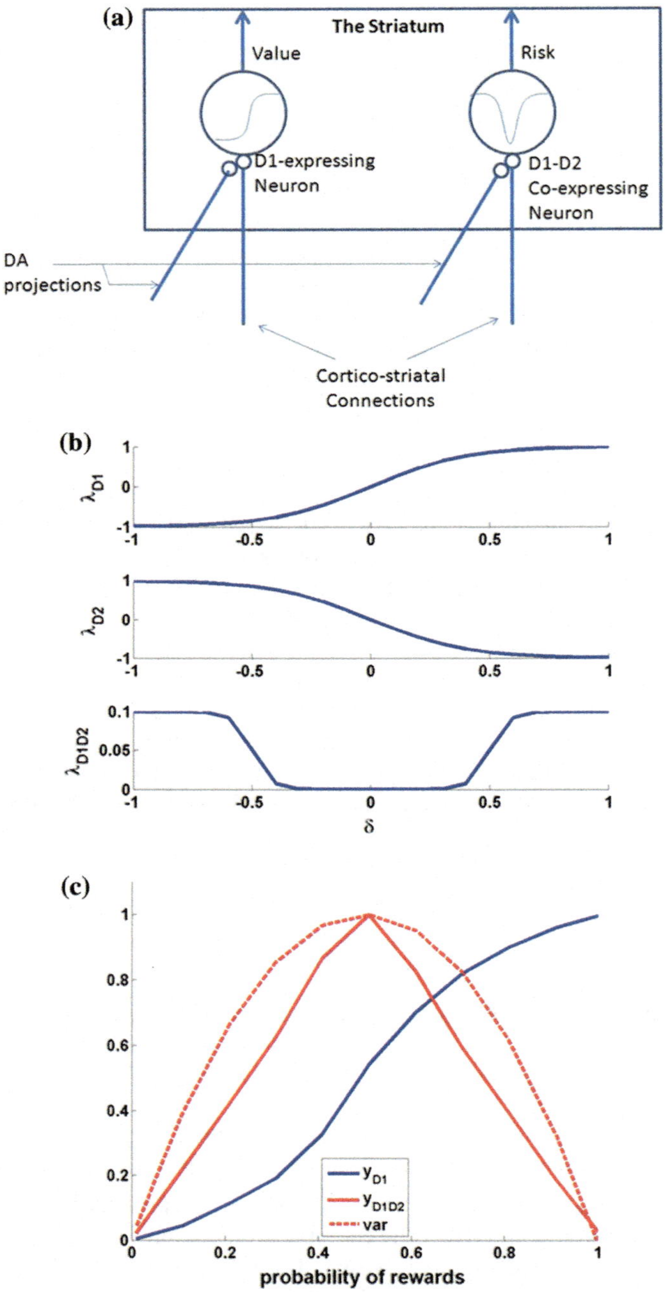
(a)
The Striatum
Value
Risk
D1-expressing
Neuron
D1-D2
Co-expressing
Neuron
DA
projections
Cortico-striatal
Connections
(b)
λ_{D1}
λ_{D2}
λ_{D1D2}
δ
(c)
y_{D1}
y_{D1D2}
var
probability of rewards

◂**Fig. 12.1** **a** Schematic of the cellular correlate model for the value and the risk computation in the striatum, **b** D1, D2, and D1D2 gain functions, **c** the output activity of D1R MSN (y_{D1}), D1R–D2R-co-expressing MSN (y_{D1D2}), and normalized variance computed analytically (var) = p* (1 − p); here p is the probability associated with rewards, i.e., with probability p, reward = 1, else reward = 0. The resemblance of var to y_{D1D2} shows the ability of D1R–D2R-co-expressing MSN to perform risk computation. Published in (Balasubramani et al., 2015a)

$$\lambda_{\mathrm{D1D2}}^{\mathrm{Str}} = \lambda_{h-\mathrm{D1}}^{\mathrm{Str}} + \lambda_{h-\mathrm{D2}}^{\mathrm{Str}} \tag{12.13}$$

Note that the characteristic of $\lambda_{h-\mathrm{D1}}^{\mathrm{Str}}$ and $\lambda_{h-\mathrm{D2}}^{\mathrm{Str}}$ as a function of δ depends on the constants c_1, c_2, c_3 of the Eq. (12.10). Response of such a neuron is given as,

$$h = w_{\mathrm{D1D2}} s \tag{12.14}$$

and the change in corresponding weight, $\delta w_{h,}$ is given as,

$$\Delta w_{\mathrm{D1D2}} = \eta_{\mathrm{D1D2}} \lambda_{\mathrm{D1D2}}^{\mathrm{Str}}(\delta)\, s \tag{12.15}$$

where η_{D1D2} is the learning rate. Thus, the unified network model proposes that (D1R-expressing) striatal MSNs with δ-dependent λ^{Str} functions that are of increasing sigmoidal shape are capable of computing value. Similarly (D1R–D2R-co-expressing), striatal neurons with δ-dependent λ^{Str} functions that are of 'U' shaped can compute risk (Fig. 12.1a). Just as D1R-expressing MSNs can be regarded as cellular level substrates for value computation in the striatum, D1R–D2R-co-expressing MSNs could be cellular level substrates for risk computation (Fig. 12.1b). The gain expression for risk coding MSNs $\left(\lambda_{h-\mathrm{D1}}^{\mathrm{Str}}, \lambda_{h-\mathrm{D2}}^{\mathrm{Str}}\right)$ uses a logarithmic sigmoid function that is unipolar, while the gain expression of other D1R and D2R MSNs $\left(\lambda_{\mathrm{D1}}^{\mathrm{Str}}, \lambda_{\mathrm{D2}}^{\mathrm{Str}}\right)$ uses a tangent-sigmoid function that is bipolar. The above cellular substrates for value and risk computation are put in a network model of BG (Fig. 12.2) to show that the network, including serotonin, is capable of reward–punishment–risk-based decision making (Balasubramani et al., 2014, 2015). This network model captures the anatomical details of the BG and represents the following nuclei: the striatum, STN, GPe, and GPi. The training of the cortico-striatal connections by nigrostriatal DA correlate (δ) also occurs as described in the earlier section. It models, in an elementary form, the action of DA in switching between DP and IP, via the differential action of DA on the D1-, D2-, and D1–D2-co-expressing receptors (R) of striatal MSNs. The model also proposes different DA signals for the updating of cortico-striatal weights and the switching in GPi (Chakravarthy & Balasubramani, 2014). Some of the key properties of the STN–GPe system such as their bidirectional connectivity facilitating oscillations and 'Exploratory' behavior are also captured. The model framework is adapted from the classical models of the BG as described in (Albin, Young, & Penney, 1989; Bar-Gad & Bergman, 2001; DeLong, 1990).

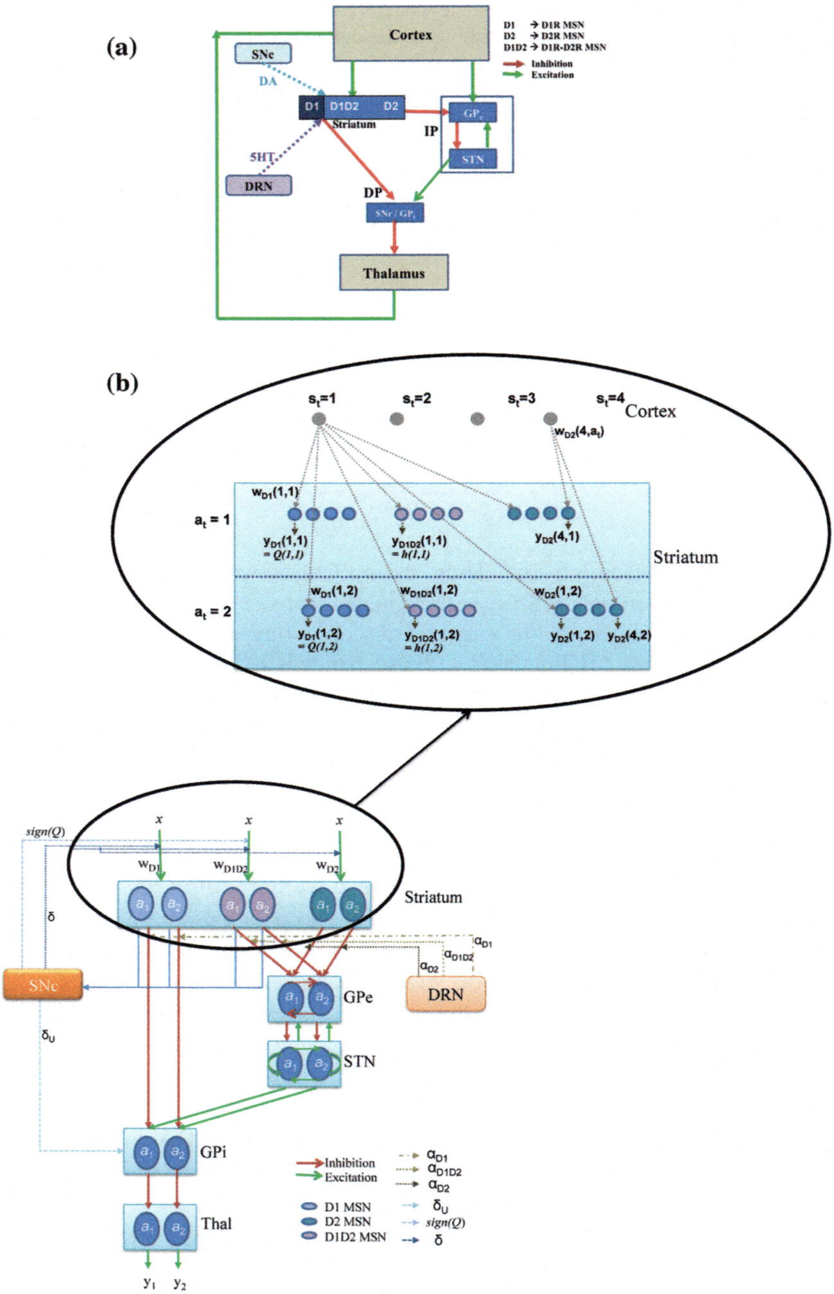
(a)
Cortex
D1 → D1R MSN
D2 → D2R MSN
D1D2 → D1R-D2R MSN
Inhibition
Excitation
SNc
DA
D1 D1D2 D2
Striatum
IP
5HT
DRN
STN
DP
Thalamus
(b)
Cortex
Striatum
a_t = 1
a_t = 2
Striatum
GPe
DRN
SNc
STN
GPi
Thal
y_1 y_2
sign(Q)
δ
δ_U
Inhibition
Excitation
D1 MSN
D2 MSN
D1D2 MSN

◀**Fig. 12.2** **a** Schematic of the BG showing the direct (DP) and indirect (IP) pathways; **b** the schematic flow of the signal in the network model. Here *s* denotes the state; *a* denotes the action; with the subscript denoting the index *i*. Since most of the experiments in the study simulate two possible actions for any state, we depict the same in the above figure for a state s_i; The D1, D2, D1D2 represent the D1R-, D2R-, D1R–D2R MSNs, respectively, and *w* denotes subscript corresponding cortico-striatal weights. The schematic also has the representation of DA forms: (1) the δ affecting the cortico-striatal connection weights (Houk et al., 2007; Schultz, 1998), (2) the δ_U affecting the action selection at the GPi (Chakravarthy & Balasubramani, 2014), (3) the Q affecting the D1/D2 MSNs (Schultz, 2010b); and 5-HT forms represented by α_{D1}, α_{D2}, and α_{D1D2} modulating the D1R, D2R, and the D1R–D2R-co-expressing neurons, respectively. The inset details the notations used in model section for representing cortico-striatal weights (*w*) and responses (*y*) of various kinds of MSNs (D1R-expressing, D2R-expressing, and D1R–D2R-co-expressing) in the striatum, with a sample cortical state size of 4, and maximum number of action choices available for performing selection in every state as 2. Adapted from (Balasubramani et al., 2014; Balasubramani, Chakravarthy, Ravindran, et al., 2015; Balasubramani, Chakravarthy, Ali, et al., 2015)

Serotonin's influence on decision making extends to various functions such as risk sensitivity, timescale of reward prediction, and punishment sensitivity. Therefore, the next section deals with application of the described unified model representing 5-HT to control the risk prediction error, and DA controlling the reward prediction error, to the distinct experiments dealing with various representative functions of 5-HT.

12.3 Results

12.3.1 Reward–Punishment Sensitivity

Several studies find that tryptophan levels mediate differential learning and decision making through rewards and punishments. Let's take this instance of reversal learning task-based study performed by (Cools et al., 2008; Robinson et al., 2012) in which the subjects were presented with two stimuli, one associated with reward and the other with punishment. On each trial, subjects had to predict whether the highlighted stimulus would lead to reward or punishment response. The subjects were tested in either a balanced or a depleted tryptophan levels (drink), on their association of the stimulus to the corresponding action at any time. Results showed that performance did not vary significantly with cases in both balanced and tryptophan-depleted conditions. Errors were fewer in the tryptophan-depleted than balanced conditions in both cases. Specifically, errors were fewer for punishment prediction trials compared to reward prediction trials in tryptophan-depleted conditions. Thus the experiment suggests that tryptophan depletion selectively enhances punishment prediction relative to reward prediction.

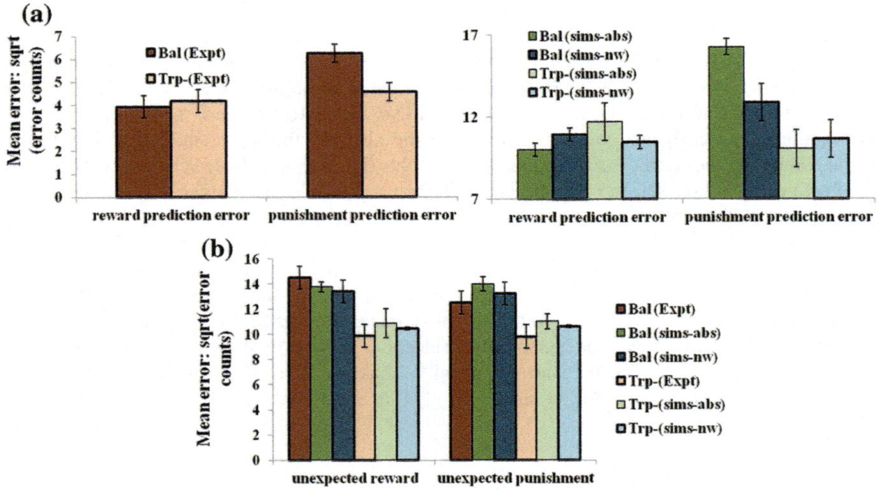

Fig. 12.3 Mean number of errors in non-switch trials **a** as a function of 'α' and outcome trial type; balanced (5-HT param $\alpha = 0.5$ in model I; $\alpha_{D1D2} = 1$, $\alpha_{D1} = 1$, $\alpha_{D2} = 5$ in model II) and tryptophan depletion ($\alpha = 0.3$ in model I; $\alpha_{D1D2} = 1$, $\alpha_{D1} = 1$, $\alpha_{D2} = 2.25$ in model II). Error bars represent standard errors of the difference as a function of 'α' in simulation for size 'N' = 100 (Sims). **b** Experimental error percentages adapted from Cools et al. (2008). Error bars represent standard errors as a function of drink in experiment (Expt). The results in **b** were reported after the exclusion of the trials from the acquisition stage of each block. Adapted from (Balasubramani et al., 2014; Balasubramani, Chakravarthy, Ravindran, et al., 2015)

The reconciled model (Fig. 12.2) has effectively captured the reward–punishment sensitivity controlling property of serotonin (Balasubramani et al., 2014; Balasubramani, Chakravarthy, Ravindran, et al., 2015) (Fig. 12.3).

12.3.2 Serotonin and Timescale of Reward Prediction

In another model, it is argued that 5-HT plays a role in the scaling of future rewards (Doya, 2002; Tanaka et al., 2007). In order to verify the hypothesis that 5-HT corresponds to the discount factor, γ (as in Eq. 12.4), Tanaka et al. (2007) conducted an fMRI experiment in which subjects performed a multistep delayed reward choice task. Subjects had to choose between a white square leading to a small early reward and a yellow square leading to a large but delayed reward (Tanaka et al., 2007). They were tested in: (1) tryptophan-depleted, (2) control, and (3) excess tryptophan conditions.

The model (Fig. 12.2) of (Balasubramani et al. 2014; Balasubramani, Chakravarthy, Ravindran, et al., 2015) has captured the timescale of reward prediction functional control of serotonin (Fig. 12.4).

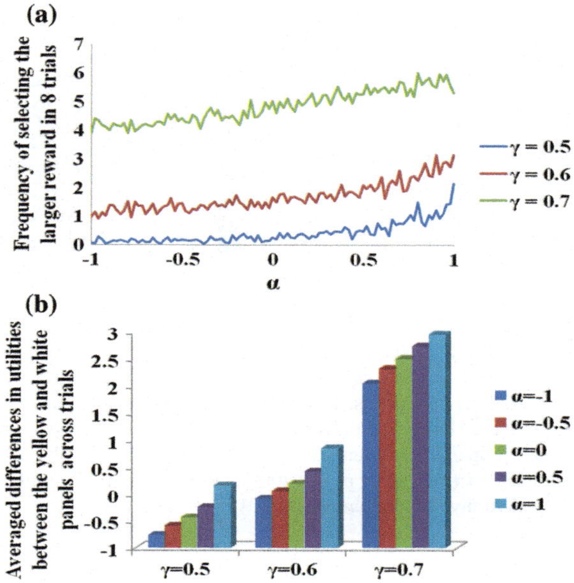

Fig. 12.4 **a** Selection of the long-term reward as a function of α. Increasing γ increased the frequency of selecting the larger and more delayed reward. Increasing α also gave similar results for a fixed γ. **b** Differences in the utilities (U) between the yellow and white panels averaged across trials for the states, s_t, as a function of γ and α. Here $N = 2000$. Adapted from (Balasubramani et al., 2014)

12.3.3 Serotonin and Risk Sensitivity

Long et al. (2009) studies risk sensitivity under conditions of tryptophan depletion. In this experiment, a monkey was required to saccade to one of two given targets. One target was associated with a guaranteed juice reward (safe) and the other with a variable juice volume (risky). A nonlinear risk sensitivity toward juice rewards by adopting risk-seeking behavior for small juice rewards and risk aversive behavior for the larger ones (Long et al., 2009) was observed in the monkeys. They showed that when 5-HT levels are reduced in the brain using Rapid Tryptophan Depletion (RTD), monkeys preferred risky over safer alternatives (Long et al., 2009). Tryptophan acts as a precursor to 5-HT, and therefore reduction in tryptophan causes reduction in 5-HT.

The model (Fig. 12.2) of Chakravarthy and colleagues (Balasubramani et al., 2014; Balasubramani, Chakravarthy, Ravindran, et al., 2015) has successfully explained risk sensitivity property of serotonin (Fig. 12.5).

12.4 Discussion

The various functional manifestations of 5-HT at different scales of analysis made it very difficult for devising a unified theory to explain its multiple roles (Dayan & Huys, 2015). Thus, most existing computational models focus on one or a few of

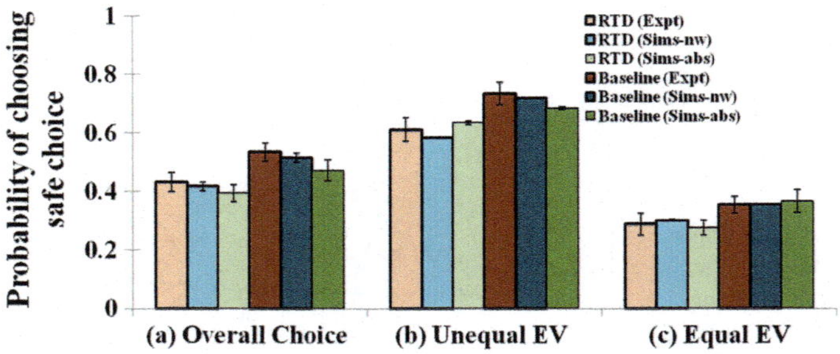

Fig. 12.5 Comparison between the experimental and simulated results for the **a** overall choice, **b** unequal EV, **c** equal EV, under RTD (with 5-HT param $\alpha = 1.658$ in model I; $\alpha_{D1D2} = 0.0012$, $\alpha_{D1} = 1$, $\alpha_{D2} = 1$ in model II) and baseline (control, with 5-HT param $\alpha = 1.985$ in model I; 1.32 in model II) conditions. Error bars represent the standard error (SE) with size 'N' = 100. The experiment (Expt) and the simulation (Sims) result of any case did not reject the null hypothesis, which proposes no difference between means, with P value > 0.05. Here the experimental results are adapted from Long et al. (2009). Adapted from (Balasubramani et al., 2014; Balasubramani, Chakravarthy, Ravindran, et al., 2015)

5-HT's functions. Many of these models employ high-level RL model such as TD learning (Daw et al., 2002; Doya, 2002). This chapter presents a model of serotonin in controlling risk-based utility, for reconciling major functions such as punishment prediction and response inhibition, timescale of reward prediction, and risk sensitivity (Balasubramani et al., 2012, 2014; Balasubramani, Chakravarthy, Ravindran, et al., 2015; Balasubramani, Chakravarthy, Ali, et al., 2015).

The starting point of our model was to understand the contributions of 5-HT to BG function (Boureau & Dayan, 2011; Tanaka et al., 2009). We use the notion of risk, since 5-HT is shown to be associated with risk sensitivity through the following instances. On presentation of the choices with risky and safe rewards, the reduction of 5-HT levels favors the selection of risky choices in comparison to the baseline levels (Long et al., 2009). The nonlinearity in risk-based decision making —risk aversion in the case of the gains and risk seeking in the case of losses—is postulated to be affected by central 5-HT levels (Murphy et al., 2009). Negative affective behavior such as depression, anxiety, and impulsivity which are caused by the reduction of the central 5-HT levels is argued to be a risky choice selection in a risk-based decision-making framework (Dayan & Huys, 2008). Based on the putative link between 5-HT function and risk sensitivity, we have extended the classical RL approach of policy execution using the utility function (Eq. 12.7) instead of value function. In the utility function, the weightage (α) that combines value and risk is proposed to represent 5-HT functioning in BG. Using this formulation, we show that three different experimental paradigms instantiating diverse theories of serotonin function in the BG can be explained under a single framework. Recent work by Bogacz and colleagues also supports this idea at the level of striatal synaptic plasticity (Mikhael & Bogacz, 2016).

12.4.1 Significance of Sign(Q_t)

The sign(Q_t) term presented in the modified formulation of utility function (Eq. 12.7) denotes the preference for risk in a given context of the experiment. At high mean reward values, humans are found to be risk-averse, whereas at low mean reward values they are risk-seeking (Kahneman & Tversky, 1979). In neuroeconomic experiments, this risk preference is statistically determined, for example, by maximizing the log likelihood of the decisions (d'Acremont et al., 2009). Though this method estimates the risk preference subjectively, it is derived from decisions made throughout the experiment. The use of sign(Q_t) in our model takes into account the variation of the subjective risk preference, according to the expected cumulative reward outcomes observed *within* an experiment.

12.4.2 5-HT-DA Interaction in the 'Risk' Component of Decision Making

The risk part of the utility function (Eq. 12.7) has three components: α, sign(Q_t), and $\sqrt{h_t}$. While 'α' represents 5-HT, the remaining two components are dependent on 'δ' or DA. Thus, the proposed model of risk computation postulates a complex interaction between DA and 5-HT. In neurobiology, complex interactions are indeed seen between DA and 5-HT (Di Matteo, Di Giovanni, Pierucci, & Esposito, 2008; Di Matteo, Pierucci, et al., 2008) at the cellular level. The 5-HT afferents from the DRN differentially modulate the DA neurons in SNc and Ventral Tegmental Area (VTA) (Gervais & Rouillard, 2000). The 5-HT projections act via specific receptor subtypes in the DA neurons. Action of 5-HT 1A, 5-HT 1B, 5-HT 2A, 5-HT 3, 5-HT 4 agonists facilitate dopaminergic release, whereas 5-HT 2C agonists inhibit the same. Selective 5-HT reuptake inhibitors are known to reduce the spontaneous activity of DA neurons in VTA (Alex & Pehek, 2007; Di Giovanni, Di Matteo, Pierucci, & Esposito, 2008; Di Mascio, Di Giovanni, Di Matteo, Prisco, & Esposito, 1998). 5-HT neurons in DRN also receive dense DA innervations from midbrain DA neurons (Ferre, Cortes, & Artigas, 1994) and express D2R (Suzuki, Hurd, Sokoloff, Schwartz, & Sedvall, 1998).

12.4.3 Main Finding of the DA-5-HT-Based BG Network Model for Utility-Based Decision Making

Ideally, a convincing model of value computation in the striatum must go beyond an abstract lumped representation and demonstrate how value may be computed by neural substrates of the striatum. There is a strong evidence for the existence of DA-modulated plasticity in cortico-striatal connections, an effect that is necessary to account for value computation in the MSNs of the striatum [see review by (Kötter & Wickens, 1998)]. The idea that MSNs are probably cellular substrates for value computation has found its place in the recent modeling literature (Morita, Morishima, Sakai, & Kawaguchi, 2012). Starting from the fact that the effect of DA on the D1R-expressing MSNs of the striatum is to increase the firing rate, it has been shown in a computational model of the BG that the D1R-expressing MSNs are capable of computing value (Krishnan et al., 2011). The present study presented a model of co-expressing D1R–D2R MSNs' gain function as an addition of the gain functions of D1R and the D2R MSNs. As a result, the D1D2R MSNs acquire a 'U'-shaped gain function. A few experiments provide support for such a representation, for instance, the study by Allen, Maher, Wani, Betts and Chase (2011) on neurons co-expressing D1-like and D2-like receptors in *C. elegans* (Allen et al., 2011). Here the D1R and D2R of a co-expressing neuron have antagonistic effects on neurotransmitter (acetylcholine) release. Specifically, they propose that the D1R–D2R-co-expressing neurons could simply be a combination of D1R and D2R neurons. Even studies on rodents and in vitro striatal cultures have shown the antagonistic nature of the D1 and the D2 receptor components of a co-expressing neuron (Hasbi et al., 2011). They report that these co-expressing neurons activate the CAMKII and BDNF machinery, each of which is known to play opposing roles in synaptic plasticity—long-term potentiation and long-term depression (Surmeier et al., 2007). We follow such a perspective of simple addition of the antagonistic D1 and the D2 neuronal gain functions to model the D1R–D2R MSN in our modeling study. In the BG, the ventral striatal neurons are known to be specially involved in risk processing (Stopper & Floresco, 2011). In this regard, we further hypothesize that D1R–D2R MSNs in those nuclei (Stopper & Floresco, 2011) would specifically contribute to risk computation as observed in experimental work by Stopper et al. (2011). We also predict that the selective loss of these co-expressing neurons would make the subject less sensitive to risk and therefore show risk-seeking behavior. Then the chapter continues toward realizing action selection through network dynamics of the BG. The underlying stochasticity in the softmax rule used in our early study (Balasubramani et al., 2014) is achieved indirectly by the chaotic dynamics of the STN–GPe loop (Kalva et al., 2012).

12.4.4 Striatal DA and 5-HT

The DA signals used in our model are a function of reward–value, and TD in value–utility (Fig. 12.2). The existence of different forms could be possible because of the following:

(1) Distinct sets of dopamine neurons are known to project to striatum. For instance, structures such as the striosome and matrisome are proposed to receive different DA modulatory signals [See the section 'Modularity of dopamine signals' in (Amemori, Gibb, & Graybiel, 2011)]. Some other studies

find that all the SNc DA neurons innervate both the striosomes and matrisomes, but each neuron's activity might favor any one of the compartment (Matsuda et al., 2009).

(2) Similarly, dopaminergic neurons from different regions dorsal–ventral of SNc–VTA might represent different computational quantities [See section 'Modularity of dopamine signals' in (Amemori et al., 2011)].

(3) Moreover, certain DAergic signals are known to specifically modulate between trials, while some other are proposed to act like a teaching signal within a trial (Stauffer, Lak, & Schultz, 2014; Tai, Lee, Benavidez, Bonci, & Wilbrecht, 2012).

A review by Schultz (Schultz, 2013) and other studies (Lak, Stauffer, & Schultz, 2014; Stauffer et al., 2014) states that the dopamine neurons are known to reflect various reward attributes such as the magnitude, probability, and delay. In fact, the above-mentioned attributes also get reflected when dopamine neurons can inform the first derivative of value or the utility function, as a common neuronal implementation (Stauffer et al., 2014).

Our model proposes that the δ and sign(Q) (Fig. 12.2) affect the computation of utility function by the MSNs. It must be noted that δ affects all the three kinds of MSNs (D1R, D2R, and the D1R–D2R MSNs) presynaptically as investigated through many experimental studies [Refer (Kötter & Wickens, 1998; Reynolds & Wickens, 2002)]. But the sign(Q) correlate of DA is proposed to affect the responses of D1R–D2R MSNs.

The receptors 5-HT 1, 2A, 2C, and 6 (Di Matteo, Pierucci, et al., 2008; Ward & Dorsa, 1996) are most abundantly expressed in the striatum. None of these receptors show preferential co-localization to any striatal proteins, such as substance P, dynorphin (neurons that contribute to the striato-nigral direct pathway), or enkephalin (contributing to the IP). However, a differential expression indeed exists —5-HT2C is highly expressed in the patches, and 5-HT2A in the matrix (Eberle-Wang, Mikeladze, Uryu, & Chesselet, 1997). These 5-HT receptors are more likely to be co-expressed even along with the D1R–D2R MSNs which form a substantial portion of the striatum according to certain experimental studies (Bertran-Gonzalez, Hervé, Girault, & Valjent, 2010; Paolo Calabresi, Picconi, Tozzi, Ghiglieri, & Di Filippo, 2014; Hasbi et al., 2010, 2011; Nadjar et al., 2006; Perreault et al., 2010). It is true that 5-HT's specificity in expression along with a particular type of MSN is still not clear.

In order to investigate the possibility that 5-HT modulation of MSNs may not be limited only to D1R–D2R MSNs, but could have a differential action on the three pools of MSNs (D1R, D2R, and D1R–D2R), we have conducted additional simulations and obtained quite revealing results. On varying different subsets of $\{\alpha_{D1}, \alpha_{D2}, and\ \alpha_{D1D2}\}$, the following inferences are made:

- The modulation of α_{D1} alone [$\alpha_{D2} = 1$, $\alpha_{D1D2} = 1$] is not able to consistently model the behavior of a balanced (high α_{D1}) or the reduced tryptophan (low α_{D1}) conditions in any experiment. Similar is the case of modulating α_{D2} [$\alpha_{D1} = 1$, $\alpha_{D1D2} = 1$] alone.

- The joint modulation of α_{D1} and α_{D2} [α_{D1D2} = 1] was not able to explain any of the experiments satisfactorily.
- α_{D1D2} is found to be able to explain the results of the experiment by Cools et al. (2008) better only when optimized along with α_{D2}. The joint modulation of α_{D2} and α_{D1D2} [α_{D1} = 1] achieves best fit for all the experiments.
- α_{D1} is not found to be as sensitive as α_{D1D2} and α_{D2} in all the experiments, though a nonzero α_{D1} is preferred.
- In summary, α_{D1} representation of 5-HT can be fixed at 1, while the others α_{D1D2} and α_{D2} can be varied and optimized to explain different 5-HT-based experimental results.

The optimization of fixed 5-HT values might also be related to the tonic modulation exerted by DRN during reward processing (Alex & Pehek, 2007; Jiang, Ashby Jr., Kasser, & Wang, 1990; Nakamura, 2013).

12.4.5 The Co-expressing D1R–D2R MSNs

There have been varied reports of the proportion of co-expressing D1R–D2R MSNs in the striatum. These neurons were not modeled in any of the earlier studies (Ashby, Turner, & Horvitz, 2010; Frank et al., 2004; Humphries & Prescott, 2010; Krishnan et al., 2011), though present in significant proportion to D1R- and D2R-expressing MSNs (Bertran-Gonzalez et al., 2010; Calabresi et al., 2014; Hasbi et al., 2010, 2011; Nadjar et al., 2006; Perreault et al., 2010). Such unpopular nature of the D1R–D2R MSNs in the striatum might be due to the following: The existence of co-expressing D1R–D2R MSNs have been under debate over years. Many studies supported distinct populations of the striatal MSNs projecting in striato-nigral and striato-pallidal pathways including neurochemical and genetic ontology analysis in mice (Araki, Sims, & Bhide, 2007), transgenic mice engineered using bacterial artificial chromosome (BAC) with enhanced green fluorescent protein (Bertler & Rosengren, 1966; Matamales et al., 2009; Shuen, Chen, Gloss, & Calakos, 2008; Valjent, Bertran-Gonzalez, Hervé, Fisone, & Girault, 2009), biochemical and imaging assays including in situ hybridization (ISH) combined with retrograde axonal tracing (Gerfen et al., 1990; Catherine Le Moine & Bloch, 1995; Le Moine, Normand, & Bloch, 1991), fluorescence-activated cell sorting (FACS) of MSNs or translating ribosome affinity purification approach (TRAP) (Heiman et al., 2008; Lobo, Karsten, Gray, Geschwind, & Yang, 2006). These studies report that D1Rs are present in striato-nigral MSNs and are Substance P positive, whereas the D2R is enriched with enkephalin and is striato-pallidal in nature [Classical models of the BG: (Albin et al., 1989; DeLong, 1990)].

However, some of these highly sensitive studies are under debate (Bertran-Gonzalez et al., 2010; Calabresi et al., 2014) due to the following reasons: The developmental regulation of D1R and D2R mRNAs as analyzed in the genetic ontology studies with mice (Araki et al., 2007) results from intrinsic genetic programs that control the receptors' expression, whereas the actual dopaminergic neuron's innervation in a projection area (here, the striatum) is found to control the D1R and D2R expressions (Jung & Bennett, 1996). Furthermore, the genetically engineered bacterial artificial chromosome (BAC) mice are found to have alterations in comparison with wild-type mice in terms of behavioral, electrophysiological, and molecular characterization. Even highly advanced optogenetics and other imaging techniques that support segregation of the pathways are questioned for their ability to monitor the subcortical activity accurately in the behaving animals [See the reviews by (Bertran-Gonzalez et al., 2010; Calabresi et al., 2014)].

Meanwhile, there are many other findings questioning the strict segregation of the direct and the IPs. See review by (Bertran-Gonzalez et al., 2010; Calabresi et al., 2014) for more details. These studies report various modes of cross-talk existing between the 'classical' dichotomous projections from the striatum. Studies also report co-expression of the D1R and the D2R in a MSN to be a medium for cross-talk. They even propose the receptors' heteromerization to such an extent that these co-expressing MSNs would have their downstream effects completely different from that of the neurons solely expressing the D1R or the D2R.

The studies reporting co-expression of D1R–D2R in the MSNs analyze components such as calcium and brain-derived neurotrophic factor (BDNF) (Hasbi et al., 2009; Rashid, O'Dowd, Verma, & George, 2007), using techniques such as reverse transcription polymerase chain reaction (RT-PCR) that is reviewed in (Surmeier & Kitai, 1993), co-immunoprecipitation (Lee et al., 2004), or fluorescence resonance energy transfer (FRET) using fluorophore-labeled antibodies (Hasbi et al., 2009). Some quantitative measures regarding the proportion of D1R–D2R MSNs in the striatum include nearly 17% in the nucleus accumbens shell, and 6% in the caudate–putamen, when estimated using BAC transgenic mice (Bertran-Gonzalez et al., 2008). Though there have been doubts regarding the accurate neuronal labeling in BAC transgenic mice, the proportions have been confirmed by the later studies too (Matamales et al., 2009). Similarly, a quantitative FRET in situ showed that more than 90% of the D1R–D2R-co-expressing neuronal bodies in the nAcc and nearly 25% of them were found in the caudate–putamen (Perreault et al., 2010). Hence, these studies favor the presence of D1R–D2R MSNs in significant levels in the striatum.

A few studies report the projection of D1R–D2R-co-expressing neurons to GPi (Perreault et al., 2010, 2011). Though our present study accounts for their projection to GPe alone, this study suggests that the D1R–D2R-co-expressing neurons targeting the pallidum mainly contribute to risk computation. Those D1R–D2R MSNs that project to SNc may be utilized for the TD in utility computation. These projections of the D1R–D2R-co-expressing neurons toward both the indirect and the direct pathways support the study that DA D1R containing neurons may not solely project onto the direct pathway. This is because some of the D1R containing MSNs are known to also project to the IP (Calabresi et al., 2014). Those D1R neurons could be co-expressing D2R, since D1R–D2R-co-expressing MSNs are capable of invading both the direct and the IPs (Bertran-Gonzalez et al., 2010;

Calabresi et al., 2014; Hasbi et al., 2010, 2011; Nadjar et al., 2006; Perreault et al., 2010). Similarly, the D2R MSN need not just solely project to the IP. The study of (Calabresi et al., 2014) shows that D1R–D2R MSNs are one of the means by which the direct and the IPs interact. Such a notion is also implemented in our modeling study, and hence these D1R–D2R-co-expressing MSNs might play a major role in the cross-talk between the direct and the IPs.

Moreover, DA D1R and D2R are also shown to form heteromeric complexes with unique functional properties and phenotype (Hasbi et al., 2011; Perreault, Fan, Alijaniaram, O'Dowd, & George, 2012). These heteromers are found to have increased sensitivity following repeated increases in DA transmission. The up-regulated state of these heteromers persisted after DA agonist removal, identifying these heteromeric complexes as therapeutic targets in DA-related disorders, such as schizophrenia and drug addiction. These heteromers are also predicted to significantly influence cognition, learning, and memory (Perreault et al., 2011, 2012). We would expect that there might be differences between the co-expressing neurons and the heteromers, but in the absence of more data, this study has used the simple model of addition of D1R and D2R MSN's gain functions to represent the D1R–D2R-co-expressing neurons.

12.5 Future Work

The co-expressing D1R–D2R MSNs are experimentally shown to significantly contribute to both the direct and the IPs of the BG (Bertran-Gonzalez et al., 2010; Calabresi et al., 2014; Hasbi et al., 2010, 2011; Nadjar et al., 2006; Perreault et al., 2010). These two distinct pools of D1R–D2R MSNs—one following DP that controls exploitation, and the other following IP that controls exploration (Chakravarthy & Balasubramani, 2013; Chakravarthy, Joseph, & Bapi, 2010; Kalva et al., 2012)—might be used for modeling the nonlinearity in risk sensitivity based on outcomes (risk aversion during gains and risk seeking during losses) (Kahneman & Tversky, 1979). The inherent opponency between the DP and IP (Albin, 1998; DeLong, 1990) would facilitate the projections of the corresponding D1R–D2R MSNs for showing contrasting risk-sensitive behavior. Each of the neuronal pools computing the risk function should then be weighed by appropriate sensitivity coefficients [representing neuromodulators DA and 5-HT (Balasubramani et al., 2014)] to capture the nonlinear risk-sensitive behavior (Kahneman & Tversky 1979) based on the reward–punishment outcomes. This is simplified in the present modeling study by considering the D1R–D2R MSNs to IP alone, multiplied by a (α sign(Q)) term. Moreover, the increased magnitude of risk associated with an action is experimentally found to enhance exploration in action selection (Cohen, McClure, & Angela, 2007; Daw, O'Doherty, Dayan, Seymour, & Dolan, 2006; Frank, Doll, Oas-Terpstra, & Moreno, 2009). This is implemented in our network model by routing the co-expressing D1R–D2R MSN activity to the IP that controls the exploration of the BG dynamics (Chakravarthy & Balasubramani, 2013;

Chakravarthy et al., 2010; Kalva et al., 2012). Expanding the framework to include the D1R–D2R MSNs projections to GPi (in the DP) would be done in our future work.

Projections from GPe to GPi are found in the primates (Gerfen & Wilson, 1996; Kawaguchi, Wilson, & Emson, 1990; Mink, 1996). GPe projections to GPi are thought to be more focused, compared to the more diffuse projections of STN to GPi. These GPe–GPi connections bypass the GPe–STN–GPi connectivity. The former are thought to perform a *focused suppression of GPi response to a particular action*, whereas the latter impose a *Global NoGo* influence (Mink, 1996; Parent & Hazrati, 1995). Though the functional significance of these connections is not known, not accounting for this connectivity (STN–GPe–GPi) is a limitation of our modeling study. However, since we do not differentiate a global/local NoGo signal in our study, the proposed minimal model adapted from classical BG models (Albin et al., 1989; Bar-Gad & Bergman, 2001; DeLong, 1990; Mink, 1996) is able to capture the required experimental results at the neural network level.

Further investigation should examine in more detail DA–5-HT interactions based on the specific receptor type distribution in the BG. This study only deals with the theoretical principles underlying DA–5-HT interactions in the BG, which can be then expanded to understand the detailed influence of the same interactions in the cortex, SNc, and Raphe nucleus. Apart from analyzing the details of the interactions in various regions of the brain, attempts to include other major neuromodulators like acetylcholine (ACh) and norepinephrine (NE) are also desired. This could be realized by including a self-organized map (SOM) model of the striatum which captures its topologically ordered arrangement of the striosomes and matrisomes (Stringer, Rolls, Trappenberg, & De Araujo, 2002) and is controlled by the ACh mediated tonically active interneurons. The model is expected to analyze ACh influence in the selection of striosome–matrisome pairs and the plasticity of cortico-striatal connections (Ding et al., 2011; Spehlmann & Stahl, 1976). Specific investigation of how the neuromodulator NE affects the STN–GPe system and the BG dynamics is also of special interest. Neuromodulator NE has been compared to the inverse temperature parameter of Eq. (12.7) and is thought to specifically affect the exploration dynamics of the BG action selection machinery (Aston-Jones & Cohen, 2005; Doya, 2002). In our earlier study, we have showed that the STN lateral connections can also influence the BG exploration dynamics significantly (Chakravarthy & Balasubramani, 2013). Control of response inhibition through STN is thought to be established through the NE activity in STN, and a dysfunction in such control could be related to impulse control disorders (ICD) (Economidou, Theobald, Robbins, Everitt, & Dalley, 2012; Swann et al., 2013). The impact of DA and NE activities on STN functioning should be tested in future, paving way to a comprehensive computational understanding of the roles of all the four major neuromodulators (DA, 5-HT, NE, ACh) in the BG dynamics.

STN also receives extensive norepinephrine (NE) afferents (Parent & Hazrati, 1995; Wang et al., 1996). And since many studies report that the dynamics of STN–GPe is strongly controlled by the neuromodulator NE (Belujon, Bezard, Taupignon, Bioulac, & Benazzouz, 2007; Delaville, Zapata, Cardoit, & Benazzouz, 2012), future work should explore the possible role of NE in the BG dynamics. Particularly, NE is expected to control the lateral connection strengths in STN–GPe, and the gain of cortical input (Aston-Jones & Cohen, 2005; Cohen et al., 2007; Dayan & Yu, 2006) to striatum and STN. The control of response inhibition through STN is thought to be established through the NE activity in STN, and a dysfunction in such control could be related to ICD (Economidou et al., 2012; Swann et al., 2013). A detailed model of STN–GPe dynamics and the effect of NE on the same could help us better understand the role of the STN–GPe system in impulsivity and design better deep brain stimulation protocols to cure impulsivity (Frank et al., 2007).

Although DA, 5-HT, and NE along with the STN–GPe dynamics figure prominently in the experimental studies on action selection dynamics and their reaction times, computational models that closely resemble the neurobiological data supporting all those factors do not exist. Our model is the first to include the contribution of both DA and 5-HT in behavioral measures mediated by the BG dynamics.

References

Abbott, P. D. a. L. F. (2001). *Theoretical neuroscience: Computational and mathematical modeling of neural systems*. Cambridge, Massachusetts, London, England: The MIT Press.

Aghajanian, G. K., & Marek, G. J. (2000). Serotonin model of schizophrenia: Emerging role of glutamate mechanisms. *Brain Research Reviews, 31*(2), 302–312.

Albin, R. L. (1998). Fuch's corneal dystrophy in a patient with mitochondrial DNA mutations. *Journal of Medical Genetics, 35*(3), 258–259.

Albin, R. L., Young, A. B., & Penney, J. B. (1989). The functional anatomy of basal ganglia disorders. *Trends in Neurosciences, 12*(10), 366–375.

Alex, K. D., & Pehek, E. A. (2007). Pharmacologic mechanisms of serotonergic regulation of dopamine neurotransmission. *Pharmacology & Therapeutics, 113*(2), 296–320. https://doi.org/10.1016/j.pharmthera.2006.08.004.

Allen, A. T., Maher, K. N., Wani, K. A., Betts, K. E., & Chase, D. L. (2011). Coexpressed D1-and D2-like dopamine receptors antagonistically modulate acetylcholine release in Caenorhabditis elegans. *Genetics, 188*(3), 579–590.

Amemori, K., Gibb, L. G., & Graybiel, A. M. (2011). Shifting responsibly: The importance of striatal modularity to reinforcement learning in uncertain environments. *Frontiers in Human Neuroscience, 5,* 47. https://doi.org/10.3389/fnhum.2011.00047.

Angiolillo, P. J., & Vanderkooi, J. M. (1996). Hydrogen atoms are produced when tryptophan within a protein is irradiated with ultraviolet light. *Photochemistry and Photobiology, 64*(3), 492–495.

Araki, K. Y., Sims, J. R., & Bhide, P. G. (2007). Dopamine receptor mRNA and protein expression in the mouse corpus striatum and cerebral cortex during pre-and postnatal development. *Brain Research, 1156,* 31–45.

Ashby, F. G., Turner, B. O., & Horvitz, J. C. (2010). Cortical and basal ganglia contributions to habit learning and automaticity. *Trends in Cognitive Sciences, 14*(5), 208–215.

Aston-Jones, G., & Cohen, J. D. (2005). An integrative theory of locus coeruleus-norepinephrine function: Adaptive gain and optimal performance. *Annual Review of Neuroscience, 28,* 403–450.

Azmitia, E. C. (1999). Serotonin neurons, neuroplasticity, and homeostasis of neural tissue. *Neuropsychopharmacology, 21*(2 Suppl), 33S–45S. https://doi.org/10.1016/S0893-133X(99)00022-6.

Azmitia, E. C. (2001). Modern views on an ancient chemical: Serotonin effects on cell proliferation, maturation, and apoptosis. *Brain Research Bulletin, 56*(5), 413–424.

Balasubramani, P. P., Chakravarthy, S., Ravindran, B., & Moustafa, A. A. (2014). An extended reinforcement learning model of basal ganglia to understand the contributions of serotonin and dopamine in risk-based decision making, reward prediction, and punishment learning. *Frontiers in Computational Neuroscience, 8,* 47.

Balasubramani, P. P., Chakravarthy, S., Ravindran, B., & Moustafa, A. A. (2015a). A network model of basal ganglia for understanding the roles of dopamine and serotonin in reward-punishment-risk based decision making. *Name. Frontiers in Computational Neuroscience, 9,* 76.

Balasubramani, P. P., Chakravarthy, V. S., Ali, M., Ravindran, B., & Moustafa, A. A. (2015b). Identifying the Basal Ganglia network model markers for medication-induced impulsivity in Parkinson's Disease patients. *PLoS ONE, 10*(6), e0127542.

Balasubramani, P. P., Ravindran, B., & Chakravarthy, S. (2012). *Understanding the role of serotonin in basal ganglia through a unified model.* Paper presented at the International Conference on Artificial Neural Networks, Lausanne, Switzerland.

Bar-Gad, I., & Bergman, H. (2001). Stepping out of the box: Information processing in the neural networks of the basal ganglia. *Current Opinion in Neurobiology, 11*(6), 689–695.

Bell, C. (2001). Tryptophan depletion and its implications for psychiatry. *The British Journal of Psychiatry, 178*(5), 399–405. https://doi.org/10.1192/bjp.178.5.399.

Bell, D. E. (1995). Risk, return and utility. *Management Science, 41,* 23–30.

Belujon, P., Bezard, E., Taupignon, A., Bioulac, B., & Benazzouz, A. (2007). Noradrenergic modulation of subthalamic nucleus activity: Behavioral and electrophysiological evidence in intact and 6-hydroxydopamine-lesioned rats. *The Journal of Neuroscience, 27*(36), 9595–9606.

Bertler, A., & Rosengren, E. (1966). Possible role of brain dopamine. *Pharmacological Reviews, 18*(1), 769–773.

Bertran-Gonzalez, J., Bosch, C., Maroteaux, M., Matamales, M., Herve, D., Valjent, E., et al. (2008). Opposing patterns of signaling activation in dopamine D1 and D2 receptor-expressing striatal neurons in response to cocaine and haloperidol. *Journal of Neuroscience, 28*(22), 5671–5685. https://doi.org/10.1523/JNEUROSCI.1039-08.2008.

Bertran-Gonzalez, J., Hervé, D., Girault, J.-A., & Valjent, E. (2010). What is the degree of segregation between striatonigral and striatopallidal projections? *Front Neuroanat, 4.*

Boureau, Y. L., & Dayan, P. (2011). Opponency revisited: Competition and cooperation between dopamine and serotonin. *Neuropsychopharmacology, 36*(1), 74–97. https://doi.org/10.1038/npp.2010.151.

Buhot, M.-C. (1997). Serotonin receptors in cognitive behaviors. *Current Opinion in Neurobiology, 7*(2), 243–254.

Calabresi, P., Maj, R., Pisani, A., Mercuri, N. B., & Bernardi, G. (1992). Long-term synaptic depression in the striatum: Physiological and pharmacological characterization. *Journal of Neuroscience, 12*(11), 4224–4233.

Calabresi, P., Picconi, B., Tozzi, A., Ghiglieri, V., & Di Filippo, M. (2014). Direct and indirect pathways of basal ganglia: A critical reappraisal. *Nature Neuroscience, 17*(8), 1022–1030.

Chakravarthy, V. S., & Balasubramani, P. P. (2013). Basal Ganglia System as an engine for exploration. In J. R. Jaeger D. (Ed.), *Encyclopedia of Computational Neuroscience*. Berlin Heidelberg: SpringerReference (http://www.springerreference.com/). Springer-Verlag.

Chakravarthy, V. S., & Balasubramani, P. P. (2014). *Basal Ganglia System as an engine for exploration.* Berlin Heidelberg: SpringerReference (http://www.springerreference.com/). Springer-Verlag.

Chakravarthy, V. S., Joseph, D., & Bapi, R. S. (2010). What do the basal ganglia do? A modeling perspective. *Biological Cybernetics, 103*(3), 237–253. https://doi.org/10.1007/s00422-010-0401-y.

Chao, M. Y., Komatsu, H., Fukuto, H. S., Dionne, H. M., & Hart, A. C. (2004). Feeding status and serotonin rapidly and reversibly modulate a Caenorhabditis elegans chemosensory circuit. *Proceedings of the National Academy of Science U S A, 101*(43), 15512–15517. https://doi.org/10.1073/pnas.0403369101.

Cohen, J. D., McClure, S. M., & Angela, J. Y. (2007). Should I stay or should I go? How the human brain manages the trade-off between exploitation and exploration. *Philosophical Transactions of the Royal Society B: Biological Sciences, 362*(1481), 933–942.

Cools, R., Nakamura, K., & Daw, N. D. (2011). Serotonin and dopamine: Unifying affective, activational, and decision functions. *Neuropsychopharmacology, 36*(1), 98–113. https://doi.org/10.1038/npp.2010.121.

Cools, R., Robinson, O. J., & Sahakian, B. (2008). Acute tryptophan depletion in healthy volunteers enhances punishment prediction but does not affect reward prediction. *Neuropsychopharmacology, 33*(9), 2291–2299. https://doi.org/10.1038/sj.npp.1301598.

d'Acremont, M., Lu, Z. L., Li, X., Van der Linden, M., & Bechara, A. (2009). Neural correlates of risk prediction error during reinforcement learning in humans. *Neuroimage, 47*(4), 1929–1939. https://doi.org/10.1016/j.neuroimage.2009.04.096.

Dalley, J. W., Everitt, B. J., & Robbins, T. W. (2011). Impulsivity, compulsivity, and top-down cognitive control. *Neuron, 69*(4), 680–694.

Daw, N. D., Kakade, S., & Dayan, P. (2002). Opponent interactions between serotonin and dopamine. *Neural Network, 15*(4–6), 603–616.

Daw, N. D., O'Doherty, J. P., Dayan, P., Seymour, B., & Dolan, R. J. (2006). Cortical substrates for exploratory decisions in humans. *Nature, 441*(7095), 876–879. https://doi.org/10.1038/nature04766.

Dayan, P., & Huys, Q. (2015). Serotonin's many meanings elude simple theories. *Elife, 4.*

Dayan, P., & Huys, Q. J. (2008). Serotonin, inhibition, and negative mood. *PLoS Computational Biology, 4*(2), e4.

Dayan, P., & Yu, A. J. (2006). Phasic norepinephrine: A neural interrupt signal for unexpected events. *Network: Computation in Neural Systems, 17*(4), 335–350.

Delaville, C., Zapata, J., Cardoit, L., & Benazzouz, A. (2012). Activation of subthalamic alpha 2 noradrenergic receptors induces motor deficits as a consequence of neuronal burst firing. *Neurobiology of Diseases, 47*(3), 322–330.

DeLong, M. R. (1990). Primate models of movement disorders of basal ganglia origin. *Trends in Neurosciences, 13*(7), 281–285.

Di Giovanni, G., Di Matteo, V., Pierucci, M., & Esposito, E. (2008). Serotonin–dopamine interaction: Electrophysiological evidence. *Progress in Brain Research, 172,* 45–71.

Di Mascio, M., Di Giovanni, G., Di Matteo, V., Prisco, S., & Esposito, E. (1998). Selective serotonin reuptake inhibitors reduce the spontaneous activity of dopaminergic neurons in the ventral tegmental area. *Brain Research Bulletin, 46*(6), 547–554.

Di Matteo, V., Di Giovanni, G., Pierucci, M., & Esposito, E. (2008a). Serotonin control of central dopaminergic function: Focus on in vivo microdialysis studies. *Progress in Brain Research, 172,* 7–44.

Di Matteo, V., Pierucci, M., Esposito, E., Crescimanno, G., Benigno, A., & Di Giovanni, G. (2008b). Serotonin modulation of the basal ganglia circuitry: Therapeutic implication for Parkinson's disease and other motor disorders. *Progress in Brain Research, 172,* 423–463.

Ding, Y., Won, L., Britt, J. P., Lim, S. A. O., McGehee, D. S., & Kang, U. J. (2011). Enhanced striatal cholinergic neuronal activity mediates l-DOPA–induced dyskinesia in parkinsonian mice. *Proceedings of the National Academy of Sciences, 108*(2), 840–845.

Divac, I., Fonnum, F., & Storm-Mathisen, J. (1977). High affinity uptake of glutamate in terminals of corticostriatal axons. *Nature, 266*(5600), 377–378.

Doya, K. (2002). Metalearning and neuromodulation. *Neural Network, 15*(4–6), 495–506.

Eberle-Wang, K., Mikeladze, Z., Uryu, K., & Chesselet, M. F. (1997). Pattern of expression of the serotonin2C receptor messenger RNA in the basal ganglia of adult rats. *Journal of Comparative Neurology, 384*(2), 233–247.

Economidou, D., Theobald, D. E., Robbins, T. W., Everitt, B. J., & Dalley, J. W. (2012). Norepinephrine and dopamine modulate impulsivity on the five-choice serial reaction time task through opponent actions in the shell and core sub-regions of the nucleus accumbens. *Neuropsychopharmacology, 37*(9), 2057–2066.

Ferre, S., Cortes, R., & Artigas, F. (1994). Dopaminergic regulation of the serotonergic raphe-striatal pathway: Microdialysis studies in freely moving rats. *Journal of Neuroscience, 14*(8), 4839–4846.

Fox, S. H., Chuang, R., & Brotchie, J. M. (2009). Serotonin and Parkinson's disease: On movement, mood, and madness. *Movement Disorders, 24*(9), 1255–1266. https://doi.org/10.1002/mds.22473.

Frank, M. J. (2005). Dynamic dopamine modulation in the basal ganglia: A neurocomputational account of cognitive deficits in medicated and nonmedicated Parkinsonism. *Journal of Cognitive Neuroscience, 17*(1), 51–72. https://doi.org/10.1162/0898929052880093.

Frank, M. J., Doll, B. B., Oas-Terpstra, J., & Moreno, F. (2009). Prefrontal and striatal dopaminergic genes predict individual differences in exploration and exploitation. *Nature Neuroscience, 12*(8), 1062–1068.

Frank, M. J., Samanta, J., Moustafa, A. A., & Sherman, S. J. (2007a). Hold your horses: Impulsivity, deep brain stimulation, and medication in parkinsonism. *Science, 318*(5854), 1309–1312.

Frank, M. J., Seeberger, L. C., & O'Reilly R, C. (2004). By carrot or by stick: Cognitive reinforcement learning in Parkinsonism. *Science, 306*(5703), 1940–1943. https://doi.org/10.1126/science.1102941.

Gerfen, C. R., Engber, T. M., Mahan, L. C., Susel, Z., Chase, T. N., Monsma, F., et al. (1990). D1 and D2 dopamine receptor-regulated gene expression of striatonigral and striatopallidal neurons. *Science, 250*(4986), 1429–1432.

Gerfen, C. R., & Wilson, C. J. (1996). Chapter II The basal ganglia. *Handbook of Chemical Neuroanatomy, 12,* 371–468.

Gervais, J., & Rouillard, C. (2000). Dorsal raphe stimulation differentially modulates dopaminergic neurons in the ventral tegmental area and substantia nigra. *Synapse, 35*(4), 281–291. https://doi.org/10.1002/(sici)1098-2396(20000315)35: 4 < 281::aid-syn6 > 3.0.co;2-a.

Gillette, R. (2006). Evolution and function in serotonergic systems. *Integrative and Comparative Biology, 46*(6), 838–846. https://doi.org/10.1093/icb/icl024.

Halford, J. C., Harrold, J. A., Lawton, C. L., & Blundell, J. E. (2005). Serotonin (5-HT) drugs: Effects on appetite expression and use for the treatment of obesity. *Current Drug Targets, 6*(2), 201–213.

Hasbi, A., Fan, T., Alijaniaram, M., Nguyen, T., Perreault, M. L., O'Dowd, B. F., et al. (2009). Calcium signaling cascade links dopamine D1-D2 receptor heteromer to striatal BDNF production and neuronal growth. *Proceedings of the National Academy of Science U S A, 106* (50), 21377–21382. https://doi.org/10.1073/pnas.0903676106.

Hasbi, A., O'Dowd, B. F., & George, S. R. (2010). Heteromerization of dopamine D2 receptors with dopamine D1 or D5 receptors generates intracellular calcium signaling by different mechanisms. *Current Opinion in Pharmacology, 10*(1), 93–99. https://doi.org/10.1016/j.coph.2009.09.011.

Hasbi, A., O'Dowd, B. F., & George, S. R. (2011). Dopamine D1-D2 receptor heteromer signaling pathway in the brain: Emerging physiological relevance. *Molecular Brain, 4,* 26. https://doi.org/10.1186/1756-6606-4-26.

He, Q., Xue, G., Chen, C., Lu, Z., Dong, Q., Lei, X., … Chen, C. (2010). Serotonin transporter gene-linked polymorphic region (5-HTTLPR) influences decision making under ambiguity and risk in a large Chinese sample. *Neuropharmacology, 59*(6), 518–526.

Heiman, M., Schaefer, A., Gong, S., Peterson, J. D., Day, M., Ramsey, K. E., … Surmeier, D. J. (2008). A translational profiling approach for the molecular characterization of CNS cell types. *Cell, 135*(4), 738–748.

Hernandez-Echeagaray, E., Starling, A. J., Cepeda, C., & Levine, M. S. (2004). Modulation of AMPA currents by D2 dopamine receptors in striatal medium-sized spiny neurons: Are dendrites necessary? *European Journal of Neuroscience, 19*(9), 2455–2463. https://doi.org/10.1111/j.0953-816X.2004.03344.x.

Houk, J. C., Bastianen, C., Fansler, D., Fishbach, A., Fraser, D., Reber, P. J., ... Simo, L. S. (2007). Action selection and refinement in subcortical loops through basal ganglia and cerebellum. *Philosophical Transactions of the Royal Society of London. Series B, Biological Sciences, 362*(1485), 1573–1583. https://doi.org/10.1098/rstb.2007.2063.

Humphries, M. D., & Prescott, T. J. (2010). The ventral basal ganglia, a selection mechanism at the crossroads of space, strategy, and reward. *Progress in Neurobiology, 90*(4), 385–417. https://doi.org/10.1016/j.pneurobio.2009.11.003.

Jiang, L. H., Ashby, C. R., Jr., Kasser, R. J., & Wang, R. Y. (1990). The effect of intraventricular administration of the 5-HT ₃ receptor agonist 2-methylserotonin on the release of dopamine in the nucleus accumbens: An in vivo chronocoulometric study. *Brain Research, 513*(1), 156–160.

Jung, A. B., & Bennett, J. P. (1996). Development of striatal dopaminergic function. I. Pre-and postnatal development of mRNAs and binding sites for striatal D1 (D1a) and D2 (D2a) receptors. *Developmental Brain Research, 94*(2), 109–120.

Kahneman, D., & Tversky, A. (1979). Prospect theory: An analysis of decision under risk. *Econometrica, 47,* 263–292.

Kalva, S. K., Rengaswamy, M., Chakravarthy, V. S., & Gupte, N. (2012). On the neural substrates for exploratory dynamics in basal ganglia: A model. *Neural Netw, 32,* 65–73. https://doi.org/10.1016/j.neunet.2012.02.031.

Kawaguchi, Y., Wilson, C. J., & Emson, P. C. (1990). Projection subtypes of rat neostriatal matrix cells revealed by intracellular injection of biocytin. *The Journal of Neuroscience, 10*(10), 3421–3438.

Kötter, R., & Wickens, J. (1998). Striatal mechanisms in Parkinson's disease: New insights from computer modeling. *Artificial Intelligence in Medicine, 13*(1), 37–55.

Kravitz, E. A. (2000). Serotonin and aggression: Insights gained from a lobster model system and speculations on the role of amine neurons in a complex behavior. *Journal of Comparative Physiology. A, Sensory, Neural, and Behavioral Physiology, 186*(3), 221–238.

Krishnan, R., Ratnadurai, S., Subramanian, D., Chakravarthy, V. S., & Rengaswamy, M. (2011). Modeling the role of basal ganglia in saccade generation: Is the indirect pathway the explorer? *Neural Network, 24*(8), 801–813. https://doi.org/10.1016/j.neunet.2011.06.002.

Kuhnen, C. M., Samanez-Larkin, G. R., & Knutson, B. (2013). Serotonergic genotypes, neuroticism, and financial choices. *PLoS ONE, 8*(1), e54632.

Lak, A., Stauffer, W. R., & Schultz, W. (2014). Dopamine prediction error responses integrate subjective value from different reward dimensions. *Proceedings of the National Academy of Sciences, 111*(6), 2343–2348.

Le Moine, C., & Bloch, B. (1995). D1 and D2 dopamine receptor gene expression in the rat striatum: Sensitive cRNA probes demonstrate prominent segregation of D1 and D2 mRNAs in distinct neuronal populations of the dorsal and ventral striatum. *Journal of Comparative Neurology, 355*(3), 418–426.

Le Moine, C., Normand, E., & Bloch, B. (1991). Phenotypical characterization of the rat striatal neurons expressing the D1 dopamine receptor gene. *Proceedings of the National Academy of Sciences, 88*(10), 4205–4209.

Lee, S. P., So, C. H., Rashid, A. J., Varghese, G., Cheng, R., Lanca, A. J., . . . George, S. R. (2004). Dopamine D1 and D2 receptor Co-activation generates a novel phospholipase C-mediated calcium signal. *Journal of Biological Chemistry, 279*(34), 35671–35678. https://doi.org/10.1074/jbc.m401923200.

Lobo, M. K., Karsten, S. L., Gray, M., Geschwind, D. H., & Yang, X. W. (2006). FACS-array profiling of striatal projection neuron subtypes in juvenile and adult mouse brains. *Nature Neuroscience, 9*(3), 443–452.

Long, A. B., Kuhn, C. M., & Platt, M. L. (2009). Serotonin shapes risky decision making in monkeys. *Social Cognitive Affective Neuroscience, 4*(4), 346–356. https://doi.org/10.1093/scan/nsp020.

Lopez-Ibor, J. (1992). Serotonin and psychiatric disorders. *International Clinical Psychopharmacology*.

Markowitz, H. (1952). Portfolio selection. *The Journal of Finance, 7*(1), 77–91. https://doi.org/10.2307/2975974.

Matamales, M., Bertran-Gonzalez, J., Salomon, L., Degos, B., Deniau, J.-M., Valjent, E., . . . Girault, J.-A. (2009). Striatal medium-sized spiny neurons: identification by nuclear staining and study of neuronal subpopulations in BAC transgenic mice. *PLoS One, 4*(3), e4770.

Matsuda, W., Furuta, T., Nakamura, K. C., Hioki, H., Fujiyama, F., Arai, R., et al. (2009). Single nigrostriatal dopaminergic neurons form widely spread and highly dense axonal arborizations in the neostriatum. *The Journal of Neuroscience, 29*(2), 444–453.

McGeorge, A. J., & Faull, R. L. (1989). The organization of the projection from the cerebral cortex to the striatum in the rat. *Neuroscience, 29*(3), 503–537.

Mink, J. W. (1996). The basal ganglia: Focused selection and inhibition of competing motor programs. *Progress in Neurobiology, 50*(4), 381.

Morita, K., Morishima, M., Sakai, K., & Kawaguchi, Y. (2012). Reinforcement learning: Computing the temporal difference of values via distinct corticostriatal pathways. *Trends in Neurosciences, 35*(8), 457–467.

Moyer, J. T., Wolf, J. A., & Finkel, L. H. (2007). Effects of dopaminergic modulation on the integrative properties of the ventral striatal medium spiny neuron. *Journal of Neurophysiology, 98*(6), 3731–3748. https://doi.org/10.1152/jn.00335.2007.

Murphy, S. E., Longhitano, C., Ayres, R. E., Cowen, P. J., Harmer, C. J., & Rogers, R. D. (2009). The role of serotonin in nonnormative risky choice: The effects of tryptophan supplements on the "reflection effect" in healthy adult volunteers. *Journal of Cognitive Neuroscience, 21*(9), 1709–1719. https://doi.org/10.1162/jocn.2009.21122.

Nadjar, A., Brotchie, J. M., Guigoni, C., Li, Q., Zhou, S.-B., Wang, G.-J., ... Bezard, E. (2006). Phenotype of striatofugal medium spiny neurons in parkinsonian and dyskinetic nonhuman primates: a call for a reappraisal of the functional organization of the basal ganglia. *The Journal of Neuroscience, 26*(34), 8653–8661.

Nakamura, K. (2013). The role of the dorsal raphé nucleus in reward-seeking behavior. *Frontiers in Integrative Neuroscience, 7*.

Parent, A., & Hazrati, L.-N. (1995). Functional anatomy of the basal ganglia. II. The place of subthalamic nucleus and external pallidium in basal ganglia circuitry. *Brain Research Reviews, 20*(1), 128–154.

Perreault, M. L., Fan, T., Alijaniaram, M., O'Dowd, B. F., & George, S. R. (2012). Dopamine D1-D2 receptor heteromer in dual phenotype GABA/glutamate-coexpressing striatal medium spiny neurons: Regulation of BDNF, GAD67 and VGLUT1/2. *PLoS ONE, 7*(3), e33348. https://doi.org/10.1371/journal.pone.0033348.

Perreault, M. L., Hasbi, A., Alijaniaram, M., Fan, T., Varghese, G., Fletcher, P. J., . . . George, S. R. (2010). The dopamine D1-D2 receptor heteromer localizes in dynorphin/enkephalin neurons: increased high affinity state following amphetamine and in schizophrenia. *Journal of Biological Chemistry, 285*(47), 36625–36634. https://doi.org/10.1074/jbc.m110.159954.

Perreault, M. L., Hasbi, A., O'Dowd, B. F., & George, S. R. (2011). The dopamine d1-d2 receptor heteromer in striatal medium spiny neurons: Evidence for a third distinct neuronal pathway in Basal Ganglia. *Frontiers in Neuroanatomy, 5*, 31. https://doi.org/10.3389/fnana.2011.00031.

Preuschoff, K., Bossaerts, P., & Quartz, S. R. (2006). Neural differentiation of expected reward and risk in human subcortical structures. *Neuron, 51*(3), 381–390. https://doi.org/10.1016/j.neuron.2006.06.024.

Rashid, A. J., O'Dowd, B. F., Verma, V., & George, S. R. (2007). Neuronal Gq/11-coupled dopamine receptors: An uncharted role for dopamine. *Trends in Pharmacological Sciences, 28*(11), 551–555.

Reynolds, J. N., Hyland, B. I., & Wickens, J. R. (2001). A cellular mechanism of reward-related learning. *Nature, 413*(6851), 67–70.

Reynolds, J. N., & Wickens, J. R. (2002). Dopamine-dependent plasticity of corticostriatal synapses. *Neural Network, 15*(4–6), 507–521.

Robinson, O. J., Cools, R., & Sahakian, B. J. (2012). Tryptophan depletion disinhibits punishment but not reward prediction: Implications for resilience. *Psychopharmacology (Berl), 219*(2), 599–605. https://doi.org/10.1007/s00213-011-2410-5.

Rogers, R. D. (2011). The roles of dopamine and serotonin in decision making: Evidence from pharmacological experiments in humans. *Neuropsychopharmacology, 36*(1), 114–132. https://doi.org/10.1038/npp.2010.165.

Schultz, W. (1998). Predictive reward signal of dopamine neurons. *Journal of Neurophysiology, 80*(1), 1–27.

Schultz, W. (2013). Updating dopamine reward signals. *Current Opinion in Neurobiology, 23*(2), 229–238.

Seymour, B., Daw, N. D., Roiser, J. P., Dayan, P., & Dolan, R. (2012). Serotonin selectively modulates reward value in human decision-making. *Journal of Neuroscience, 32*(17), 5833–5842. https://doi.org/10.1523/JNEUROSCI.0053-12.2012.

Shuen, J. A., Chen, M., Gloss, B., & Calakos, N. (2008). Drd1a-tdTomato BAC transgenic mice for simultaneous visualization of medium spiny neurons in the direct and indirect pathways of the basal ganglia. *The Journal of Neuroscience, 28*(11), 2681–2685.

So, C. H., Verma, V., Alijaniaram, M., Cheng, R., Rashid, A. J., O'Dowd, B. F., et al. (2009). Calcium signaling by dopamine D5 receptor and D5-D2 receptor hetero-oligomers occurs by a mechanism distinct from that for dopamine D1-D2 receptor hetero-oligomers. *Molecular Pharmacology, 75*(4), 843–854. https://doi.org/10.1124/mol.108.051805.

Spehlmann, R., & Stahl, S. (1976). Dopamine acetylcholine imbalance in Parkinson's disease: Possible regenerative overgrowth of cholinergic axon terminals. *The Lancet, 307*(7962), 724–726.

Stauffer, W. R., Lak, A., & Schultz, W. (2014). Dopamine reward prediction error responses reflect marginal utility. *Current Biology, 24*(21), 2491–2500.

Stopper, C. M., & Floresco, S. B. (2011). Contributions of the nucleus accumbens and its subregions to different aspects of risk-based decision making. *Cognitive Affective Behavioral Neuroscience, 11*(1), 97–112. https://doi.org/10.3758/s13415-010-0015-9.

Stringer, S., Rolls, E., Trappenberg, T., & De Araujo, I. (2002). Self-organizing continuous attractor networks and path integration: Two-dimensional models of place cells. *Network: Computation in Neural Systems, 13*(4), 429–446.

Surmeier, D., & Kitai, S. (1993). D 1 and D 2 dopamine receptor modulation of sodium and potassium currents in rat neostriatal neurons. *Progress in Brain Research, 99*, 309–324.

Surmeier, D. J., Ding, J., Day, M., Wang, Z., & Shen, W. (2007). D1 and D2 dopamine-receptor modulation of striatal glutamatergic signaling in striatal medium spiny neurons. *Trends in Neurosciences, 30*(5), 228–235. https://doi.org/10.1016/j.tins.2007.03.008.

Sutton, R. S., & Barto, A. G. (1998). *Reinforcement learning: An introduction. Adaptive computations and machine learning*. Bradford: MIT Press.

Suzuki, M., Hurd, Y. L., Sokoloff, P., Schwartz, J. C., & Sedvall, G. (1998). D3 dopamine receptor mRNA is widely expressed in the human brain. *Brain Research, 779*(1–2), 58–74.

Swann, A. C., Lijffijt, M., Lane, S. D., Cox, B., Steinberg, J. L., & Moeller, F. G. (2013). Norepinephrine and impulsivity: Effects of acute yohimbine. *Psychopharmacology (Berl), 229*(1), 83–94.

Tai, L.-H., Lee, A. M., Benavidez, N., Bonci, A., & Wilbrecht, L. (2012). Transient stimulation of distinct subpopulations of striatal neurons mimics changes in action value. *Nature Neuroscience, 15*(9), 1281–1289.

Tanaka, S. C., Schweighofer, N., Asahi, S., Shishida, K., Okamoto, Y., Yamawaki, S., et al. (2007). Serotonin differentially regulates short- and long-term prediction of rewards in the ventral and dorsal striatum. *PLoS ONE, 2*(12), e1333. https://doi.org/10.1371/journal.pone.0001333.

Tanaka, S. C., Shishida, K., Schweighofer, N., Okamoto, Y., Yamawaki, S., & Doya, K. (2009). Serotonin affects association of aversive outcomes to past actions. *Journal of Neuroscience, 29*(50), 15669–15674. https://doi.org/10.1523/JNEUROSCI.2799-09.2009.

Tops, M., Russo, S., Boksem, M. A., & Tucker, D. M. (2009). Serotonin: Modulator of a drive to withdraw. *Brain and Cognition, 71*(3), 427–436. https://doi.org/10.1016/j.bandc.2009.03.009.

Valjent, E., Bertran-Gonzalez, J., Hervé, D., Fisone, G., & Girault, J.-A. (2009). Looking BAC at striatal signaling: Cell-specific analysis in new transgenic mice. *Trends in Neurosciences, 32* (10), 538–547.

Wang, R., Macmillan, L., Fremeau, R., Jr., Magnuson, M., Lindner, J., & Limbird, L. (1996). Expression of α2-adrenergic receptor subtypes in the mouse brain: Evaluation of spatial and temporal information imparted by 3 kb of 5′ regulatory sequence for the α2A AR-receptor gene in transgenic animals. *Neuroscience, 74*(1), 199–218.

Ward, R. P., & Dorsa, D. M. (1996). Colocalization of serotonin receptor subtypes 5-HT2A, 5-HT2C, and 5-HT6 with neuropeptides in rat striatum. *Journal of Comparative Neurology, 370*(3), 405–414.

Zhong, S., Israel, S., Xue, H., Ebstein, R. P., & Chew, S. H. (2009a). Monoamine oxidase A gene (MAOA) associated with attitude towards longshot risks. *PLoS ONE, 4*(12), e8516.

Zhong, S., Israel, S., Xue, H., Sham, P. C., Ebstein, R. P., & Chew, S. H. (2009b). A neurochemical approach to valuation sensitivity over gains and losses. *Proceedings of the Royal Society B: Biological Sciences, 276*(1676), 4181–4188.

Chapter 13
Modeling Serotonin's Contributions to Basal Ganglia Dynamics in Parkinson's Disease with Impulse Control Disorders

Pragathi Priyadharsini Balasubramani, V. Srinivasa Chakravarthy, Balaraman Ravindran and Ahmed A. Moustafa

Abstract Impulsivity involves irresistibility in execution of actions and is prominent in medication condition of Parkinson's disease (PD) patients. In this chapter, we model a probabilistic reversal learning task in PD patients with and without impulse control disorder (ICD) to understand the basis of their neural circuitry responsible for displaying ICD in PD condition. The proposed model is of the basal ganglia (BG) action selection dynamics, and it predicts the dysfunction of both dopaminergic (DA) and serotonergic (5HT) neuromodulator systems to account for the experimental results. Furthermore, the BG is modeled after utility function framework with DA controlling reward prediction and 5HT controlling the loss and risk prediction, respectively. The striatal model has three pools of medium spiny neurons (MSNs) including those with D1 receptor (R) alone, D2R alone, and co-expressing D1R–D2R neurons. Some significant results modeled are increased reward sensitivity during ON medication and an increased punishment sensitivity during OFF medication in patients. The lower reaction times (RT) in ICD subjects compared to that of the non-ICD category of the PD ON patients are also explained. Other modeling predictions include a significant decrease in the sensitivity to loss and risk in the ICD patients.

13.1 Introduction

The network model explained in the Chap. 12 is extended to model behavior of Parkinson's disease (PD), PD with impulsivity in this chapter. The reaction time behavior of subjects as modeled by an action selection paradigm in this study is compared against experimental literature. Impulsivity is a multi-factorial problem, is assessed depending on the accuracy of goal-directed action performance, as well as the ability to exert inhibitory control over action impulses that could possibly interfere with goal-directed action (Ahlskog, 2010; Wylie, Ridderinkhof, Bashore, & van den Wildenberg, 2010). Some popular action selection paradigms used to test for impulsivity include a simple stimulus–response task, delayed stimulus–response task, and their modifications through contingency devaluations

V. S. Chakravarthy and A. Moustafa, *Computational Neuroscience Models of the Basal Ganglia*, Cognitive Science and Technology,
https://doi.org/10.1007/978-981-10-8494-2_13

(Dougherty, Mathias, Marsh, & Jagar, 2005; Nombela, Rittman, Robbins, & Rowe, 2014). Impulsive behaviors are generally marked in experimental measures by shorter reaction times (RT) and reduced accuracy indicating their compromised inhibitory control of non-optimal actions and higher delay discounting of rewards (Dalley, Everitt, & Robbins, 2011; Dalley, Mar, Economidou, & Robbins, 2008; Evenden, 1999).

A subset of PD patients with impulse control disorder (ICD) suffers from a lack of inhibitory control over some inappropriate hedonic drives potentially associated with harmful consequences; they form 14% of ON medication PD patients, PD ON, and are mostly undergoing medication with DA agonists (Bugalho & Oliveira-Maia, 2013). Their characteristics include pathological gambling, binge eating, overuse of dopaminergic medications among others. Optimal therapy and medications are required to control for withdrawal symptoms in motor and the non-motor domains (Djamshidian, Averbeck, Lees, & O'Sullivan, 2011). Cortical structures such as prefrontal cortex and orbitofrontal cortex, subcortical structures like basal ganglia (BG) are important correlates of impulsivity (Dalley et al. 2008; Ray, Antonelli, & Strafella, 2011). Dysfunction of neuromodulators such as dopamine and serotonin, and their receptors DA D2, and 5HT 1, 2, 6, in the frontostriatal circuitry has been identified for impulsivity as well (Averbeck, O'Sullivan, & Djamshidian, 2014; Bugalho & Oliveira-Maia, 2013; Dalley et al., 2008), especially in PD (Dalley et al., 2008; Dalley et al., 2011). Therefore, a unified and complete modeling approach to PD, and ICD in PD, should include 5HT along with DA system for their better understanding and toward their better therapeutics. The network dynamics of the BG as delineated in the previous chapter, modulated through DA and 5HT based utility dynamics, is applied to an experiment invoking reward-based decisions, assessing their accuracy as well as RT, conducted on healthy controls and PD patients with and without ICD. The model has been able shown to propose distinctive neural correlates contributing to ICD in PD patients (Balasubramani, Chakravarthy, Ali, Ravindran, & Moustafa, 2015).

13.2 Probabilistic Learning, Parkinson's Disease, and Impulse Control Disorder

Using the lumped model of the BG policy to model a probabilistic learning task.

The simulation studies presented so far in the previous chapter are performed under controlled conditions. This section simulates a study that relates to reward/punishment learning of PD condition.

13.2.1 Experiment Summary

Bodi et al. (2009) used a probabilistic classification task for assessing reward/punishment learning under the different medication conditions of PD patients. The medications used in the study were a mix of DA agonists (pramipexole and ropinirole) and L-DOPA. The task was as follows: one of four random fractal images (I1–I4) was presented. In response to each image, the subject had to press on one of two buttons—A or B—on a keypad. Stimuli I1 and I2 were always associated with reward (+25 points), while I3 and I4 were associated with loss/punishment (−25 points). The probability of reward or punishment outcome depended on the button (A or B) that the subject pressed in response to viewing an image (Bodi et al., 2009). There are 160 trials administered in 4 blocks. Experiments were performed on healthy controls, never-medicated (PD OFF), and recently medicated PD (PD ON) patients. The study (Bodi et al., 2009) showed that the never-medicated patients were more sensitive to punishment than the recently medicated patients and healthy controls. On the other hand, the recently medicated patients outperformed the never-medicated patients and healthy controls on reward learning tasks. The optimal decision is the selection of A for I1 and I3, and B for I2 and I4.

13.2.2 Simulation

We adapt the version of abstract extended reinforcement learning model described in the previous chapter here. The immediate reward case of the experiment is expressed by Eq. (3) of Chap. 12, with which the value update, Eq. (13.2), and the risk update, Eq. (5) of Chap. 12, are made for a (state, action) pair. The states here are four images, and the action is categorized as either A or B. The utility for a particular (state, action) pair is constructed using Eq. (7). The measure of change in utility is calculated by the following Eq. (13.1).

$$\delta_{\mathrm{U}}(t) = U_t(s_t, a_t) - U_{t-1}(s_t, a_{t-1}) \tag{13.1}$$

where 'U' is the utility represented in Eq. (12.7) of Chap. 12. The change in utility, Eq. (13.1), now controls the action selection dynamics set out by the following Eq. (13.2).

$$\begin{aligned}
&\text{if } \delta_{\mathrm{U}_i} > \delta_{\mathrm{hi}}; \text{Go} \\
&\text{elseif } \delta_{\mathrm{U}_i} < \delta_{\mathrm{lo}}; \text{NoGo} \\
&\text{else Explore}\{\text{if rand} > \varepsilon; \\
&\quad \text{if } \delta_{\mathrm{U}_i} > \delta_m; \text{Go} \\
&\quad \text{else NoGo} \\
&\text{else Select random action}\}
\end{aligned} \tag{13.2}$$

where

$$\delta_m = (\delta_{\mathrm{hi}} + \delta_{\mathrm{lo}})/2$$

$$\varepsilon = \exp((\delta_{\mathrm{U}_i} - \delta_m)^2/\sigma^2)$$

The Go-Explore-NoGo (GEN) policy-based BG action selection dynamics has been discussed earlier in the Chap. 5. The PD condition is modeled by equations in the section of Chap. 5 with parameters, $\delta_{\mathrm{Lim}} = 0$, and $\delta_{\mathrm{Med}} = 0.15$. The simulation is run for 160 trials.

13.2.3 Results

The modeling study finds that α (5HT) takes a lower value in PD compared to the healthy controls to represent the overall reduction of 5HT levels. The model of healthy controls shows almost equal sensitivity to rewards and punishments. The PD ON model shows an increased sensitivity to reward compared to that of punishment, whereas PD OFF shows the opposite trend (Fig. 13.1).

13.3 Applying the Network Model of BG to Probabilistic Learning Task

Bodi et al. (2009) experiment is again modeled here using the network model as described in the previous chapter. The +25 reward is represented as reward 'r = 1' and the −25 punishment as 'r = −1'. The weights for the D1R, D2R, and the D1R–D2R neurons are initialized randomly between 0 and 1.

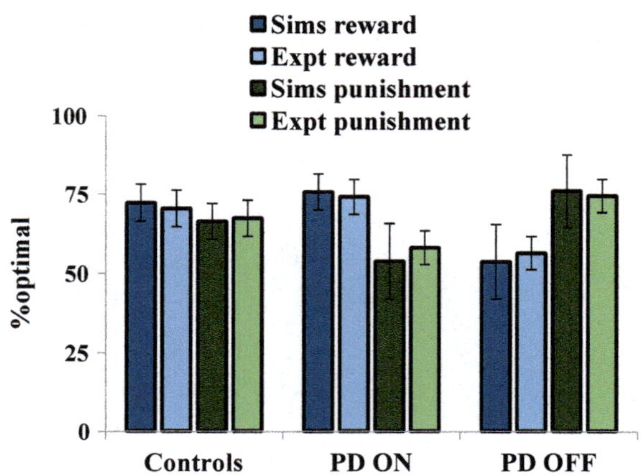

Fig. 13.1 Accuracy of decisions depicted for various subject categories of experimental data and their model simulation (run for 100 instances)

13.3.1 Results

In the model, the healthy controls show high sensitivity to rewards, punishments, and risk (α_{D1}, α_{D2}, α_{D1D2}). The PD ON patients show an increased sensitivity to reward compared to that of punishment, whereas the PD OFF patients show the opposite trend. The parameters of the model that best represent the experiment are: [α_{D1}, α_{D2}, α_{D1D2}] = [1, 1, 0.2] for the healthy controls; [δ_{Lim}, α_{D1}, α_{D2}, α_{D1D2}] = [0.001, 1, 0.99, 0.001] for PD OFF; and [δ_{Lim}, δ_{Med}, α_{D1}, α_{D2}, α_{D1D2}] = [0.001, 0.021, 1, 0.2, 0.001] for PD ON. The results are put forth in Fig. 13.2 and are consistent across other model extensions detailed in this chapter.

The results substantiate both the differential modulation of 5HT in the MSNs that differentiating various PD conditions as well—(1) The differential modulation of 5HT in the D1R–D2R MSNs with $\alpha_{D1D2} = 0.2$ (in healthy controls) and $\alpha_{D1D2} < 0.2$ (in PD) (Fig. 13.2) is noticed. (2) The activity of 5HT in the D2R MSNs is significantly lowered specifically in the PD ON condition (PD ON $\alpha_{D2} = 0.2$ compared to $\alpha_{D2} > 0.2$ in PD OFF and healthy controls). Many neurobiological experimental studies have observed lowered 5HT levels in PD conditions compared to the healthy controls (Bedard et al., 2011; Fahn, Libsch, & Cutler, 1971; Halliday, Blumbergs, Cotton, Blessing, & Geffen, 1990). This is captured in our modeling study with a smaller α value observed to modulate both the D2R and the D1R–D2R MSNs. (3) The PD ON condition is reported to have lowered 5HT levels than the OFF-medicated PD condition. This is shown by reduced 5HT release and increased DA release from the serotonergic neurons in the presence of L-DOPA (Reed, Nijhout, & Best, 2012; Tan, Salgado, & Fahn, 1996). This is specifically reflected by a significant decrease in the level of α_{D2} affecting the D2R MSNs in our modeling study.

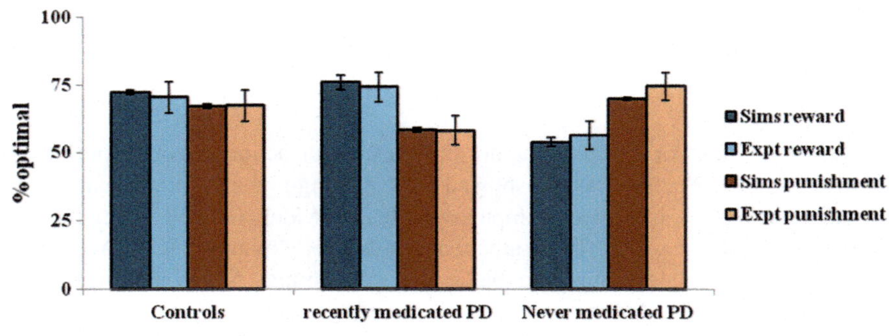

Fig. 13.2 Reward–punishment sensitivity obtained by simulated (Sims)-PD and healthy control models to explain the experiment (Expt) of Bodi et al. (2009). Error bars represent the standard error (SE) with $N = 100$ (N = number of simulation instances). The simulations match the experimental value distribution closely and are not found to be significantly different ($p > 0.05$)

13.4 Analyzing the Reaction Times and Impulsivity

This experimental setup is the same as the previous section and has been earlier performed with PD patients and healthy control subjects as described in (Piray, Zeighami, Bahrami, Eissa, Hewedi, & Moustafa, 2014), but the present modeling study extends the same experimental setup to analyze the subject's RT. The experimental results suggest ICD patients are more sensitive to rewards in the medication condition and had lower RT than non-ICD. The non-ICD patients had no significant difference between reward and punishment learning, and similar was the case of healthy controls. The PD OFF patients showed higher efficiency toward punishment learning.

13.4.1 Modeling Results

The network model described in the previous section is now applied to the experimental data. The experimental and simulated RT as well as task accuracy is provided in Fig. 13.3 for all subject groups. The modeling results suggest optimal parameters to DA as δ (viz. δ_{Lim} and δ_{Med}) and 5HT—(α_{D1}, α_{D2}, α_{D1D2}) for representing ICD behavior in the PD. Particularly, an increased reward sensitivity in PD ON with a significantly reduced 5HT modulation of the striatal D2R (α_{D2}) and the D1R–D2R (α_{D1D2}) MSNs represents PD ON ICD condition. The parameters of the model that best represent the experiment are: $[\alpha_{D1}, \alpha_{D2}, \alpha_{D1D2}] = [1, 0.185, 0.997]$ for the healthy controls; $[\delta_{Lim}, \alpha_{D1}, \alpha_{D2}, \alpha_{D1D2}] = [0.001, 1, 0.99, 0.033]$ for PD OFF; and $[\delta_{Lim}, \delta_{Med}, \alpha_{D1}, \alpha_{D2}, \alpha_{D1D2}] = [0.001, 0.06, 1, 0.046, 0.001]$ for PD ON ICD, [0.001, 0.06, 1, 0.916, 0.160] for PD ON non-ICD.

13.5 Discussion

The developed network model was not only tested for action selection problems, but also for their RT. The haste displayed on executing actions, for premature and inaccurate responses, is termed as impulsivity. ICD is widely noticed during the ON medication condition of PD. There are many models for explaining ICD: According to one model, ICD is thought to be a kind of habitual behavior (Bugalho & Oliveira-Maia, 2013). Another one makes use of the opponency between the BG pathways as mediated dopamine to explain ICD (Frank, 2005; Frank, Samanta, Moustafa, & Sherman, 2007a; Frank, Scheres, & Sherman, 2007b). One another model employs an Actor–Critic approach to find abnormal evaluation-related computations in ventral striatum to explain impulsivity in PD ON (Piray, Zeighami, Bahrami, Eissa, Hewedi, & Moustafa, 2014). There exists other models which uses

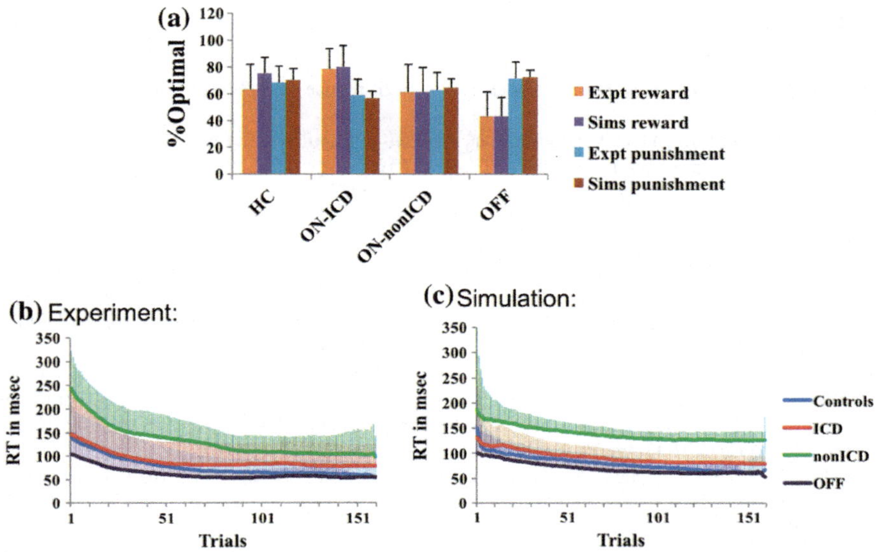

Fig. 13.3 Comparing action selection accuracy and reaction times between experiment and simulation for various subject categories. **a** Accuracy is presented as run for 100 instances. RTs are shown for (**b**) the experimental data, and (**c**) for simulation ($p > 0.05$, ANOVA, with reward valence, punishment valence, and RT as factors of analysis). Refer to (Balasubramani et al., 2015) for more details on parameters, and Chap. 12 for details on the network model methods

matching law (Evenden, 1999). Some relate them to an increased reward discounting and myopicity of reward predictions (Doya, 2002; Tanaka et al., 2007).

We show that such effects can be captured in the proposed model by the risk sensitivity term (α_{D1D2}). Earlier models of impulsivity in PD take only the deficiency of DA into consideration (Piray et al., 2014), leaving behind other potential salient factors such as 5HT as indicated in this study. The reduced learning from the punishments in PD ON ICD patients was captured by prior models using an explicitly negative prediction learning rate (Piray et al., 2014), whereas this study takes the nonlinearity in reward–punishment behavior through the *sign*() term to differentiate reward–punishment learning among subject groups.

Our model shows the importance of modulating both DA and 5HT in the BG to effectively explain ICD and PD behavior for the probabilistic task; the clamping of DA models the marked reduced DA availability in PD (Evans et al., 2006; Steeves et al., 2009). Our model also predicts a lower levels of 5HT in the BG for both PD OFF and PD ON as found through many experimental studies (Bedard et al., 2011; Fahn et al., 1971; Fahn, Snider, Prasad, Lane, & Makadon, 1975; Halliday et al., 1990). Specifically, based on model, a lowered sensitivity of 5HT in D2R and D1R–D2R MSNs is observed to mark ICD behavior. The model also shows a higher sensitivity of 5HT in the D2R MSNs for PD OFF behavior.

References

Ahlskog, J. E. (2010). Think before you leap Donepezil reduces falls? *Neurology, 75*(14), 1226–1227.

Averbeck, B., O'Sullivan, S., & Djamshidian, A. (2014). Impulsive and compulsive behaviors in Parkinson's disease. *Annual Review of Clinical Psychology, 10,* 553–580.

Balasubramani, P. P., Chakravarthy, V. S., Ali, M., Ravindran, B., & Moustafa, A. A. (2015). Identifying the basal ganglia network model markers for medication-induced impulsivity in Parkinson's disease patients. *PLoS ONE, 10*(6), e0127542.

Bedard, C., Wallman, M. J., Pourcher, E., Gould, P. V., Parent, A., & Parent, M. (2011). Serotonin and dopamine striatal innervation in Parkinson's disease and Huntington's chorea. *Parkinsonism and Related Disorders, 17*(8), 593–598. https://doi.org/10.1016/j.parkreldis.2011.05.012.

Bodi, N., Keri, S., Nagy, H., Moustafa, A., Myers, C. E., Daw, N., … Gluck, M. A. (2009). Reward-learning and the novelty-seeking personality: a between- and within-subjects study of the effects of dopamine agonists on young Parkinson's patients. *Brain, 132*(Pt 9), 2385–2395. https://doi.org/10.1093/brain/awp094.

Bugalho, P., & Oliveira-Maia, A. J. (2013). Impulse control disorders in Parkinson's disease: Crossroads between neurology, psychiatry and neuroscience. *Behavioural Neurology, 27*(4), 547–557.

Dalley, J. W., Everitt, B. J., & Robbins, T. W. (2011). Impulsivity, compulsivity, and top-down cognitive control. *Neuron, 69*(4), 680–694.

Dalley, J. W., Mar, A. C., Economidou, D., & Robbins, T. W. (2008). Neurobehavioral mechanisms of impulsivity: Fronto-striatal systems and functional neurochemistry. *Pharmacology, Biochemistry and Behavior, 90*(2), 250–260.

Djamshidian, A., Averbeck, B. B., Lees, A. J., & O'Sullivan, S. S. (2011). Clinical aspects of impulsive compulsive behaviours in Parkinson's disease. *Journal of the Neurological Sciences, 310*(1), 183–188.

Dougherty, D. M., Mathias, C. W., Marsh, D. M., & Jagar, A. A. (2005). Laboratory behavioral measures of impulsivity. *Behavior Research Methods, 37*(1), 82–90.

Doya, K. (2002). Metalearning and neuromodulation. *Neural Network, 15*(4–6), 495–506.

Evans, A. H., Pavese, N., Lawrence, A. D., Tai, Y. F., Appel, S., Doder, M., … Piccini, P. (2006). Compulsive drug use linked to sensitized ventral striatal dopamine transmission. Annals of *Neurology, 59*(5), 852–858.

Evenden, J. L. (1999). Varieties of impulsivity. *Psychopharmacology (Berl), 146*(4), 348–361.

Fahn, S., Libsch, L. R., & Cutler, R. W. (1971). Monoamines in the human neostriatum: Topographic distribution in normals and in Parkinson's disease and their role in akinesia, rigidity, chorea, and tremor. *Journal of the Neurological Sciences, 14*(4), 427–455.

Fahn, S., Snider, S., Prasad, A. L., Lane, E., & Makadon, H. (1975). Normalization of brain serotonin by L-tryptophan in levodopa-treated rats. *Neurology, 25*(9), 861–865.

Frank, M. J. (2005). Dynamic dopamine modulation in the basal ganglia: A neurocomputational account of cognitive deficits in medicated and nonmedicated Parkinsonism. *Journal of Cognitive Neuroscience, 17*(1), 51–72. https://doi.org/10.1162/0898929052880093.

Frank, M. J., Samanta, J., Moustafa, A. A., & Sherman, S. J. (2007a). Hold your horses: Impulsivity, deep brain stimulation, and medication in parkinsonism. *Science, 318*(5854), 1309–1312. https://doi.org/10.1126/science.1146157.

Frank, M. J., Scheres, A., & Sherman, S. J. (2007b). Understanding decision-making deficits in neurological conditions: Insights from models of natural action selection. *Philosophical Transactions of the Royal Society B: Biological Sciences, 362*(1485), 1641–1654.

Halliday, G. M., Blumbergs, P. C., Cotton, R. G., Blessing, W. W., & Geffen, L. B. (1990). Loss of brainstem serotonin- and substance P-containing neurons in Parkinson's disease. *Brain Research, 510*(1), 104–107.

Nombela, C., Rittman, T., Robbins, T. W., & Rowe, J. B. (2014). Multiple modes of impulsivity in Parkinson's disease. *PLoS ONE, 9*(1), e85747.

Piray, P., Zeighami, Y., Bahrami, F., Eissa, A. M., Hewedi, D. H., & Moustafa, A. A. (2014). Impulse control disorders in Parkinson's disease are associated with dysfunction in stimulus valuation but not action valuation. *The Journal of Neuroscience, 34*(23), 7814–7824.

Ray, N., Antonelli, F., & Strafella, A. P. (2011). Imaging impulsivity in Parkinson's disease and the contribution of the subthalamic nucleus. *Parkinson's Dis.* 2011:594860. doi: 10.4061/2011/594860.

Reed, M. C., Nijhout, H. F., & Best, J. A. (2012). Mathematical insights into the effects of levodopa. *Frontiers in integrative neuroscience, 6,* 21. doi: 10.3389/fnint.2012.00021

Steeves, T., Miyasaki, J., Zurowski, M., Lang, A., Pellecchia, G., Van Eimeren, T., … Strafella, A. (2009). Increased striatal dopamine release in Parkinsonian patients with pathological gambling: a [11C] raclopride PET study. *Brain, 132*(5), 1376–1385.

Tan, A., Salgado, M., & Fahn, S. (1996). Rapid eye movement sleep behavior disorder preceding Parkinson's disease with therapeutic response to levodopa. *Movement Disorders, 11*(2), 214–216.

Tanaka, S. C., Schweighofer, N., Asahi, S., Shishida, K., Okamoto, Y., Yamawaki, S., et al. (2007). Serotonin differentially regulates short- and long-term prediction of rewards in the ventral and dorsal striatum. *PLoS ONE, 2*(12), e1333. https://doi.org/10.1371/journal.pone.0001333.

Wylie, S. A., Ridderinkhof, K. R., Bashore, T. R., & van den Wildenberg, W. P. (2010). The effect of Parkinson's disease on the dynamics of on-line and proactive cognitive control during action selection. *Journal of Cognitive Neuroscience, 22*(9), 2058–2073.

Chapter 14
An Oscillatory Neural Network Model for Birdsong Learning and Generation: Implications for the Role of Dopamine in Song Learning

M. Maya, V. Srinivasa Chakravarthy and B. Ravindran

Abstract We present a model of bird song learning and production in which the motor control pathway is modeled by a trainable network of oscillators and the Anterior Forebrain Pathway (AFP) is modeled as a stochastic system. Song learning in many species of birds is divided into two phases. In the first phase, the sensory phase, the male bird listens to the tutor song of another male bird in the colony and memorizes some aspect of the tutor song. In the second phase, the motor learning phase, the bird establishes the songs learnt earlier by rehearsal aided by auditory self-feedback. We hypothesize that: (1) the songbird learns only *evaluations* of songs during the sensory phase; (2) the AFP plays a role analogous to the Explorer, a key component in reinforcement learning (RL); (3) the motor pathway learns the song by combining the evaluations (value information) stored from the sensory phase, and the exploratory inputs from the AFP in a temporal stage-wise manner. Model performance on real birdsong samples is presented. Impaired song output under conditions of lesions of AFP nuclei, including the Lateral Magnocellular Nucleus of the Anterior Neostriatum (LMAN) and Area X, is studied. The model also proposes a role for dopamine signal in song learning and shows that under dopamine-deficient conditions, similar to those of Parkinson's disease, song learning is impaired.

14.1 Introduction

14.1.1 Birdsong Learning

The aesthetic and emotional appeal of the bird song is a phenomenon known throughout human history. In recent times, the bird song has evoked tremendous interest among the scientific community due to the striking similarities found between the process of song acquisition by birds and speech learning in humans (Doupe & Kuhl, 1999). It has been observed that both juvenile birds and human babies need an adult of the same species—the so-called 'tutor' bird—to develop

V. S. Chakravarthy and A. Moustafa, *Computational Neuroscience Models of the Basal Ganglia*, Cognitive Science and Technology,
https://doi.org/10.1007/978-981-10-8494-2_14

normal vocalizations (Doupe & Kuhl, 1999). Not only do birds raised in social isolation develop abnormal 'isolate' songs but birds can develop geographically restricted 'dialects' for songs (Brainard & Doupe, 2002) as in case of human languages. Both humans and birds require auditory feedback of their own speech/song during the learning phase (Konishi, 1965). During the sensory learning phase, a juvenile songbird like the zebra finch listens to and memorizes the tutor song (from the tutor); in the subsequent sensory motor learning phase, they match their actual vocalization with the memorized song and use this error feedback to correct their song (Bolhuis & Moorman, 2015; Hahnloser & Kotowicz, 2010). Electrophysiological studies reveal that the site of the tutor song memory is caudomedial nidopallium (NCM), a region in the avian brain homologous to mammalian auditory cortex (Yanagihara & Yazaki-Sugiyama, 2016).

Yet another concurrent feature of speech learning, in humans and songbirds, relates to the critical periods of learning observed in both species. In humans as well as in zebra finches, the capacity to learn generally declines with age; yet this closure of the sensitive period is not strictly dependent on age but on prior experience. Infants as well as juvenile birds have the ability to discriminate their own species' speech/song when presented with many different sounds. Experiments have shown that songbirds when presented with a variety of songs from different species prefer to learn songs of their own species (Marler & Peters, 1977). At the level of the brain circuits, it has been observed that there is a dominance of the left hemisphere for speech/song production in both humans and songbirds (Doupe & Kuhl, 1999).

The process of vocal learning from an adult is a process that is found to be unique to humans, songbirds, dolphins, bats, and across species (Brainard & Doupe, 2002). Recent research revealed that there are analogous regions between the bird's brain and the human brain at a gross level and also at the level of song/speech processing pathways (Reiner, Perkel, & Mello, 2004). Furthermore, the availability of a precisely characterized neural pathway involved in song learning makes it an attractive model system for investigating the mechanisms of motor skill acquisition and the role of various brain regions. Added to this are the facts that the songbirds are domesticated species, have relatively high breeding rates, and hence are easy to rear in the laboratory.

The neural regions involved in song learning and generation process together form the 'song system,' which consists of (a) the motor pathway that is involved in song generation (Nottebohm, Stokes, & Leonard, 1976) and (b) the Anterior Forebrain Pathway (AFP) which is involved in the song learning process but is not required for song production once the song is learnt (Bottjer, Meisner, & Arnold, 1984; Scharff & Nottebohm, 1991; Sohrabji, Nordeen, & Nordeen, 1990). Since the neural circuit in songbirds is entirely devoted to a single motor task—that of song learning—it serves as an ideal system for mapping behavior to neural activity.

A remarkable feature that has emerged out of the study of the song system is the resemblance of the AFP to the human cortical–basal ganglia circuit (Luo, Ding, & Perkel, 2001). The human basal ganglia (BG) are a set of deep brain regions implicated in many motor and cognitive functions ranging from action selection

(Berns & Sejnowski, 1995), action gating, sequence learning (Berns & Sejnowski, 1998), reward-based learning (Barto, 1995; Schultz, 1998), among other processes.

In the present work, we focus on the role of BG in reward-based learning by using songbirds as model systems. In humans, phasic dopamine signal is thought to signify the error between the total expected future reward and the reward obtained (Schultz, 1998). This error drives the learning process by motivating the subject to perform those actions that maximize the reward obtained. In song birds, a region called Area X in the AFP receives dense dopaminergic projections from the midbrain (Bottjer, 1993), which is analogous to mammalian BG regions like Str and pallidum. This dopaminergic release is found to be correlated with that of various stages in song learning in birds (Harding, 1998). Just as in mammalian brains, dopamine cells in songbirds are also found to encode an error-like signal that guides song learning (Gadagkar et al., 2016).

14.1.2 Neuroanatomy of Birdsong

There are two major pathways involved in the song learning process: the motor pathway and the Anterior Forebrain Pathway (AFP). The motor pathway runs between two sets of nuclei—the HVc and the Robust nucleus of the Arcopallium (RA). The HVc receives inputs from the auditory system (Field L) and sends glutamatergic projections to the RA and Area X, a region of the AFP. The RA is characterized by the presence of long-range inhibitory interneurons and two different sets of motor neurons innervating the syringeal muscles and the respiratory muscles (Nottebohm et al., 1976; McCasland, 1987). It has been seen that lesions of any of these two nuclei, HVc or the RA, leads to muteness or abnormal song in adult songbirds (Nottebohm et al., 1976). The song system seems to be organized in a hierarchical fashion within the motor pathway (Abarnel, Gibb, Mindlin, & Talathi, 2004; Yu & Margoliash, 1996). The HVc is thought to mediate learning at the level of motifs, and the RA seems to encode the vocalizations at the level of individual syllables (Fig. 14.1).

It has been observed that the AFP plays an important role in vocal learning in juvenile birds, but has no role in the maintenance of adult song (Bottjer et al., 1984). The AFP consists of three major sets of nuclei—Area X, Dorso-Lateral Thalamus (DLM), and Lateral Magnocellular Nucleus of the Anterior Neostriatum (LMAN). Area X consists of GABAergic neurons which resemble the spiny neurons of the mammalian striatum (Str). The electrophysiological properties of these and other neurons in Area X have led to the speculation that this region behaves in a manner analogous to that of the pallidum and Str in the mammalian BG (Carrilo & Doupe, 2004; Reiner et al., 2004). Area X receives glutamatergic projections from HVc and LMAN. It also receives dense dopaminergic projections from the midbrain region called the Ventral Tegmental Area (VTA) and has neurons containing D_1, D_2, or both dopamine (DA) receptors. Similar to the situation in medium spiny neurons in the mammalian Str, D_1 receptor activation enhances excitability and is

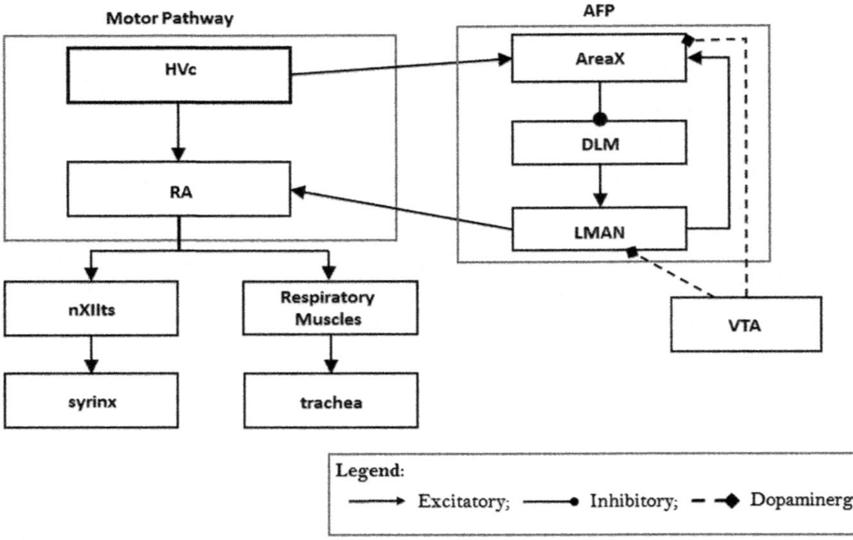

Fig. 14.1 Neuroanatomy of Birdsong. AFP—Anterior Forebrain Pathway, RA—Robust nucleus of the Archistriatum, DLM—Dorso-Lateral Thalamus, LMAN—Lateral Magnocellular Nucleus of the Anterior Neostriatum, VTA—Ventral Tegmental Area

essential for long-term potentiation (LTP), whereas D_2 receptor activation reduces excitability (Ding & Perkel, 2002). The DLM projection to LMAN is excitatory and is analogous to the thalamocortical projection in mammals. The LMAN is the output nucleus of the AFP to the motor pathway, and it projects to the RA and Area X through NMDA-mediated synapses.

The effects of lesioning studies of Area X and LMAN in juvenile zebra finches are in stark contrast to each other, yet they give an important information regarding the function of the AFP in song learning (Bottjer et al., 1984; Scharff & Nottebohm, 1991; Sohrabji et al., 1990). Area X lesions prevent the song from getting crystallized, and the bird produces highly variable songs for the rest of its life. This occurs because Area X lesions lead to reduced inhibition of LMAN, leading to increased firing of LMAN neurons, whose influence on the motor pathway seems to result in high song variability. LMAN lesions on the other hand reduce the variability in song production and cause early crystallization of the song. The AFP drives learning by perturbing the HVc–RA synaptic connections or RA activity. RA neurons have been seen to exhibit auditory responses in sleeping zebra finches (Dave & Margoliash, 2000). It has been proposed that RA neurons might be involved in the construction of an inverse model for sensorimotor learning as they can be responsive to auditory input (in sleep) and to the premotor activity (Margoliash, 2002). Hence, the AFP might be acting to regulate the activity of the RA neurons. Within the AFP, the Area X seems to control the output of the LMAN neurons through the DLM. Hence, it follows that Area X is either directly involved in the evaluation of the error between the bird's song and the tutor song or it receives this evaluation from a region in the auditory pathway (probably Field L nuclei). Based on this error, it controls the output of the LMAN to produce higher or lesser variability. Further evidence for this kind of function of the AFP comes from inactivation studies of LMAN. It was seen that reversible inactivation of LMAN during the *sensory stage*, but not during the *sensorimotor stage*, leads to reduction in song learning in the bird (Basham, Nordeen, & Nordeen, 1996). Evidence has also been found for the presence of what are known as 'song-selective' neurons in the AFP (Doupe, 1997). 'Song-selective' neurons are neurons that respond specifically to the bird's own song or to the tutor's song but not to the song of conspecifics. It has been observed that both Area X and LMAN respond selectively to the bird's own song and not to other conspecific songs or even to the bird's own song played in reverse order. The fact that this responsiveness develops during the process of learning, and is not innate, further indicates the experience-dependent learning of neurons in the AFP. It is important to note that such selectivity based on experience is also observed in human language learning (Kuhl, 1994). In the beginning, human infants can discriminate phonemes from all human languages but are gradually tuned to respond to the sounds of their native language.

14.1.3 Dopamine in Learning

In the songbird brain, dopaminergic projections are seen to originate from the midbrain regions, VTA, SNc, and the periaqueductal gray (PAG) (Appeltants, Absil, Balthazart, & Ball, 2000; Lewis, Ryan, Arnold, & Butcher, 1981). It has been observed that the major target nucleus of the VTA projections is the Area X in the AFP. This region receives much denser dopaminergic innervations compared to the other regions in the bird's brain like the motor pathway, HVc, and RA. It is not known conclusively if the same neurons innervate the HVc and RA as those of the AFP. The properties of the dopaminergic neurons innervating Area X have been well characterized. It was observed that dopaminergic neuron activity was higher during singing than during silence. It was also observed that in vitro electrical stimulation of Area X leads to the release of DA (Gale & Perkel, 2005). Further, it was observed that the role of dopamine (DA) in Area X was to reduce the excitability of the medium spiny neurons (MSNs), since there was a majority of D2 receptors in Area X (Ding & Perkel, 2002).

Studies that quantify DA activity employ assays for the enzyme, tyrosine hydroxylase (TH), which converts L-tyrosine to dihydroxyphenylalanine (DOPA), which in turn is the precursor for DA. It was observed that Area X neurons show low reactivity for TH during the initial stage of song learning, but increased reactivity during development and reached a peak by day 60 (Soha, Shimizu, & Doupe, 1996). Studies using turnover rates of DA (the rate at which DA is taken up or used by the cells) showed that the DA levels and turnover rates reduced profoundly by day 90 (Harding, 1998). Similar results were observed in other song control nuclei like LMAN and RA, and auditory neurons like Field L, though the levels of DA were much higher in Area X. This implies that DA plays a role in error-driven learning during the song learning process similar to that observed in humans and other mammals.

In mammals, it has been observed that the major area in the BG that receives dopaminergic projections is the Str from the midbrain region substantia nigra pars compacta (SNc). Some studies indicate that the SNc also projects to the GP and the subthalamic nucleus (STN) (Cossette, Lévesque, & Parent, 1999), but the projections are not as dense as those projected to the Str. Various functional roles for DA in this system have emerged from studies of animal models of Parkinson's disease (PD), a disease which occurs due to the degeneration of the dopaminergic neurons in the SNc. The role of D1 and D2 receptors in the Str has already been outlined above. It has also been observed that DA modulates the efficacy of the cortico-striatal synapses.

DA neurons have been observed to respond to novel stimuli and rewards (Schulz, 1998). It was seen that these neurons initially respond to the occurrence of a reward signal, but subsequently to cues that predict the reward signal. When a reward is not expected but is received, there is an increase in firing of dopamine neurons; when the reward is expected and is delivered, there is no change in their firing, whereas when the reward is expected but not delivered there is a dip in the activity of dopamine neurons.

14.1.4 Modeling Bird Song Learning

Song learning in birds takes place in three stages: the *sensory stage*, during which the birds listens to and memorizes the song, the *sensorimotor stage*, when the bird learns the memorized song by self-feedback, and the *crystallized stage*, when the bird has developed a stable song matching that of the tutor.

Doya & Sejnowski, (1995) proposed a reinforcement-based learning scheme of the song learning process. It was assumed that the AFP stores in it a 'song template' which it compares with the bird's own song to adaptively train the songbird. The synaptic perturbations were provided by LMAN in the AFP. Doupe et al. (2005) proposed a model based on the 'AFP comparison hypothesis.' Here it was proposed that the AFP evaluates the birdsong by producing a prediction of the feedback of the syllable, the 'efference copy.' Fiete et al. (2007) proposed a spiking neuron model involving the motor pathway and the LMAN nuclei. According to their model, the AFP acts as an adaptive critic with stochastic perturbations generated by the LMAN region.

Computational models of bird song acquisition and generation developed so far (Doya & Sejnowski, 1995; Fiete, Fee, & Seung, 2007; Troyer & Doupe, 2000) have focused on attributing a specific function to various regions within the song system in the song learning process. Yet the role of DA in the learning process has not been considered by these models. Considering the fact that the AFP functions in a manner analogous to the BG in humans, it would be interesting to study the exact role of dopamine in the AFP during learning.

A computational model for the song learning process would serve as a model system to understand motor learning. In humans, degeneration of the mesencephalic dopaminergic centers like substantia nigra pars compacta can lead to a pathological state called Parkinson's disease (PD) characterized by tremor, bradykinesia, and other motor symptoms. At the systems level, a full understanding of how the neural activity patterns are affected and how these lead to behavioral changes is currently not available. Understanding the networks subserving song learning in birds and the pathology of the same can open up new avenues to the understanding of BG function in learning.

14.1.5 Objective

We present a model of bird song production in which the motor control pathway is modeled by a trainable network of oscillators and the Anterior Forebrain Pathway (AFP) is modeled as a stochastic system. The outputs from the motor pathway coordinate the activity of the bird's vocal organ, syrinx, and that of the respiratory system which have been modeled here using a mechanistic model of the syrinx. The bird's beak is modeled as a vocal filter. The model proposes a role for dopamine in song learning, analogous to the role of dopamine in mammalian BG and motor learning. It gives a general picture of how dopamine drives learning and exploration in bird song acquisition. Simulating PD-like conditions of dopamine deficiency, the model shows impairment in song learning under such conditions. The model is trained on real birdsong data, and effects of lesioning and Parkinsonian pathology are investigated.

The chapter is organized as follows. Section 14.2 presents a detailed description of the proposed model and each of its components. Section 14.3 presents the results of simulations of the model. Model performance from real birdsong samples under (a) normal conditions and (b) pathological conditions—which include lesioning and dopamine-deficient Parkinsonian-like conditions—is presented in this chapter. Section 14.4 discusses the implications of the findings from our model. The chapter concludes with directions for future work.

14.2 Model Description

We propose a model of song learning based on reinforcement learning where the HVc–RA system is modeled by two sets of Hopf oscillators (Righetti, Buchli, & Ijspeert, 2006) which act as central pattern generator (CPG) circuits. The AFP is modeled as a random noise source which perturbs the output of the oscillatory

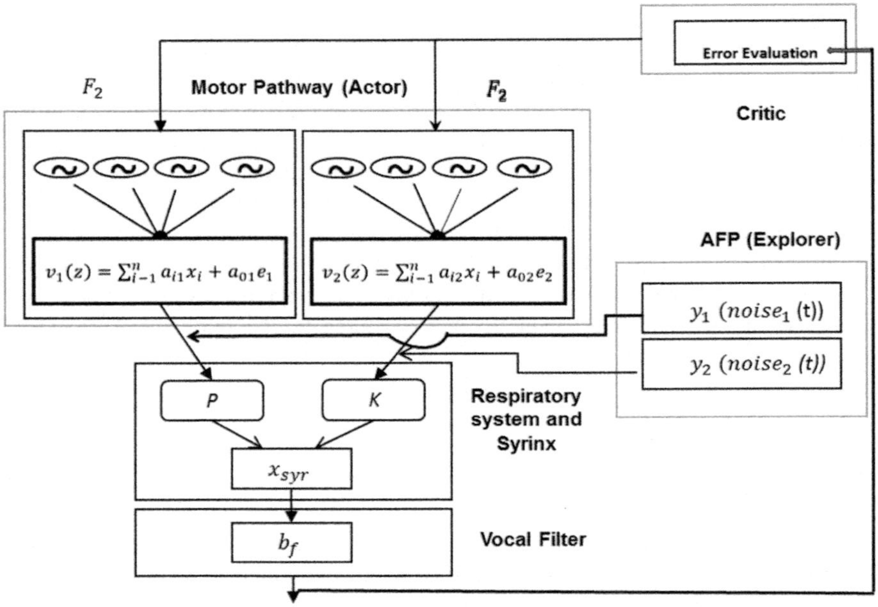

Fig. 14.2 Actor–Critic–Explorer schema of the proposed model with the various components. AFP—Anterior Forebrain Pathway. For explanation of the notations, refer text

networks. The outputs of the two networks, with the noise added from the AFP, are fed to the model of the bird vocal organ, the syrinx, which in turn is fed to a vocal filter which plays the role of analogous to the bird's peak, enhancing only certain frequency components of the incoming signal. Figure 14.2 gives a broad overview of the proposed model.

14.2.1 The Motor Pathway Model

Central pattern generators (CPGs) are systems that are capable of producing rhythmic patterns of activity in the absence of any external input (Crook & Cohen, 2003). Many rhythmic behaviors like respiration and locomotion are encoded in the form of CPGs in the brain stem. The song produced by the adult zebra finch is highly stereotyped irrespective of the presence or absence of auditory feedback. The fact that song deteriorates very little in adult songbirds after deafening indicates that the song must be generated by a mechanism that is independent of external control once it is learnt (Doupe & Kuhl, 1999). However, deafening in juvenile birds leads to impaired song learning.

The first conclusive evidence that these CPGs are located in the motor pathway was given by Vu, Mazutek, and Kuo (1994). Further evidence comes from the fact that the HVc receives inputs from auditory areas and the HVc firing patterns are modified by auditory feedback. The song 'pattern' is thought to be encoded in the HVc–RA synaptic strengths.

We use a network of Hopf oscillators to model the motor pathway (Righetti et al., 2006). During the training process, the training signal is fed as a forcing function to each of the oscillators in the network. These oscillators then embed the desired signal in the form of a stable limit cycle. Hence, the system is robust; even with small perturbations, it returns to the limit cycle pattern mimicking the stable song system in the adult bird.

The governing equations for the variables (x, y, ω, α, Φ) of the oscillators are given as:

$$\frac{dx_i}{dt} = \gamma(\mu - r_i^2)x_i - \omega_i y_i + \varepsilon F(t) + \tau \sin(R_i - \phi_i). \tag{14.1}$$

$$\frac{dy_i}{dt} = \gamma(\mu - r_i^2)y_i - \omega_i x_i. \tag{14.2}$$

$$\frac{d\omega_i}{dt} = -\varepsilon\eta\left(\frac{y_i}{r_i}\right)F(t). \tag{14.3}$$

$$\frac{d\alpha_i}{dt} = \eta x_i F(t). \tag{14.4}$$

$$\frac{d\alpha_0}{dt} = \eta c F(t). \tag{14.5}$$

$$\frac{d\phi_0}{dt} = 0. \tag{14.6}$$

$$\frac{d\phi_i}{dt} = \sin(Ri - \mathrm{sgn}(x_i)\cos^{-1}\left(-\frac{y_i}{r_i}\right) - \phi_i), \quad \forall i \neq 0 \tag{14.7}$$

$$R_i = \frac{\omega_i}{\omega_0}\mathrm{sgn}(x_0)\cos^{-1}\left(-\frac{y_0}{\sqrt{x_0^2 + y_0^2}}\right) \tag{14.8}$$

$$F(t) = P_{\mathrm{teach}}(t) - \sum_{i=1}^{n} \alpha_i x_i \tag{14.9}$$

where μ is a parameter that controls the amplitude of oscillations, ω is the intrinsic frequency of the oscillators, ε is a coupling constant, γ is the learning rate, α denotes the weights of the oscillators, α_0 denotes the baseline values of the weights, x and y denote the state variables of the oscillator, ϕ_i denotes the phase of the oscillator, $P_{\mathrm{teach}}(t)$ denotes the input signal, and $F(t)$ is the feedback error signal in the oscillator equation that denotes the difference between the training signal and its

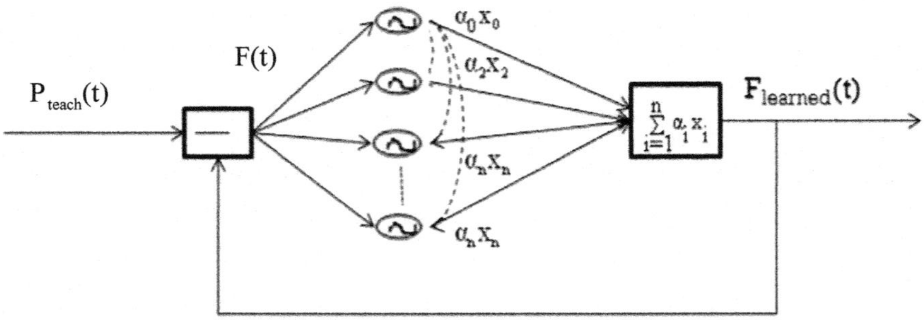

Fig. 14.3 Schematic of network used in the model showing the feedback loop and the coupling between the oscillators

reconstruction, $r = \sqrt{x^2 + y^2}$. Equations (14.1, 14.2) denote oscillator dynamics and Eqs. (14.3–14.8) denote learning dynamics of the parameters ω, α, and α_0. Note that the learning described above is a supervised form of learning. In the present model, the above learning mechanism is reformulated as a RL mechanism (Fig. 14.3).

We use the above two-network model to simulate the outputs from the motor pathway to the syringeal and respiratory muscles. The outputs of the two oscillatory networks are:

$$v_1(t) = \sum_{i=1}^{n} \alpha_{i1} x_i + \alpha_{01} c_1 \tag{14.10}$$

$$v_2(t) = \sum_{i=1}^{n} \alpha_{i2} x_i + \alpha_{02} c_2 \tag{14.11}$$

14.2.2 The Anterior Forebrain Pathway in the Model

We propose that the AFP plays the role of an 'Explorer' in the reinforcement learning framework. This pathway serves as the source of chaotic perturbations to the motor pathway. This is evident from the remodeling of the LMAN–RA synapses observed during song learning (Iyengar & Bottjer, 2002) and the reduced variability in the song following LMAN lesions (Bottjer et al., 1984). At the onset of learning, the LMAN–RA projections are diffuse and the bird produces highly variable notes (*subsong*). During the process of learning, refinement of LMAN–RA synapses takes place (*plastic song*) and at the end of *crystallized song* phase, the number of LMAN–RA synapses decreases substantially.

We incorporated two random variables (mean = 0, std = 0.25, uniform distribution) to model the role of AFP in the song learning process. In the model, the noise generated at every time instant is added to the output of the oscillator networks. The following equation governs the final output to the vocal organ:

$$u(t) = v(t) + (\chi)\zeta. \tag{14.12}$$

The term χ controls the exploratory drive to the oscillatory networks. This in turn is controlled by δ, which denotes the temporal difference (TD) error term in the RL framework (Sutton & Barto, 1998), given by

$$\delta = \tanh(s\, e_{\mathrm{avg}}(t) - \mathrm{err}(t)) \tag{14.13}$$

where s is a constant (s = 0.6), $e_{\mathrm{avg}}(t)$ is the average error of the previous training stage, and err(t) is the current error, denoted by $e_{\mathrm{r}}(t)$ in Eq. (14.25) and $e_{\mathrm{rr}}(t)$ in Eq. (14.27) (see Sect. 14.2.5 for a more detailed description of error measures). The tanh(.) is used to suppress fluctuations in error differences in Eq. (14.13). The term δ controls the exploration done by the model at each time step by controlling the value of χ.

$$\chi = c(1 - \delta) \tag{14.14}$$

where c = 0.4;

A negative value of δ implies a low error; hence, the value of χ is set to low value for the model to have less exploration.

14.2.3 The Respiratory System and Syrinx Model

The outputs from the two networks of the motor pathway control two different variables essential for song production. Network 1 influences the respiratory rhythm by controlling the bronchial pressure (P_{b}). Network 2 controls the restitution constant (K) of the bird's vocal organ, the syrinx.

$$P = P_0 + P_1 v_1; \tag{14.15a}$$

$$K = K_0 + K_1 v_2 \tag{14.15b}$$

where P_0 = 1.2 kPa, K_0 = 2.5 N/cm^3, P_1 = 1.0 kPa, K_1 = 3 N/cm^3.

The current theory of song learning in birds suggests that song is produced through oscillations of tissue folds called labia that open and close the air passage from the bronchi to the trachea (Gardner, Cecchi, Magnasco, Laje, & Mindlin, 2001). These oscillations are modeled by:

$$\frac{\mathrm{d}x_{\mathrm{syr}}}{\mathrm{d}t} = y_{\mathrm{syr}}. \tag{14.16}$$

$$M\left(\frac{\mathrm{d}y_{\mathrm{syr}}}{\mathrm{d}t}\right) + Dy_{\mathrm{syr}} + D_2 y_{\mathrm{syr}}^3 + Kx_{\mathrm{syr}} = P\left(\frac{a_0 - b_0 + 2\tau y_{\mathrm{syr}}}{x + b_0 + \tau y_{\mathrm{syr}}}\right). \tag{14.17}$$

where M is the mass (M = 0.05 g/cm^2), D and D_2 correspond to (D = 5 dyn s/cm^3, D_2 = 0.009 dyn s^3/cm^3), all of which are in per unit area; τ is the time constant (τ = 0.005 s), a_0 = 0.01 cm, b_0 = 0.02 cm.

14.2.4 The Vocal Filter Model

The oscillations produced by the vocal folds are filtered by the vocal tract so that certain high-frequency components of the incoming signal are enhanced (Hoese, Podos, Boetticher, & Nowicki, 2000). Here the filter is modeled by assuming that the trachea and the beak form two tubes of lengths L_1 and L_2 and areas A_1 and A_2, respectively (Gardner et al., 2001). The input pressure generates a fluctuation in the tracheal pressure part of which is reflected and part of which travels to the beak. Again part of the energy is reflected and part transmitted at the beak, denoted by the final output b_{f}.

$$a(t) = P_i(t) + b_{\mathrm{b}}(t - \tau_1). \tag{14.18}$$

$$b_{\mathrm{b}}(t) = r_{12} \cdot a(t - \tau_1) + t_{12} \cdot c_{\mathrm{b}}(t - \tau_2). \tag{14.19}$$

$$b_{\mathrm{f}}(t) = t_{12} \cdot a(t - \tau_1) + r_{12} \cdot c_{\mathrm{b}}(t - \tau_2). \tag{14.20}$$

$$c_{\mathrm{b}}(t) = \alpha \cdot b_{\mathrm{f}}(t - \tau_2). \tag{14.21}$$

where $P(t)$ is the input pressure, calculated by Eq. (14.22), $a(t)$ and $b_{\mathrm{f}}(t)$ are the forward traveling waves from the trachea and the beak, respectively, $b_{\mathrm{b}}(t)$ and $c_{\mathrm{b}}(t)$ denote the backward traveling (reflected) waves from the trachea and the beak, respectively, r_{12} and t_{12} denote the sound energy reflected and transmitted, respectively, at the junction of the beak and the trachea, τ_I denotes the time taken by the sound wave to traverse a tube of length L_i, and α denotes the reflection coefficient between the beak and the atmosphere.

$$P_i(t) = c_{\mathrm{p}} \frac{\mathrm{d}^2 x_{\mathrm{syr}}}{\mathrm{d}t^2} \tag{14.22}$$

where c_{p} is a constant of proportionality (c_{p} = 1), L_1 = 5 cm, L_2 = 2 cm, A_1 = 9.0 arbitrary units, A_2 = 8.0 arbitrary units, α = 0.9 (Trevisan, Eguia, & Mindlin, 2001), r_{12} and t_{12} are calculated according to the following equations:

$$r_{12} = \frac{A_1 - A_2}{A_1 + A_2} \tag{14.23}$$

$$t_{12} = 1 - r_{12} \tag{14.24}$$

14.2.5 Training Algorithm

The model is trained in a temporal stage-wise manner. The training signal received by the motor system is processed by dividing the signal into smaller segments. Each segment is processed based on the average error of the previous segment. This kind of segmentation process referred to as 'chunking' of sensory information has been observed to occur in the case of performing a sequential visuomotor task (Sakai, Kitaguchi, & Hikosaka, 2003). In the model, training is done by comparing the performance of the system with that of the previous stage using a TD learning algorithm. Training is performed if the error in a given instant is lesser than that of the previous trained segment (Fig. 14.4). The error is computed in two phases. In the first phase, the error is taken as a weighted average of errors in the peak frequency, peak amplitude, and baseline values of the signal over a small time window. In the second phase of training, the error is the average over instantaneous error values between the birdsong signal and the output signal of the model in a small time window. Figure 14.4 shows the temporal staging used for training.

The first phase error is computed using the following equation:

$$e_I(t) = e_{\text{pp}}(t) + e_{\text{fr}}(t) + e_{\text{bl}}(t) \tag{14.25}$$

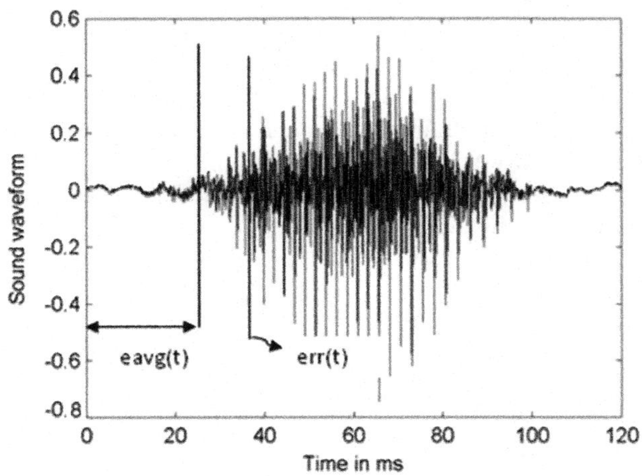

Fig. 14.4 Error computation for temporal stage-wise learning in the model

where $e_{pp}(t)$ is the error in the peak-to-peak amplitude between the two signals, and $e_{fr}(t)$ is the error in peak frequencies, $e_{bl}(t)$ is the error in the baseline values of the two signals.

The error, $e_{II}(t)$, in phase 2 of training is computed as:

$$e_{\mathrm{inst}}(t) = (F(t) - b_{\mathrm{f}}(t))^2 \tag{14.26}$$

$$e_{\mathrm{rr}}(t) = \frac{\sum_{j=(t-\mathrm{win})}^{t} e_{\mathrm{inst}}(j)}{\mathrm{win}} \tag{14.27}$$

where e_{inst} implies instantaneous value of error between the two signals and win is the size of window over which the instantaneous error is averaged.

As denoted in Fig. 14.4, the signal is segmented into chunks of smaller units, containing, say, k sample points. Then, $e_{\mathrm{avg}}(t)$ is the averaged error denoted by:

$$e_{\mathrm{avg}}(t) = \sum_{t=ik}^{(i+1)k} \frac{\mathrm{err}(t)}{k}. \tag{14.28}$$

Here, err(t) is er(t) as in Eq. (14.25) or err(t) as in Eq. (14.27), i denotes the ith chunk of the signal.

The network output error signals, F_1 and F_2, are updated based on the following equations,

$$F_1(t) = \chi_1 \zeta_1(t) + m \cdot F_1(t-1) \tag{14.29}$$

$$F_2(t) = \chi_2 \zeta_2(t) + m \cdot F_2(t-1) \tag{14.30}$$

where m denotes the momentum term in reinforcement learning (m = 0.3), χ_1 and χ_2 denote the values of the exploration variables, and ζ_1 and ζ_2 denote the noise variables as explained in Sect. 14.2.2. F_1 and F_2 denote the output error signals. They denote ($u(t) - v(t)$) as in Eq. (14.12). As $v(t) \to u(t)$, the noise variable reduces in magnitude, denoting a decrease in the exploratory drive in the model. Further, as the networks are trained with the Eqs. (14.3–14.7), for various parameters, $v_1(t) \to F_1(t)$ and $v_2(t) \to F_2(t)$.

The learning equations for the weights of the oscillators are also updated such that Eqs. (14.4) and (14.5) are modified to:

$$\frac{\mathrm{d}\alpha}{\mathrm{d}t} = \eta \delta x(t) F_j(t). \tag{14.31}$$

$$\frac{\mathrm{d}\alpha_{0j}}{\mathrm{d}t} = \eta \delta c_j F_j(t), \tag{14.32}$$

where $j = 1$ or 2 denotes network 1 or network 2

where c_1 = 1.0, c_2 = 1.0, η = 0.4.

14.3 Results

The model was tested on two different zebra finch song syllables. The results are shown in Figs. 14.5 and 14.6, respectively. Here the noise variable χ has the mean 0 and standard deviation 0.25.

14.3.1 Lesion Studies

Lesioning is a standard way to differentiate the specific contributions of a nucleus in a neural circuit. In the song system of Fig. 14.1, the efferent pathway consisting of HVc and RA is essential for both song learning and production while the recursive loop consisting of Area X and LMAN is found to be necessary for song learning and not for song production. Specifically, lesions of Area X and LMAN have distinct effects on song learning process, which will be now investigated with the help of simulations.

14.3.1.1 LMAN Lesion

LMAN lesions caused severe distortion of bird song that consisted of monotonous repetition of a single note complex. Basically, birds with early LMAN lesions manifest premature crystallization, though similar lesions in adult birds have little or no effect. Figure 14.7 from Bottjer et al. (1984) shows sonograms of songs from birds whose LMAN was lesioned at 35 days. Note that the spectral band progressively became narrower with age; at an age of about 90 days, spectrum became much narrower than that of a normal song. Notes in the song of LMAN-lesioned birds lacked the frequency modulations of a normal song; they consisted of long bouts of singing without the normal phrase structure. The song of LMAN-lesioned birds is characterized by monotonous repetitions of a single note complex and stabilizes early (Scharff & Nottebohm, 1991).

Our present model lumps Area X and LMAN together, into a single random vector with variable amplitude. LMAN has glutamatergic neurons while Area X has GABA neurons. Since LMAN directly projects to RA, and LMAN neurons are excitatory, lesioning of LMAN can be modeled as reduced amplitude of noise vector (ζ_i) in the model.

To simulate LMAN lesion, noise amplitude ζ_i is reduced from the normal 0.25 (standard deviation) to 0.05, and the song system is trained as it was in the normal case. Note that the model song corresponding to LMAN lesion has neither the frequency nor amplitude variations of a normal song note. Like the song of the LMAN-lesioned bird, the model output is confined to a narrow band of frequencies with weak harmonic component. The model song seems to consist of two repetitions of a single note complex, each lasting about 50 ms (Fig. 14.8a). Network's

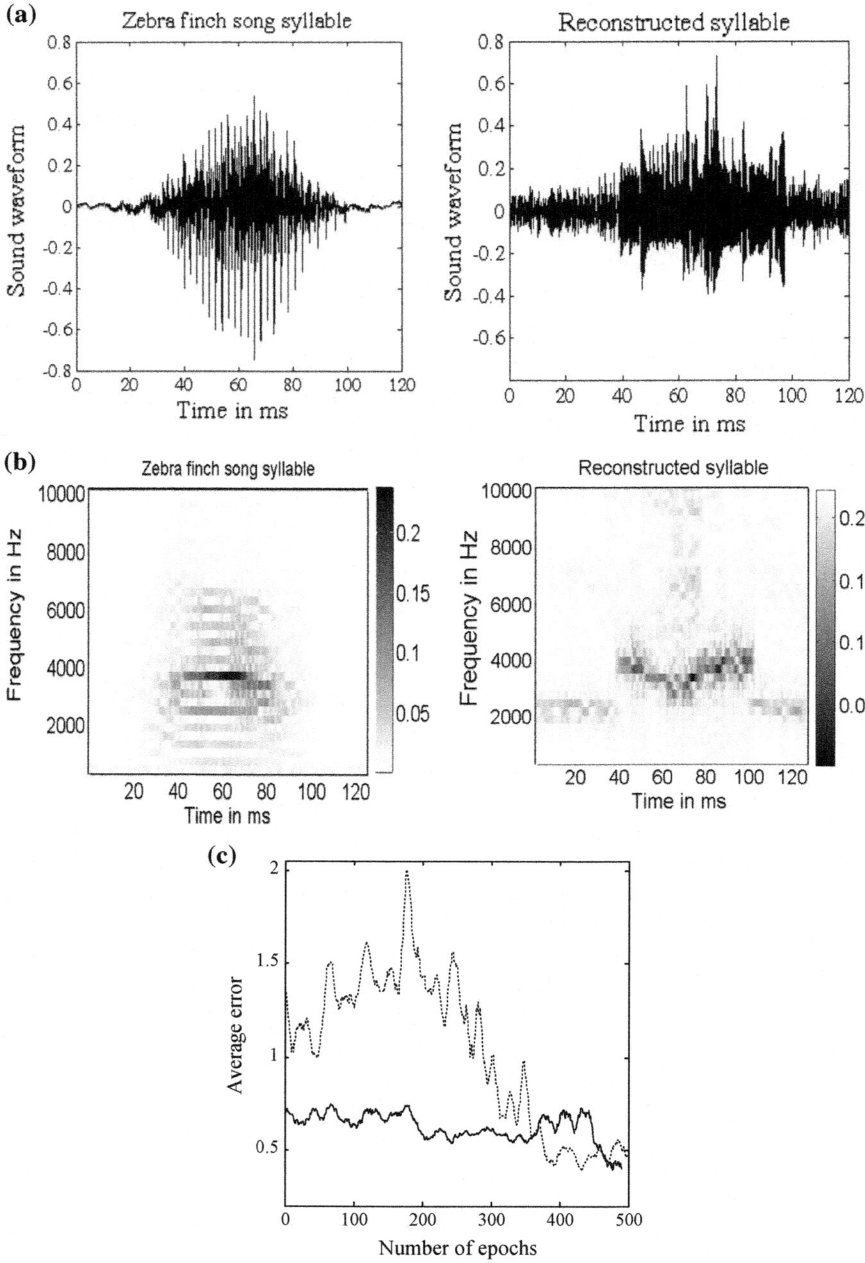
(a)
Zebra finch song syllable
Sound waveform
Time in ms
Reconstructed syllable
Sound waveform
Time in ms
(b)
Zebra finch song syllable
Frequency in Hz
Time in ms
Reconstructed syllable
Frequency in Hz
Time in ms
(c)
Average error
Number of epochs

◄**Fig. 14.5** **a** Sound pressure waveforms (normalized) of a birdsong syllable (left) and the reconstructed waveform obtained from the model (right). **b** Power spectra of an actual birdsong syllable and the reconstructed waveform obtained from the model are shown below. **c** The error plot for the different stages in the training is shown. The solid line indicates the first stage of training where the error is the weighted average of errors in the peak frequency, peak amplitude, and baseline values of the signal over a small time window. The dashed line indicates the second-stage training where the error is the average value of instantaneous error over a small time window

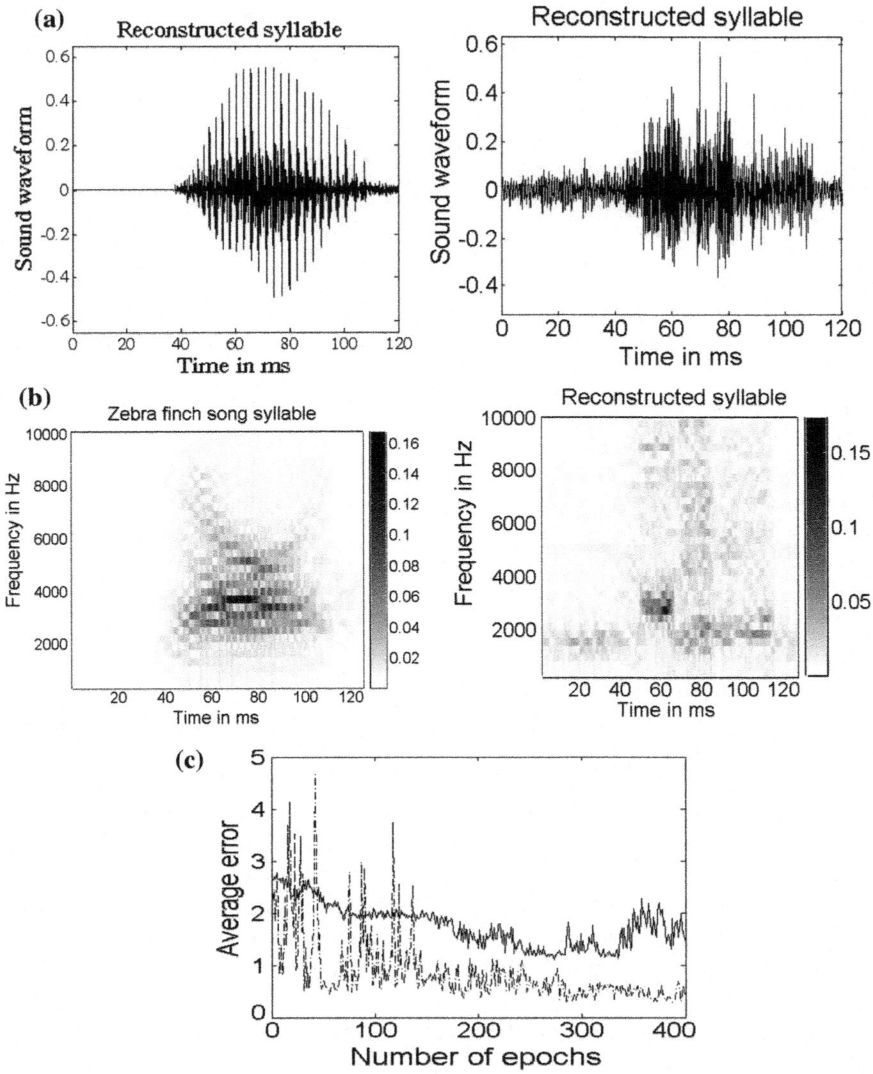

Fig. 14.6 **a** Sound pressure waveforms (normalized) of a birdsong syllable (left) and the reconstructed waveform obtained from the model (right). **b** Power spectra of an actual birdsong syllable and the reconstructed waveform obtained from the model are shown below. **c** The error plot of one of the training stages is shown

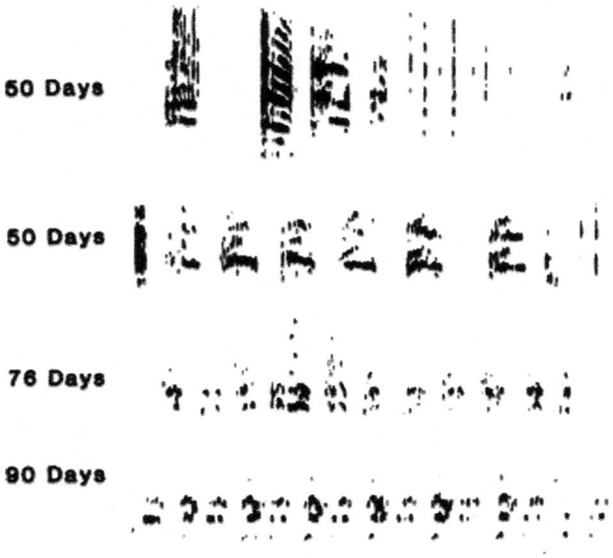

Fig. 14.7 Effect of LMAN lesions on song performance in a zebra finch From Bottjer et al. (1984)

training error drops quickly at an early stage and saturates at a high value, resembling early song stabilization (Fig. 14.8c).

Figure 14.8 shows the reconstructed syllable at the adult stage after LMAN lesion in comparison with that of the normal syllable of the zebra finch.

14.3.1.2 Area X Lesion

Contrary to effects of LMAN lesions, birds with lesions in Area X produced songs that consisted of a rambling series of long and variable notes (Scharff & Nottebohm, 1991). While LMAN lesions produced premature stabilization, songs in birds with lesions of Area X never stabilized.

We may recall that Area X is GABAergic and projects to LMAN. Thus, lesions of Area X would disinhibit LMAN, which is expected to have larger than normal activation. Therefore, we model Area X lesions by increasing the noise amplitude, ζ_i (standard deviation is 0.25 in normal case to 0.6 in case of Area X lesion).

In Fig. 14.9a, b, denote cases of different degrees of Area X damage. A denotes 43% Area X damage and B denotes 84% Area X damage. Note that the interval length and the note structure both show degradation to various degrees in both cases.

The results of the model simulations for two different trials of Area X lesions are shown in Fig. 14.10.

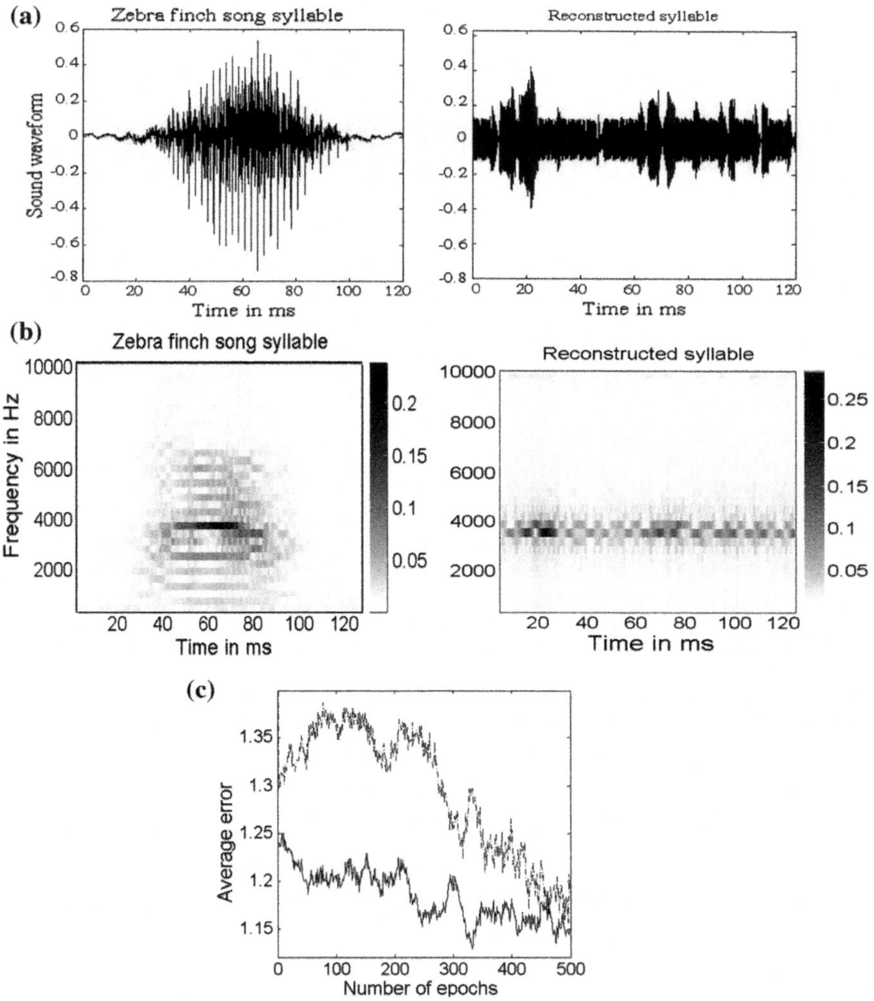

Fig. 14.8 **a** Sound pressure waveforms (normalized) of a normal birdsong syllable (left) and the reconstructed waveform obtained from the model after LMAN lesion (right). **b** Power spectra of an actual birdsong syllable and the reconstructed waveform obtained from the model after lesion are shown below. **c** The error plot of one of the training stages is shown. The solid line denotes phase 1 error, and the dotted line denotes the phase 2 error during training

14.3.2 Dopamine Depletion Studies

Significant progress seems to have been made, in the recent years, in the understanding of the role of dopaminergic system in bird song (Gadagkar et al., 2016; Kubikova & Kostal, 2009). Like in mammalian brains, sites that contain

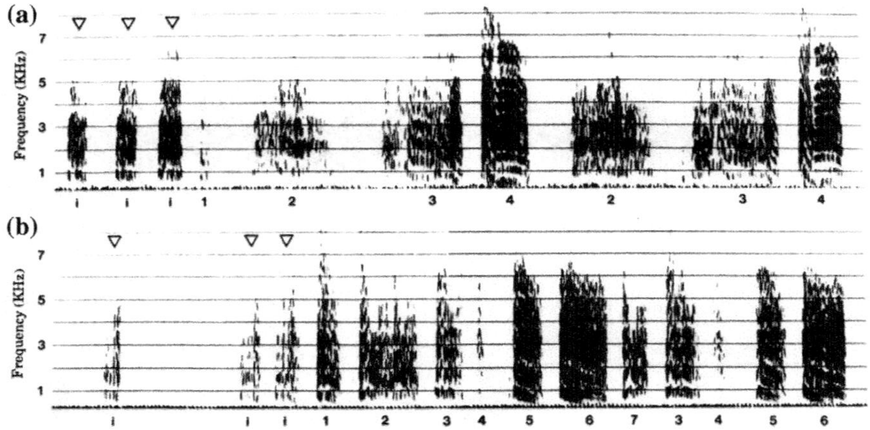

Fig. 14.9 Effect of Area X lesions on song performance in a zebra finch. From Scharff & Nottebohm, (1991)

dopaminergic cells in songbird brain are SNc, VTA, and/or the periaqueductal gray. Neurons of SNc and VTA project to the medium spiny neurons of Area X and alter the excitability of the same (Ding & Perkel, 2002). Dopaminergic projections to HVc and RA also exist (Bottjer, 1993). Although it is quite tempting to apply the current notions about the varied functions of dopamine in mammalian brains to songbird brains, there is very little experimental evidence to directly support such a view. Brainard and Doupe (2002) suggest that the AFP in songbird provides some sort of an error signal used for learning. To simulate dopamine-deficient conditions, like those in a PD-affected songbird, we constrain the upward fluctuations of δ, which signifies DA signal, as follows:

$$\delta_{\text{PD}} = \min(\delta, \text{DA}_{\text{ceil}}) \tag{14.33}$$

where min(x, a) is defined as:

$$\begin{aligned} y &= x, \quad \text{for} \quad x < a \\ &= a, \quad \text{for} \quad a \le x \end{aligned}$$

In Eq. (14.33), δ is the error signal directly calculated by Eq. (14.13), and δ_{PD} denotes a weakened DA signal. DA_{ceil} is chosen to be a value less than the maximum value of δ. Thus, lesser values of DA_{ceil} denote a greater DA loss. A similar implementation of dopamine deficiency was used recently in (Balasubramani et al., 2015; Magdoom, Subramanian, Chakravarthy, Amari, & Meenakshisundaram, 2010).

The DA_{ceil} value here was taken to be 0.5 so that DA levels as denoted by δ vary between -1 and 0.5 instead of -1 and 1, simulating reduced DA conditions. The model was trained for the same time as in previous cases (Fig. 14.11).

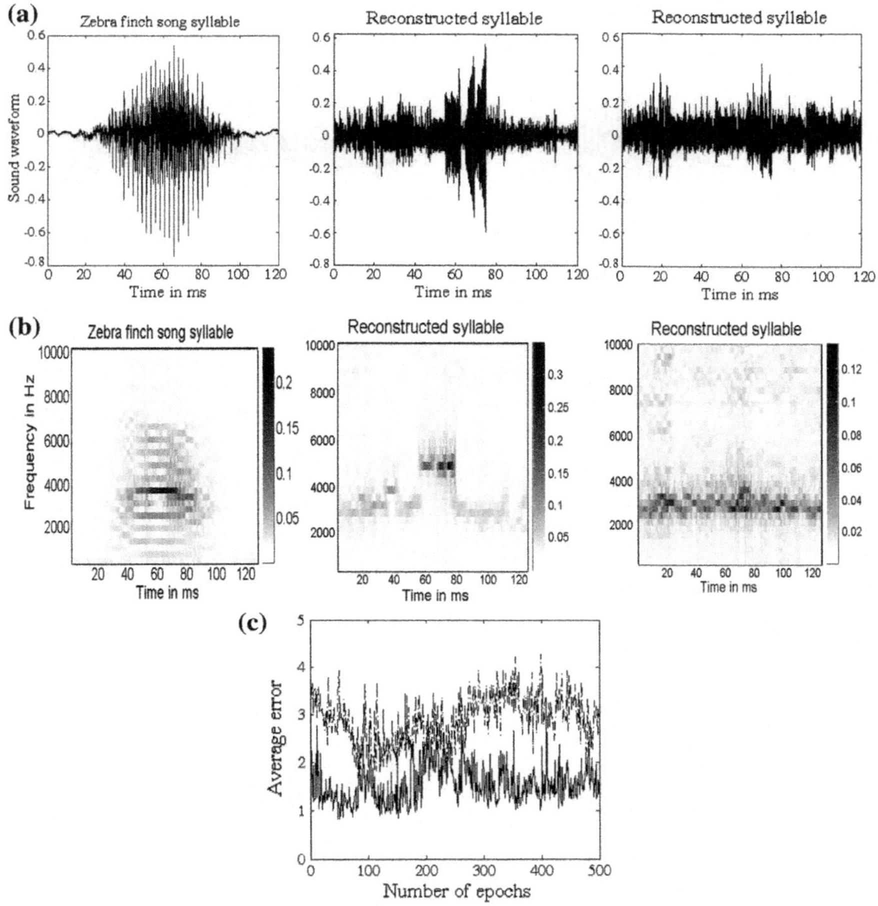

Fig. 14.10 **a** Sound pressure waveforms (normalized) of a normal birdsong syllable (left) and two instances of reconstructed waveforms (middle and right) obtained from the model after Area X lesion. Note the variability in the song produced. **b** Power spectra of an actual birdsong syllable and the reconstructed waveform after Area X lesion obtained from the model as explained in the text. **c** The error plot of one of the training stages is shown. The solid line denotes phase 1 error, and the dotted line denotes the phase 2 error during training

Correlated noise:

It is customary to model PD conditions by reduced dopamine levels. But another change also is observed in dopamine-deficient BG. Experimental studies have revealed, under dopamine-deficient conditions, prominent low-frequency periodicity (4–30 Hz) of firing and dramatically increased correlations among neurons in the GPe and the STN, though there were no significant changes in firing rates (Bergman, Wichmann, Karmon, & DeLong, 1994; Brown et al., 2001;

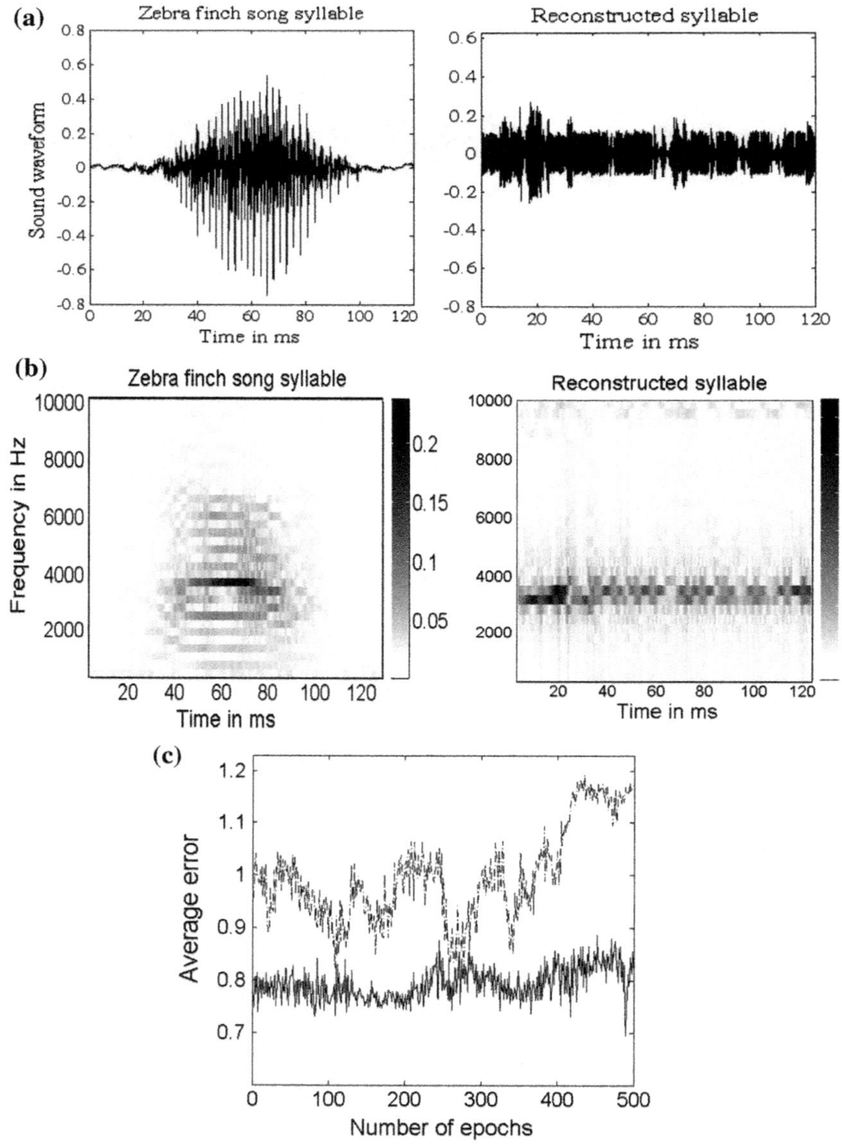

Fig. 14.11 **a** Sound pressure waveforms (normalized) of a normal birdsong syllable (left) and the reconstructed waveform obtained from the model after dopamine depletion (right). **b** Power spectra of an actual birdsong syllable and the reconstructed waveform after depletion obtained from the model are shown below. **c** The error plot of one of the training stages is shown. The solid line denotes phase 1 error, and the dotted line denotes the phase 2 error during training

Magnin, Morel, & Jeanmonod, 2000; Nini et al., 1995). Complex activity of STN–GPe system is thought to play the role of the Explorer, an important component in RL machinery and loss of such complexity has been hypothesized to contribute to motor symptoms of PD, a theme that happens to be the central theme in the line of models of basal ganglia described in this book (Chakravarthy & Balasubramani, 2014).

Therefore, we present a modeling exercise in which the noise arising out of AFP is correlated. In the model we make $\zeta_1 = \zeta_2$ to make noise values of both networks correlated (Fig. 14.12).

14.4 Discussion

We present an Actor–Critic–Explorer model of birdsong learning cast in the framework of reinforcement learning. In tune with some of the previous RL-based models on BG, we model dopamine as the temporal difference (TD) error. The value, which represents the inverse of the error between the original birdsong and its reconstruction, is thought to be computed in the critic. The location of the critic is not precisely known. In the model of Doya and Sejnowski (1995), it was assumed that the AFP stores a template of the tutor's song and uses it to generate evaluations. But there has been no clear evidence of the same though it is clear that the Area X is involved in either error evaluation or in using the error for controlling exploration. We assume that these evaluations are the only information preserved by the bird after its sensory stage. The AFP, or specifically LMAN's output to RA, is thought to be the Explorer which perturbs the motor pathway, thereby enabling it to discover the correct inputs to the syrinx necessary to produce an accurate song. Several lines of biological evidence point to the possibility that LMAN serves as an Explorer. Firstly, LMAN neural activity is more irregular and variable from trial to trial, than activity in RA (Leonardo, 2004). Secondly, LMAN activity is lower during the directed song, which is stable and stereotyped, than during the undirected song which shows more variability (Hessler & Doupe, 1999; Kao, Doupe, & Brainard, 2005).

In this preliminary model, the three modules of AFP (Area X, DLM, and LMAN) are lumped together and modeled as two uniformly distributed random variables of mean 0 and standard deviation 0.25. The motor pathway is modeled by a network of oscillators, where the oscillatory layer denotes the HVc and the output layer of the network represents RA. The song output of the model is a function of both the motor pathway and the AFP. The relative contributions of the motor pathway and the AFP vary with trials, with the contribution from AFP becoming negligible once the song is crystallized. This is incorporated in the following manner: In our model, as seen by Eq. (14.14), the variable χ controls the noise from the AFP, which in turn is controlled by δ as explained in Sect. 14.2.2. In the beginning of training, χ is high as δ is high. As training proceeds and the bird's song starts to match that of the tutor, the value of δ falls to a low value, thereby

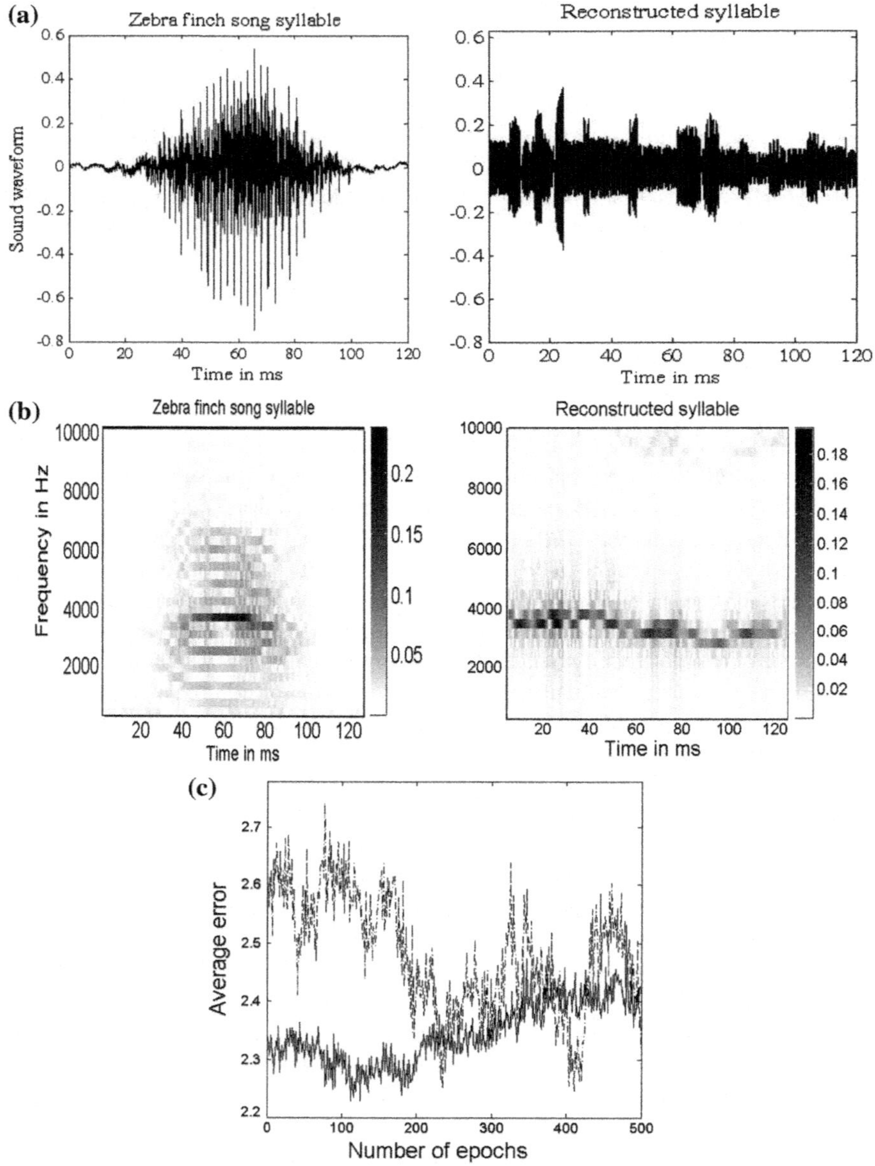

Fig. 14.12 **a** Sound pressure waveforms (normalized) of a normal birdsong syllable (left) and the reconstructed waveform obtained from the model with correlated noise (right). **b** Power spectra of an actual birdsong syllable and the reconstructed waveform with correlated noise obtained from the model are shown below. **c** The error plot of one of the training stages is shown. The solid line denotes phase 1 error, and the dotted line denotes the phase 2 error during training

reducing χ, leading to a low contribution from the AFP to the final song. This feature of the model reflects the fact that the adult song evolves to be independent of AFP.

The dopamine signal serves multiple roles in the present model. The phasic dopamine signal, $\delta(t)$, represents the TD error. It modulates the connections between the oscillatory layer and the output layer of the network that represents the motor pathway. This assumption is biologically feasible since the HVc $\rightarrow$ RA synapses are located in RA and dopamine projections from VTA do arrive at RA (Appletants, Ball, & Balthazart, 2002). The phasic dopamine also controls the level of exploration at every time step according to Eq. (14.14), which is a biologically plausible feature since there are dopaminergic projections to Area X and LMAN. Although DA projections to LMAN have been observed, the number of the projections seems to be only moderate, whereas the projections to Area X seem to be quite dense as seen by tyrosine hydroxylase immunoreactivity studies (Bottjer, 1993). The slow-varying or tonic dopamine signal, which may be thought to be represented by the average error in song reconstruction over trials, controls the amplitude of noise arising out of the AFP, on a slower timescale of trials.

The idea that AFP is probably providing exploratory inputs for training the motor pathway is not new (Doya & Sejnowski, 1995; Fiete et al., 2007). Doya and Sejnowski (1995) assume that LMAN projections to RA perturb the HVc $\rightarrow$ RA synapses, and offer, as biological justification, the fact that LMAN synapses onto RA are of NMDA-type. Expanding their original schema in a subsequent work, Doya and Sejnowski (1998) interpret Area X as the Critic and LMAN as the Explorer. In a more recent work, Fiete et al. (2007) present a model of birdsong learning which is also cast in the RL framework, but differs from Doya and Sejnowski (1995) on several issues. While both models agree that LMAN is the Explorer [referred to as *experimenter* in (Fiete et al., 2007)], the main point of departure between the two models is regarding the nature of the Explorer, or, specifically, the nature of LMAN's influence on RA.

In (Doya & Sejnowski, 1995), LMAN inputs to RA are thought to produce a change in HVc $\rightarrow$ RA weights, a change that lasts during the entire duration of the song (1–2 s). If the weight change results in an improved song, the change is reinforced; else the change is discarded. But LMAN activity is dynamic, varying over a timescale of 10–100 ms; it does not seem to supply the biological requirements of Doya and Sejnowski model. Following such a thread of arguments, Fiete et al. (2007) present a model of birdsong learning in which LMAN inputs to RA perturb, not the HVc $\rightarrow$ RA synapses, but the activities of RA neurons.

The treatment of Explorer, in the present model, is similar to that of (Fiete et al., 2007) in its broad outlines, though the mathematical form is quite different. Also, unlike both the models described above (Doya & Sejnowski, 1995; Fiete et al., 2007), we explicitly represent dopamine signal in the model. This signal is used to train HVc $\rightarrow$ RA synapses, and also to control the amplitude of LMAN activity, which represents the exploratory noise. An explicit representation of dopamine signal made it possible to model song learning in avian Parkinson's disease (PD).

Song learning under dopamine-deficient conditions was simulated in Sect. 14.3.3. DA reduction is simulated by varying two parameters: (1) constraining upward fluctuations of dopamine signal and (2) increasing the correlation (to 1) of the two outputs of the AFP. In both cases, there is a failure to learn, with reconstruction error showing non-monotonic fluctuations. In the former case of constrained dopamine fluctuations, the model song has smaller amplitude than normal song and exhibits a narrowing of spectral band. Experimental studies show that lesions of SNc–VTA do not affect song production in adult birds (Hara, Kubikova, Hessler, & Jarvis, 2007). But to our knowledge, there is no corresponding study on song learning in a juvenile bird. Studies of changes in PD speech in human subjects reveal a reduced variability in fundamental frequency, F_0 (Harel, Cannizzaro, Cohen, Reilly, & Snyder, 2004), but there is no mention of variation in harmonic content.

It would be important to compare the current understanding of the function of mammalian BG and the AFP in songbird. Knowledge in our understanding of one of these systems can probably be used to fill in the gaps in our understanding of the other. For example, the classical accounts of BG interpret the direct pathway as the 'Go' pathway, since it facilitates movement, and the indirect pathway as the 'NoGo' pathway, since it inhibits movement. Since the AFP, the avian homolog of BG, is only a single pathway, it is not clear whether it is comparable to the direct or the indirect pathway of mammalian BG. From microanatomical studies of AFP, Farries & Perkel, (2002) suggest that AFP is best considered as a mixture of the direct and indirect pathways, and Area X has both striatal and pallidal features. Extending the anatomical comparison to that of function, if we consider that the AFP's inputs to RA serve as an Explorer or an experimenter, it would be legitimate to ask if and where such an Explorer is located in mammalian BG. The question is of significance since classical treatments only speak of 'Go' and 'NoGo,' with no sign of an Explorer in the picture.

We have been developing a model of mammalian BG in which the indirect pathway serves the role of an Explorer (Chakravarthy & Balasubramani, 2014; Chakravarthy, Joseph, & Bapi, 2010; Joseph, Gangadhar, & Chakravarthy, 2010; Magdoom et al., 2010; Sridharan, Prashanth, & Chakravarthy, 2004). Dopamine signals to the Str are believed to switch cortico-striatal transmission between the direct and indirect pathways, with higher dopamine levels activating the direct pathway ('Go'), and lower dopamine levels activating the indirect pathway ('NoGo'). Between the 'Go' and 'NoGo' regimes, we posit an intermediate 'Explore' regime which corresponds to moderate dopamine levels. In this regime, the indirect pathway generates stochastic, exploratory signals, analogous to the AFP in songbird. A recent model of the role of BG in reaching movements embodies these modeling principles and also exhibits PD symptoms like tremor and bradykinesia under dopamine-deficient conditions (Magdoom et al., 2010).

Thus, mammalian BG and avian AFP seem to have deep functional similarities, in addition to the known anatomical resemblances. A more in-depth study of functional similarities might clarify and resolve points of ambiguity in both systems. The two systems indeed seem to have a deep functional similarity in that both serve as engines for learning by reinforcement. With this fundamental guiding principle, a comprehensive modeling effort that goes hand in hand with experimental investigations might soon bring about a decisive progress in our understanding of these two crucial brain circuits.

References

Abarnel, H. D. I., Gibb, L., Mindlin, G. B., & Talathi, S. (2004). Mapping neural architectures onto acoustic features of birdsong. *Journal of Neuroscience, 92*(1), 96–110.

Appeltants, D., Absil, P., Balthazart, J., & Ball, G. F. (2000). Identification of the origin of catecholaminergic inputs to HVc in canaries by retrograde tract tracing combined with tyrosine hydroxylase immunocytochemistry. *Journal of Chemical Neuroanatomy, 18,* 117–133.

Appletants, D., Ball, G. F., & Balthazart, J. (2002). The origin of catecholaminergic inputs to the song control nucleus RA in canaries. *NeuroReport, 13,* 649–653.

Balasubramani, P. P., Chakravarthy, V. S., Ali, M., Ravindran, B., & Moustafa, A. A. (2015). Identifying the basal ganglia network model markers for medication-induced impulsivity in Parkinson's Disease patients. *PLoS ONE, 10,* e0127542.

Barto, A. G. (1995). Adaptive Critics and the Basal Ganglia. In J. C. Houk, J. Davis, & D. Beiser (Eds.), *Models of information processing in the basal ganglia* (pp. 215–232). Cambridge, MA: MIT Press.

Basham, M. E., Nordeen, E. J., & Nordeen, K. W. (1996). Blockade of NMDA receptors in the anterior forebrain impairs sensory acquisition in the zebra finch (Poephila guttata). *Neurobiology of Learning and Memory, 66*(3), 295–304.

Bergman, H., Wichmann, T., Karmon, B., & DeLong, M. R. (1994). The primate subthalamic nucleus. II. Neuronal activity in the MPTP model of parkinsonism. *Journal of Neurophysiology, 72,* 507–520.

Berns, G. S., & Sejnowski, T. J. (1995). A computational model of local memory in the primate pallidal-subthalamic circuit. *Soc. Neurosc. Abstracts, 21,* 678.

Berns, G. S., & Sejnowski, T. J. (1998). A computational model of how the Basal ganglia produce sequences. *Journal of Cognitive Neuroscience, 10,* 108–121.

Bolhuis, J. J., & Moorman, S. (2015). Birdsong memory and the brain: In search of the template. *Neuroscience and Biobehavioral Reviews, 50C,* 41–55.

Bottjer, S. W. (1993). The distribution of tyrosine hydroxylase immunoreactivity in the brains of male and female zebra finches. *Journal of Neurobiology, 24,* 51–69.

Bottjer, S. W., Meisner, E. A., & Arnold, A. P. (1984). Forebrain lesions disrupt development but not maintenance of song in passerine birds. *Science, 224,* 901–903.

Brainard, M. S., & Doupe, A. J. (2002). What songbirds teach us about learning. *Nature, 417,* 351–358.

Brown, P., Olivero, A., Mazzone, P., Insola, A., Tonali, P., & Lazzaro, V. D. (2001). Dopamine dependency of oscillations in between subthalamic nucleus and pallidum in Parkinson's disease. *Journal of Neuroscience, 21,* 1033–1038.

Carrilo, G. D., & Doupe, A. J. (2004). Is the songbird Area X striatal, pallidal, or both? An anatomical study. *The Journal of Comparative Neurology, 473,* 415–437.

Chakravarthy, V. S., & Balasubramani, P. P. (2014). Basal ganglia system as an engine for exploration. In R. Jung & D. Jaeger (Eds.), *Encyclopedia of computational neuroscience.* Berlin, Heidelberg: Springer.

Chakravarthy, V. S., Joseph, D., & Bapi, R. S. (2010). What do the Basal Ganglia Do? A modeling perspective. *Biological Cybernetics.*

Cossette, M., Lévesque, M., & Parent, A. (1999). Extrastriatal dopaminergic innervation of human basal ganglia. *Neuroscience Research, 34,* 51–54.

Crook, S., & Cohen, A. (2003). Central pattern generators. The book of genesis (Internet edition).

Dave, A. S., & Margoliash, D. (2000). Song replay during sleep and computational rules for sensorimotor vocal learning. *Science, 290,* 812–816.

Ding, L., & Perkel, D. J. (2002). Dopamine modulates excitability of spiny neurons in the avian basal ganglia. *Journal of Neuroscience, 22,* 5210–5218.

Doupe, A. J. (1997). Song –and order – selective neurons in the songbird anterior forebrain and their emergence during vocal development. *Jounal of Neuroscience, 17*(3), 1147–1167.

Doupe, J. A., & Kuhl, K. P. (1999). Birdsong and human speech: Common themes and mechanisms. *Annual Review of Neuroscience, 22,* 567–631.

Doupe, A. J., Perkel, D. J., Reiner, A., & Stern, E. A. (2005). Birdbrains could teach basal ganglia research a new song. *Trends in Neurosciences, 28,* 353–363.

Doya, K., & Sejnowski, J. T. (1995). A novel reinforcement model of birdsong vocalization learning. *Advances in Neural Information Processing Systems, 7,* 101–108.

Doya, K., & Sejnowski, T. J. (1998). A computational model of birdsong learning by auditory experience and auditory feedback. *In Central auditory processing and neural modeling* (pp. 77-88). Springer, Boston, MA.

Farries, M. A., & Perkel, D. J. (2002). A telencephalic nucleus essential for song learning contains neurons with physiological characteristics of both striatum and globus pallidus. *Jounal of Neuroscience, 22,* 3776–3787.

Fiete, I. R., Fee, M. S., & Seung, H. S. (2007). Model of birdsong learning based on gradient estimation by dynamic perturbation of neural conductances. *Journal of Neurophysiology, 98,* 2038–2057.

Gadagkar, V., Puzerey, P. A., Chen, R., Baird-Daniel, E., Farhang, A. R., & Goldberg, J. H. (2016). Dopamine neurons encode performance error in singing birds. *Science, 354*(6317), 1278–1282.

Gale, S. D., & Perkel, D. J. (2005). Properties of dopamine release and uptake in the songbird basal ganglia. *Journal of Neurophysiology, 93,* 1871–1879.

Gardner, T., Cecchi, G., Magnasco, M., Laje, R., & Mindlin, G. B. (2001). Simple motor gestures for birdsongs. *Physical Review Letters, 87,* 208101.

Hahnloser, R. H., & Kotowicz, A. (2010). Auditory representations and memory in birdsong learning. *Current Opinion in Neurobiology, 20,* 332–339.

Hara, E., Kubikova, L., Hessler, N. A., & Jarvis, E. D. (2007). Role of the midbrain dopaminergic system in modulation of vocal brain activation by social context. *European Journal of Neuroscience, 25,* 3406–3416.

Harding, C. F. (1998). Changes in catecholamine levels and turnover rates in hypothalamic, vocal control and auditory nuclei in male zebra finches during development. *Journal of Neurobiology, 34,* 329–346.

Harel, B. T., Cannizzaro, M. S., Cohen, H., Reilly, N., & Snyder, P. J. (2004). Acoustic characteristics of Parkinsonian speech: A potential biomarker of early disease progression and treatment. *Journal of Neurolinguistics, 17,* 439–453.

Hessler, N. A., & Doupe, A. J. (1999). Singing-related neural activity in a dorsal forebrain-basal ganglia circuit of adult zebra finches. *Journal of Neuroscience, 19,* 10461–10481.

Hoese, W., Podos J, Boetticher, N. C., & Nowicki, S. (2000, June). Vocal tract function in birdsong production: Experimental manipulation of beak movements. *Journal of Experimental Biology, 203*(12), 1845–1855.

Iyengar, S., & Bottjer, S. J. (2002). Development of individual axon arbors in a thalamocortical circuit necessary for song learning in zebra finches. *Journal of Neuroscience, 22*(3), 901–911.

Joseph, D., Gangadhar, G., Chakravarthy, V. S. ACE (Actor—Critic—Explorer) paradigm for reinforcement learning in basal ganglia: Highlighting the role of sub-thalamic and pallidal nuclei. *Neurocomputing,* (2010, in press).

Kao, M. H., Doupe, A. J., & Brainard, M. S. (2005). Contributions of an avian basal ganglia-forebrain circuit to real-time modulation of song. *Nature, 433,* 638–643.

Konishi, M. (1965). The role of auditory feedback in the control of vocalization in the white-crowned sparrow. *Zeitschrift fur Tierpsychologie, 22,* 770–783.

Kubikova, L., & Kostal, L. (2009). Dopaminergic system in birdsong learning and maintenance. *Journal of Chemical Neuroanatomy*.

Kuhl, P. K. (1994). Learning and representation in speech and language. *Current Opinion in Neurobiology, 4*, 812–822.

Leonardo, A. (2004). Experimental test of the birdsong error-correction model. *Proceedings of the National Academy of Sciences of the United States of America, 101*, 16935–16940.

Lewis, J. W., Ryan, S. M., Arnold, A. P., & Butcher, L. L. (1981). Evidence for a catecholarninergic projection to area X in the zebra finch. *Journal of Comparative Neurology, 196*, 347–354.

Luo, M., Ding, L., & Perkel, D. J. (2001). An avian basal ganglia pathway essential for vocal learning forms a closed topographic loop. *Journal of Neuroscience, 21*, 6836–6845.

Magdoom, K. N., Subramanian, D., Chakravarthy, V. S., Amari, S.-I., & Meenakshisundaram, N. Modelling basal ganglia for understanding Parkinsonian reaching movements. *Neural Computation*, (2010, in press).

Magnin, M., Morel, A., & Jeanmonod, D. (2000). Single-unit analysis of the pallidum, thalamus and subthalamic nucleus in Parkinsonian patients. *Neuroscience, 96*, 549–564.

Margoliash, D. (2002). Evaluating theories of bird song learning: Implications for future directions. *Journal of Comparative Physiology. A, Sensory, Neural, and Behavioral Physiology, 188*, 851–866.

Marler, P., & Peters, S. (1977). Selective vocal learning in a sparrow. *Science, 198*, 519–521.

McCasland, J. S. (1987). Neuronal control of bird song production. *Jounal of Neuroscience,7*, 23–39.

Nini A., Feingold, A., Slovin, H., & Bergman, H. (1995). Neurons in the globus pallidus do not show correlated activity in the normal monkey, but phase-locked oscillations appear in the MPTP model of Parkinsonism. *Journal of Neurophysiology, 74*, 1800–1805.

Nottebohm, F., Stokes, T. M., & Leonard, C. M. (1976). Central control of song in the canary, Serinus canarinus. *Journal of Comparative Neurology, 165*, 457–486.

Reiner, A., Perkel, D. J., Mello, C. V., & Jarvis, E. D. (2004). Songbirds and the revised avian brain nomenclature. *Annals of the New York Academy of Sciences, 1016*, 77–108.

Righetti, L., Buchli, J., & Ijspeert, A. J. (2006). Dynamic Hebbian learning in adaptive frequency oscillators. *Physica D: Nonlinear Phenomena, 216*, 269–281.

Sakai, K., Kitaguchi, K., & Hikosaka, O. (2003). Chunking during human visuomotor sequence learning. *Experimental Brain Research, 152*, 229–242.

Scharff, C., & Nottebohm, F. (1991). A comparative study of the behavioral deficits following lesions of various parts of the zebra finch song system: Implications for vocal learning. *Journal of Neuroscience, 11*, 2896–2913.

Schultz, W. (1998). Predictive reward signal of dopamine neurons. *Journal of Neurophysiology, 80*, 1–27.

Soha, J. A., Shimizu, T., & Doupe, A. J. (1996). Development of the catecholaminergic innervation of the song system of the male zebra finch. *Journal of Neurobiology, 29*, 473–489 (464951).

Sohrabji, F., Nordeen, E. J., & Nordeen, K. W. (1990). Selective impairment of song learning following lesions of a forebrain nucleus in the juvenile zebra finch. *Behavioral and Neural Biology., 53*, 51–63.

Sridharan, D., Prashanth, P. S., & Chakravarthy, V. S. (2004). The role of the basal ganglia in exploratory behavior in a model based on reinforcement learning. In N. R. Pal, N. Kasabov, R. K. Mudi, S. Pal, & S. K. Parui (Eds.), *International Conference on Neural Information Processing, Lecture Notes in Computer Science (LNCS)* (Vol. 3316, pp. 70–77). Berlin: Springer.

Sutton, R. S., & Barto, A. G. (1998). *Reinforcement learning: An introduction*. Cambridge, MA: MIT Press.

Trevisan, M. A., Eguia, M. C., & Mindlin, G. B. (2001). Non-linear aspects of analysis and synthesis of analysis and synthesis of speech time series data. *Physical Review E, 63*, 026216.

Troyer, T. W., & Doupe, A. J. (2000). An associational model of birdsong sensorimotor learning: Efference copy and the learning of song syllables. *Journal of Neurophysiology, 84,* 1024–1223.
Vu, E. T., Mazutek, M. E., & Kuo, Y.-C. (1994). Identification of a forebrain motor programming network for the learned song of zebra finches. *Journal of Neuroscience, 14,* 6924–6934.
Yanagihara, S., & Yazaki-Sugiyama, Y. (2016, June 21). Auditory experience-dependent cortical circuit shaping for memory formation in bird song learning. *Nature Communications*.
Yu, A. C., & Margoliash, D. (1996). Temporal hierarchical control of singing in birds. *Science, 273*, 1871–1875.

Chapter 15
The Basal Ganglia: Summary and Future Modeling Research

V. Srinivasa Chakravarthy and Ahmed A. Moustafa

Abstract The pivotal idea of the Go-Explore-NoGo (GEN) approach to BG modeling, expounded in this book, is the hypothesis that the complex dynamics of the STN–GPe loop can introduce certain randomness in the action selection mechanisms that occur downstream in other BG areas as well as the cortex, thalamus, and other subcortical structures. The indirect pathway, that consists of the STN–GPe loop, in addition to its classical role as a source of movement inhibition, also serves as a source of randomness, and therefore, in the jargon of reinforcement learning, can control the levels of exploration in action selection. In this chapter, we summarize how the Go-Explore-NoGo approach can account for several functions of the basal ganglia. We also provide a few ideas on how to use the Go-Explore-NoGo approach to simulate behavioral performance in other basal ganglia-related functions (e.g., attention, working memory, episodic memory) and disorders (e.g., schizophrenia).

For more details on the GEN approach, the reader is referred to Chap. 5 in this book. Modeling studies described in various prior chapters in this book have further shown that as the tonic dopamine levels in the striatum are varied continuously, action selection exhibits three regimes for three different ranges of dopamine. For high dopamine levels, the model tends to choose the optimal (highest value) output (analogous to 'Go'); for low dopamine levels, no action is selected (analogous to 'NoGo'); but for intermediate levels of dopamine the model selects a nonoptimal action, a selection which randomly varies from trial to trial. It is this new facet of the model that emerges naturally out of the dynamics of STN–GPe, without explicitly introducing stochastic elements, that is being proposed to underlie exploratory action selection dynamics of the BG. Thus, we have incorporated in our modeling approach the so-called GEN approach according to which the action selected dynamics of BG can be expressed in terms of the three regimes: Go, Explore, and NoGo.

Our approach to modeling BG described in Chap. 5 is built around value-based decision making. In Chap. 12, we extended this approach to utility-based decision making. Utility is a linear sum of value and risk, where risk is defined as the

V. S. Chakravarthy and A. Moustafa, *Computational Neuroscience Models of the Basal Ganglia*, Cognitive Science and Technology,
https://doi.org/10.1007/978-981-10-8494-2_15

expected value of the square of the temporal difference error. This modeling expansion accommodated another neuromodulator, serotonin, which is known to have a multitude of functions in the BG. Specifically, we hypothesized that the coefficient of the risk term in the utility function, the risk sensitivity, α, is the neural correlate of the action of serotonin in the BG. This model, which combines the roles of both dopamine and serotonin in BG, accounts for several functions of serotonin. Currently, there are three prominent theories of the role of serotonin in BG: (1) risk-sensitive decision making, where serotonin controls risk assessment, (2) temporal reward prediction, where serotonin controls timescale of reward prediction, and (3) reward/punishment sensitivity, in which the punishment prediction error depends on serotonin levels. Our model reconciles all the three functions of serotonin in a single framework.

The above-described model of the action of dopamine and serotonin has been recently extended to model bipolar disorder (Balasubramani & Chakravarthy, 2017). Bipolar disorder is characterized by mood swings—oscillations between manic and depressive states. These swings (oscillations) mark the length of an episode in a patient's mood cycle (period) and can vary from hours to years. Our modeling study uses decision-making framework to investigate the role of basal ganglia network in generating bipolar oscillations. In this model, the basal ganglia system performs a simple two-arm bandit task (Bourdaud, Chavarriaga, Galán, & del R Millan, 2008; Daw, O'Doherty, Dayan, Seymour, & Dolan, 2006), consisting of probabilistic positive and negative rewards as outcomes of each of the arms (states); they correspond to the model's positive and negative mood states, respectively. The outcomes are probabilistic with probability = 0.5. We chose this task as it is simple enough for observing the oscillatory effect in decision-making framework between positive and negative states. Here, we study the utility function approach in more detail, focusing on the influence of subjective reward and risk sensitivity in the overall choice selection dynamics between positive and negative mood states. In the first model (A), we build value, risk, and utility functions from classic RL strategies for positive and negative states and use softmax policy (Sutton & Barto, 1998) to choose between actions. Then, in model (B), we extend the concepts to a more detailed network model of the BG, with abstract activities of D1 receptor-expressing medium spiny neurons (D1 MSNs) of striatum for computing value, and the D1 and D2 receptor co-expressing MSNs for computing risk function. The direct and indirect pathways of BG, encompassing STN, GPe, GPi, and thalamus, implement the selection strategy to choose between utilities of positive and negative states. In both the models, we associate putative mechanisms driving the selection dynamics with bipolar-like oscillations between mood states. The bipolar oscillations, it must be noted, are obviously different from the pathological oscillations of STN–GPe dynamics that are exhibited in diseases such as Parkinson's (Gillies, Willshaw, & Li, 2002; Weinberger, Hutchison, Lozano, Hodaie, & Dostrovsky, 2009; Willshaw & Li, 2002). While the STN–GPe oscillations are in the range of Hertz, the bipolar oscillations span over months and years. Finally, a reduced dynamical system model (C) consisting of a simple two-variable system, that captures the essential dynamics of both the above models, is presented and the correspondences between the key parameters of different models are discussed.

Since it is generally considered that the central function of BG is action selection, several computational BG models address action, which is often restricted to discrete or binary action selection. However, several motor functions of the BG involve operating in continuous spaces—reaching, gait, saccades, spatial navigation, among others. Therefore, there is an obvious need to build BG models that operate in continuous action spaces, a topic that has been often neglected in prior models of the basal ganglia (Joel, Niv, & Ruppin, 2002). In Chap. 5, we extended the discrete action selection framework to continuous spaces. The discrete values associated with discrete actions now give place to a value function defined over continuous state spaces. The GEN approach applied to such continuous value function can be reduced, for computational convenience, to a three-step algorithm, which can be interpreted as a form of stochastic hill-climbing over the value landscape, a process that is analogous to simulated annealing (Kirkpatrick, Gelatt, & Vecchi, 1983). The above-mentioned general framework has been extended to explain a wide variety of BG functions including reaching, gait, precision grip, song learning, and action selection. Extensions of the above-mentioned framework that substitute the value function with a utility function, which is a combination of value and risk, have been used to model the joint functions of dopaminergic and serotonergic systems in the BG.

We now explain how the aforementioned modeling approach can be used to explain several motor and some non-motor functions of BG. This is a guide for computational modeling neuroscientists who aim to develop models of the basal ganglia.

15.1 Applying the BG Model to Various Behavioral Processes

In Chap. 10, we describe a model of the role of BG in generating reaching movements. The model consists of a cortical loop including the motor cortex, proprioceptive cortex, a layer of neurons representing the spinal cord and a simple arm model closing the loop. The cortical loop is first trained using unsupervised learning. Once the cortical loop is trained, the activation of a specific part of the motor cortex places the arm at a specific point in its workspace. Now the BG and the prefrontal cortex are added to the cortical loop to introduce goal-oriented reaching. The distance between the target and the BG system is represented as a value function in the BG system: The goal information is projected to the striatum from the prefrontal cortex, and information about the actual location of the hand comes from the sensory cortical projections to the striatum. The cortico-basal ganglia dynamics drives the hand up the value gradient following the GEN policy taking the hand closer to the goal. Once the hand approaches the goal sufficiently closely, the final muscle activations of the hand at that state are used to train the prefrontal to motor cortical connections. Thus, the muscle activations required to reach a given target are discovered by the BG dynamics by a slow search process and transferred progressively to the cortical pathway by. Thus, the model's architecture and function are consistent with the classical perspective of the role of BG in motor learning that motor learning first occurs in the BG and gradually transferred to the cortex. This gradual transfer from BG to the cortex is implemented by two coefficients that represent the level of influence of the BG and the prefrontal cortex on the movement generated by the model. Parkinson's disease is simulated by suppressing the dopamine signal represented by the temporal difference error, and also by altering the connectivity of the STN–GPe system so as to increase the synchronization levels of that system. The simulated Parkinson's disease model shows both akinetic rigidity and tremor for various parameter settings.

Chapter 10 describes a model of song learning in birds, specifically in zebra finches. Birds have a BG homolog, known as the Anterior Forebrain Pathway (AFP), although it is slightly different from mammalian BG in detailed anatomy. The AFP consists of a chain of modules including Area X, Dorso-Lateral Thalamus (DLM), and Lateral Magnocellular Nucleus of the Anterior Neostriatum (LMAN). Then there is the motor pathway, analogous to the sensory-motor cortical pathway mammalian brains, consisting of two regions: hyperstriatum ventrale pars caudalis (HVc) and Robust nucleus of the Archistriatum (RA). Song learning in many species of birds is divided into two phases. In the first phase, the sensory phase, the male bird listens to the tutor song of another male bird in the colony and memorizes some aspect of the tutor song. In the second phase, the motor learning phase, the bird establishes the songs learnt earlier by rehearsal aided by auditory self-feedback. We hypothesize that: (1) the songbird learns only *evaluations* of songs during the sensory phase; (2) the AFP plays a role analogous to the Explorer, a key component in reinforcement learning (RL); (3) the motor pathway learns the song by combining the evaluations (value information) stored from the sensory phase, and the exploratory inputs from the AFP in a temporal stage-wise manner. Although there are differences in anatomical features between the avian BG homolog and mammalian BG, at an abstract level, there are strong functional similarities. Here too, as in the earlier cases that describe motor learning in mammalian brains, there is a BG homolog that computes the value and also supplies exploratory drive to the main pathway—the motor pathway. Thus, the BG homolog in the avian case discovers the correct output by a slow and gradual search process and transfers the results of the search to the motor pathway for rapid learning.

In Chap. 7, we develop modeling strategies to elucidate gait abnormalities in PD specifically freezing of gait, which is characterized by sudden and paroxysmal cessation of locomotion. It is often triggered by certain environmental contexts which include approaching narrow doorways/passages, turns, and also during movement-related scenarios such as movement initiation and dual-tasking. The models are driven by the hypothesis that FOG is not a motor problem, but a problem of the evaluation of space in PD subjects, as the contextual features which usually trigger freezing include turning environments, narrow doorways, obstacles among others. We model the cortico-BG network to understand the role of these contexts on freezing behavior. We propose the concept of 'value of space' which represents a value function an agent (simulated entity) builds while exploring a given environment. This value function is used by the motor machinery to perform appropriate actions in different scenarios, like narrow doorways and winding passages. In this work, we model the cortico-BG-spinal system to explain the Parkinsonian gait impairment. The spinal areas are modeled as central pattern generators which produce the necessary rhythmic output for sustaining locomotion. We study the role of higher level areas on lower level mechanisms leading to gait regulation and its breakdown in PD freezers (subjects who show freezing). We further extend this model to study the role of cognition in freezing. Cognitive tasks which are often coupled with normal motor processes (dual-tasking) evoke freezing in PD. We attempt to understand freezing by modeling multiple cortico-BG loops, viz. motor and cognitive loops and the effect of their interaction on gait movements. We also model the influence of turning on gait, another factor known to trigger freezing. The modeling efforts suggest that freezing is perhaps not exclusively dependent on dopamine insufficiency in PD, and there is a plausible role of other neuromodulators, especially serotonin and norepinephrine which could precipitate this episodic behavior.

Chapter 8 presents a model of precision grip performance in normal and PD conditions. Precision grip refers to the task of holding a small object by gripping it between the thumb and the index finger. Precision grip performance involves a threshold effect: Precision grip succeeds only if the grip force exceeds a certain threshold, known as the slip force. Otherwise the object slips and falls. The problem, therefore, lends itself naturally to be treated using risk-based decision-making methods. Close to the slip force, even a small fluctuation in grip force can lead to slippage and fall. This large variability in the outcome close to the slip force is associated with high risk. Therefore, grip force must be maintained sufficiently higher than the threshold value of slip force, so as to minimize risk. The utility-based decision-making approach of Chap. 12 is ideally suited to model the grip force changes in normal and PD conditions. Experimental studies on grip force generation in PD patients show an increase in grip force during ON medication and an increase in the variability of the grip force during OFF medication (Fellows, Noth, & Schwarz, 1998; Ingvarsson, Gordon, & Forssberg, 1997). Since there are grip force changes in PD conditions, the role of BG in precision grip performance is strongly indicated. The model consists of two components: (1) the sensory-motor loop component and (2) the Basal Ganglia component. The sensory-motor loop component converts a reference position and a reference grip force into lift force and grip force profiles, respectively. The lift force and grip force work together in lifting a load. The sensory-motor loop component also includes a plant model that represents the interaction between two fingers involved in PG and the object to be lifted. The proposed model is able to account for the precision grip results from normal and PD patients accurately (Fellows et al., 1998; Ingvarsson et al., 1997).

We had also extended the approach to BG modeling developed in this book to model the role of BG in spatial navigation. This work was published elsewhere and not discussed in this book (Sukumar, Rengaswamy, & Chakravarthy, 2012). It is generally thought that the hippocampus is the primary subcortical structure involved in spatial navigation. However, there is evidence that BG and hippocampus actually cooperate in driving spatial navigation (Packard & Knowlton, 2002). In fact, more recent evidence suggests the involvement of cerebellum in spatial navigation, particularly in integrating vestibular signals with other sensory signals like visual, auditory, or proprioceptive signals that aid navigation (Rochefort, Lefort, & Rondi-Reig, 2013). Elsewhere we had described a model of the combined roles of BG and hippocampus in spatial navigation (Sukumar et al., 2012). The model describes the BG as subserving cue-based navigation, while the hippocampus subserves place-based navigation. The model was applied to a standard spatial navigation set up like the Morris water maze. Place is coded by the spatial context represented by a set of poles uniformly placed around the rim of the circular water maze. When the simulated animal reaches the hidden platform, it is given a positive reward; when it hits the surrounding wall it is given a punishment. On exploration of the maze, the simulated animal constructs a value landscape of the ambience in terms of the spatial context represented by the array of poles. The model was able to account for the experimental results of using the Morris water maze (Devan & White, 1999). Under Parkinsonian conditions, the model also exhibited impaired spatial learning characterized by longer escape latencies (Miyoshi et al., 2002).

Our BG modeling approach was also extended to model the role of BG in saccade generation (Krishnan, Ratnadurai, Subramanian, Chakravarthy, & Rengaswamy, 2011). Although there is extensive evidence supporting the role of BG in saccade generation, there are very few computational models simulating this function (Dominey & Arbib, 1992). One of the output ports of BG, the substantia nigra pars reticulate (SNr), influences saccades via the Superior Colliculus (SC), a key nucleus that controls eye movements. Hikosaka and colleagues have shown that the caudate–SNr–SC pathway can be trained to produce reward-based saccadic movements (Isoda & Hikosaka, 2008). Saccades were found to be prolonged with reduced velocities in MPTP monkeys (Kori et al., 1995). As with the other BG models, our BG model for saccade generation was cast within the reinforcement learning (RL) framework, with the dopamine representing the temporal difference error, the striatum serving as the critic, and the indirect pathway playing the role of the Explorer. The model captures experimentally observed performance on standard saccade tasks like feature and conjunction searches, directional selectivity, and a successive saccade task. Under Parkinson's conditions, the model also exhibited longer saccade reaction times and prolonged saccades, consistent with patient performance.

In this book, we have primarily focused on motor functions of the basal ganglia, in normal and pathological conditions. However, the motor deficits in pathology of the basal ganglia are not confined to the musculoskeletal system and extend also to motor apparatus that controls speech. Motor deficits in PD include those of respiration, phonation, and articulation and therefore affect speech production (Harel, Cannizzaro, & Snyder, 2004). Speech motor deficits in PD are manifested as dysarthria, a disorder of spoken communication due to central or peripheral nervous system damage, and are associated with disturbance in the muscular control of speech. It is characterized by a monotony of pitch and loudness, reduced stress, variable rate, imprecise consonants, and a generally poor level of articulation (Pinto et al., 2004). In Chap. 14, we described a model of song learning and production in songbirds. An interesting modeling challenge for the future is to extend the song learning model described in this book, to model speech impairment in PD patients. The fact that the basal ganglia have a role in speech production, anticipates a role for this subcortical circuit in general language functions. BG models of sequence learning can be extended to model language functions, since language processing is an instance of sequential information processing. Thus, modeling effort of the basal ganglia, starting from modeling of motor functions, to modeling sequence processing, proceeding to modeling language function, is a logical path along with modeling can progress. Such a manner of progression is meaningful even from the point of view of the so-called motor theories of language origins (Allott, 1992). According to this theory, language had originated by a new combination of pre-existing neural motor elements, by a redirection of neural programs which were hitherto occupied with movement coordination, to be extended to the muscles of the mouth, throat, and chest among others in order to produce speech sounds. Thus, the modeling framework laid out in this book, to explain motor functions of the basal ganglia, can perhaps be effectively extended to model also the language functions of the basal ganglia.

The Go-Explore-NoGo approach to BG modeling can also be expanded to cognitive processes, such as attention, working memory, and episodic memory. In attentional processes, one needs the Go mechanism to select which information in the environment to pay attention to. But in addition, we also learn not to pay attention to certain cues in the environment, such as task-irrelevant information, which require the NoGo mechanism (Moustafa, 2015). Impairment of the NoGo mechanism of attention can lead to excessive attentional performance to several cues in the environment, which is a key feature of schizophrenia (Bergman, O'Brien, Osgood, & Cornblatt, 1995; Morris, Griffiths, Le Pelley, & Weickert, 2013; Pankow et al., 2016). Impaired dopamine projections to the basal ganglia underlie attentional deficits in schizophrenia (Mehler-Wex, Riederer, & Gerlach, 2006). In addition, choosing which environmental cues to pay attention to requires one to search for important cues in the environment; such search is based on the Explore mechanism, as what we pay attention to in the environment is not always fixed. Accordingly, future modeling work should incorporate the Go-Explore-NoGo approach to simulate attentional task performance in healthy people and also patients with schizophrenia. The same mechanism can also be used to simulate attentional impairment in ADHD, a disorder that also involves degeneration of dopamine projections to the basal ganglia (Mehler-Wex et al., 2006). The same exact mechanism can also be used to simulate working memory, as both attentional and working memory processes rely on similar neural mechanisms (LaBar, Gitelman, Parrish, & Mesulam, 1999; Mayer et al., 2007). The same Go-Explore-NoGo policy can be used to simulate episodic memory. Some studies have investigated memory retrieval vs. memory suppression, that is, to understand the Go mechanism to retrieve memory or NoGo mechanism to suppress memory retrieval (Anderson & Green, 2001; Benoit & Anderson, 2012; Levy & Anderson, 2008). The importance of inhibiting memory retrieval (NoGo) is linked to trauma-related disorders, such as post-traumatic stress disorder (PTSD), when it is potentially important not to remember negative life events. While it has been reported that different parts of the prefrontal cortex play a role in memory retrieval and memory suppression (Depue, Curran, & Banich, 2007), to our knowledge it is not known whether the basal ganglia play a role in these processes as well. Future modeling work can also incorporate the Go-Explore-NoGo mechanism to simulate episodic memory in healthy individuals and also patients with PTSD.

15.2 Clinical Applications

Throughout the book, the models described have been used to describe basal ganglia function in both normal and disease conditions, particularly under Parkinsonian conditions. But the power of the models does not stop at explaining the impaired function in disease conditions; our models can be extended to explain the effect of various therapeutic measures employed to treat the disease. Two prominent therapeutic modalities for Parkinson's disease are drugs and deep brain stimulation (DBS); both of these approaches have been included in the models described in this book.

The Izhikevich spiking network model (Mandali, Rengaswamy, Chakravarthy, & Moustafa, 2015) was used to study impulsivity in PD OFF and ON conditions using cognitive tasks, such as the Iowa gambling task (IGT) and probabilistic learning task (PLT) (Mandali & Chakravarthy, 2016; Mandali, Chakravarthy, Rajan, Sarma, & Kishore, 2016) (see Chap. 11). The model showed that the IGT performance was poor in PD condition (both OFF and ON medication) compared to healthy condition, with worse performance in ON medication condition. The simulated PD ON condition does not learn from its action outcomes (rewards/ punishments) and wanders among the decks, which is reflected in the negative IGT score. Physiologically this negative behavior is attributed to excess DA levels in the striatum (Frank, Samanta, Moustafa, & Sherman, 2007). In the model, striatal weights were positively updated even in punishment situation due to dopaminergic medication (δ_{med}), leading to the selection of wrong choice. Stimulation decreased the IGT performance compared to ON condition. We then changed the position of electrode within the STN nucleus and observed a significant change in IGT score. Apart from position, another parameter thought to influence cognition was observed to be current amplitude. With the above results, one can consider the possibility that stimulation current when applied to topographical areas within the STN might lead to inhibition/facilitation of the corresponding panel selection depending on the current amplitude.

In the probabilistic learning task, the model's ability to differentiate between a low and high rewarding choice in each of the physiological and pathological conditions (PD OFF, PD medicated conditions (L-DOPA and dopamine agonist)) was tested. This behavior in PD OFF condition was also experimentally observed where PD patients under medication tend to learn more from rewards than punishments (Frank, Seeberger, & O'reilly, 2004; Hazy, Frank, & O'Reilly, 2007). This can be accounted for by the medication term ($\delta_{\text{med}} = 3$) which prevents the dip selection of punitive choices. The model's performance in the dopamine agonist condition did not yield good accuracy either in reward learning but performed better than L-DOPA condition in punishment learning. We then studied the performance of the model in DBS condition and observed that the performance was dependent on the electrode position. As the position of electrode was changed, the model switched from reward-based to punishment-based learning. Based on the experimental data that an increase in STN activity was observed during high-conflict conditions, we analyzed the reaction time for each of the conditions in low-conflict and high-conflict cases in each of the five conditions. We observed that the model in normal conditions took more time to make a choice during high-conflict case compared to that in low-conflict in both correct and error trials. The impulsivity behavior observed clinically due to dopamine agonist medication (Ondo & Lai, 2008; Voon et al., 2007) was captured by the model wherein it was observed that a lower reaction time for high-conflict case. We observed that the reaction times were different for different electrode positions and a lower reaction time was obtained for high-conflict case during both correct and error trials for a specific electrode position (Pos 3) in the DBS condition.

One of the aims of building computational models is to understand the mechanisms of pathophysiology and aid in their diagnosis and treatment. It is of utmost importance to translate the developed models such that they can be used clinically to provide patient/symptom-specific intervention. There is a need to build computational models with reasonable details which can predict the differences in pathophysiological features within patients that could lead to their symptomatic differences. These differences in the estimated model's parameters within subjects would then correspond to various cellular parameters such as neurotransmitter levels and anatomical architecture, which can be used to optimally target and improve patient's clinical condition. The development of comprehensive, multiscale, and closed-loop models to better understand and deliver DBS in a clinical setting has strong analogies with the problem of understanding drug effects on behavior.

Even in the domain of modeling pharmaceutical effects of neurological drugs, there is a need to develop comprehensive, multiscale models of the brain that capture drug effects from the molecular level to behavior. Such an approach to modeling drug action has been identified as a whole new field known as computational neuropharmacology (Aradi & Érdi, 2006). Though computational models of drug action often confine themselves to modeling drug–receptor interactions (Jorgensen, 2004), its ultimate aim is to understand drug's effect on symptom manifestation; its effect at the molecular is only instrumental in ensuring the desirable effect at the level of the symptom. Our modeling approach used in this book can be also used to simulate drug effects on the treatment of various basal ganglia-related disorders.

References

Allott, R. (1992). Language origin: A multidisciplinary approach. In J. Wind (Ed.), (pp. 105–119). Dordrecht: Kluwer.

Anderson, M. C., & Green, C. (2001). Suppressing unwanted memories by executive control. *Nature, 410*(6826), 366–369. https://doi.org/10.1038/35066572.

Aradi, I., & Érdi, P. (2006). Computational neuropharmacology: Dynamical approaches in drug discovery. *Trends in Pharmacological Sciences, 27*(5), 240–243.

Balasubramani, P. P., & Chakravarthy, S. (2017). Bipolar oscillations between positive and negative mood states in a computational model of Basal Ganglia. *bioRxiv*, 205310.

Benoit, R. G., & Anderson, M. C. (2012). Opposing mechanisms support the voluntary forgetting of unwanted memories. *Neuron, 76*(2), 450–460. https://doi.org/10.1016/j.neuron.2012.07.025.

Bergman, A., O'Brien, J., Osgood, G., & Cornblatt, B. (1995). Distractibility in schizophrenia. *Psychiatry Research, 57*(2), 131–140.

Bourdaud, N., Chavarriaga, R., Galán, F., & del R Millan, J. (2008). Characterizing the EEG correlates of exploratory behavior. *IEEE Transactions on Neural Systems and Rehabilitation Engineering, 16*(6), 549–556.

Daw, N. D., O'Doherty, J. P., Dayan, P., Seymour, B., & Dolan, R. J. (2006). Cortical substrates for exploratory decisions in humans. *Nature, 441*(7095), 876–879.

Depue, B. E., Curran, T., & Banich, M. T. (2007). Prefrontal regions orchestrate suppression of emotional memories via a two-phase process. *Science, 317*(5835), 215–219. https://doi.org/10.1126/science.1139560.

Devan, B. D., & White, N. M. (1999). Parallel information processing in the dorsal striatum: Relation to hippocampal function. *Journal of Neuroscience, 19*(7), 2789–2798.

Dominey, P. F., & Arbib, M. A. (1992). A cortico-subcortical model for generation of spatially accurate sequential saccades. *Cerebral Cortex, 2*(2), 153–175.

Fellows, Noth, J., & Schwarz, M. (1998). Precision grip and Parkinson's disease. *Brain, 121*(9), 1771–1784. https://doi.org/10.1093/brain/121.9.1771.

Frank, M. J., Samanta, J., Moustafa, A. A., & Sherman, S. J. (2007). Hold your horses: Impulsivity, deep brain stimulation, and medication in Parkinsonism. *Science, 318*(5854), 1309–1312.

Frank, M. J., Seeberger, L. C., & O'reilly, R. C. (2004). By carrot or by stick: Cognitive reinforcement learning in Parkinsonism. *Science, 306*(5703), 1940–1943.

Gillies, A., Willshaw, D., & Li, Z. (2002). Subthalamic-pallidal interactions are critical in determining normal and abnormal functioning of the basal ganglia. *Proceedings of the Royal Society of London B: Biological Sciences, 269*(1491), 545–551. https://doi.org/10.1098/rspb.2001.1817.

Harel, B., Cannizzaro, M., & Snyder, P. J. (2004). Variability in fundamental frequency during speech in prodromal and incipient Parkinson's disease: A longitudinal case study. *Brain and Cognition, 56*(1), 24–29.

Hazy, T. E., Frank, M. J., & O'Reilly R, C. (2007). Towards an executive without a homunculus: computational models of the prefrontal cortex/basal ganglia system. *Philosophical Transactions of the Royal Society of London. Series B, Biological Sciences.*

Ingvarsson, P. E., Gordon, A. M., & Forssberg, H. (1997). Coordination of manipulative forces in Parkinson's disease. *Experimental Neurology, 145*(2), 489–501.

Isoda, M., & Hikosaka, O. (2008). Role for subthalamic nucleus neurons in switching from automatic to controlled eye movement. *The Journal of Neuroscience, 28*(28), 7209–7218. https://doi.org/10.1523/jneurosci.0487-08.200828/28/7209. [pii].

Joel, D., Niv, Y., & Ruppin, E. (2002). Actor–critic models of the basal ganglia: New anatomical and computational perspectives. *Neural Networks, 15*(4), 535–547.

Jorgensen, W. L. (2004). The many roles of computation in drug discovery. *Science, 303*(5665), 1813–1818.

Kirkpatrick, S., Gelatt, C. D., Jr., & Vecchi, M. P. (1983). Optimization by simulated annealing. *Science, 220*(4598), 671–680. https://doi.org/10.1126/science.220.4598.671.

Kori, A., Miyashita, N., Kato, M., Hikosaka, O., Usui, S., & Matsumura, M. (1995). Eye movements in monkeys with local dopamine depletion in the caudate nucleus. II. Deficits in voluntary saccades. *Journal of Neuroscience, 15*(1), 928–941.

Krishnan, R., Ratnadurai, S., Subramanian, D., Chakravarthy, V. S., & Rengaswamy, M. (2011). Modeling the role of basal ganglia in saccade generation: Is the indirect pathway the explorer? *Neural Networks, 24*(8), 801–813.

LaBar, K. S., Gitelman, D. R., Parrish, T. B., & Mesulam, M. (1999). Neuroanatomic overlap of working memory and spatial attention networks: A functional MRI comparison within subjects. *Neuroimage, 10*(6), 695–704. https://doi.org/10.1006/nimg.1999.0503.

Levy, B. J., & Anderson, M. C. (2008). Individual differences in the suppression of unwanted memories: The executive deficit hypothesis. *Acta Psychologica (Amst), 127*(3), 623–635. https://doi.org/10.1016/j.actpsy.2007.12.004.

Mandali, A., & Chakravarthy, V. S. (2016). Probing the Role of Medication, DBS Electrode Position, and Antidromic Activation on Impulsivity Using a Computational Model of Basal Ganglia. *Frontiers in Human Neuroscience, 10.*

Mandali, A., Chakravarthy, V. S., Rajan, R., Sarma, S., & Kishore, A. (2016). Electrode position and current amplitude modulate impulsivity after subthalamic stimulation in Parkinsons disease—A computational study. *Frontiers in Physiology, 7.*

Mandali, A., Rengaswamy, M., Chakravarthy, S., & Moustafa, A. A. (2015). A spiking basal ganglia model of synchrony, exploration and decision making. *Frontiers in Neuroscience, 9,* 191.

Mayer, J. S., Bittner, R. A., Nikolic, D., Bledowski, C., Goebel, R., & Linden, D. E. (2007). Common neural substrates for visual working memory and attention. *Neuroimage, 36*(2), 441–453. https://doi.org/10.1016/j.neuroimage.2007.03.007.

Mehler-Wex, C., Riederer, P., & Gerlach, M. (2006). Dopaminergic dysbalance in distinct basal ganglia neurocircuits: Implications for the pathophysiology of Parkinson's disease, schizophrenia and attention deficit hyperactivity disorder. *Neurotoxicity Research, 10*(3–4), 167–179.

Miyoshi, E., Wietzikoski, S., Camplessei, M., Silveira, R., Takahashi, R. N., & Da Cunha, C. (2002). Impaired learning in a spatial working memory version and in a cued version of the water maze in rats with MPTP-induced mesencephalic dopaminergic lesions. *Brain Research Bulletin, 58*(1), 41–47.

Morris, R., Griffiths, O., Le Pelley, M. E., & Weickert, T. W. (2013). Attention to irrelevant cues is related to positive symptoms in schizophrenia. *Schizophrenia Bulletin, 39*(3), 575–582. https://doi.org/10.1093/schbul/sbr192.

Moustafa, A. A. (2015). On and off switches in the brain. *Frontiers in Behavioral Neuroscience, 9,* 114. https://doi.org/10.3389/fnbeh.2015.00114.

Ondo, W. G., & Lai, D. (2008). Predictors of impulsivity and reward seeking behavior with dopamine agonists. *Parkinsonism & Related Disorders, 14*(1), 28–32.

Packard, M. G., & Knowlton, B. J. (2002). Learning and memory functions of the basal ganglia. *Annual Review of Neuroscience, 25,* 563–593.

Pankow, A., Katthagen, T., Diner, S., Deserno, L., Boehme, R., Kathmann, N., … Schlagenhauf, F. (2016). Aberrant salience is related to dysfunctional self-referential processing in psychosis. *Schizophrenia Bulletin, 42*(1), 67–76. https://doi.org/10.1093/schbul/sbv098.

Pinto, S., Ozsancak, C., Tripoliti, E., Thobois, S., Limousin-Dowsey, P., & Auzou, P. (2004). Treatments for dysarthria in Parkinson's disease. *Lancet Neurol, 3*(9), 547–556. https://doi.org/10.1016/s1474-4422(04)00854-3S1474442204008543. [pii].

Rochefort, C., Lefort, J. M., & Rondi-Reig, L. (2013). The cerebellum: A new key structure in the navigation system. *Front Neural Circuits, 7,* 35. https://doi.org/10.3389/fncir.2013.00035.

Sukumar, D., Rengaswamy, M., & Chakravarthy, V. S. (2012). Modeling the contributions of basal ganglia and hippocampus to spatial navigation using reinforcement learning. *PLoS ONE, 7*(10), e47467.

Sutton, R. S., & Barto, A. G. (1998). *Reinforcement learning : An introduction*. Cambridge, MA: MIT Press.

Voon, V., Thomsen, T., Miyasaki, J. M., de Souza, M., Shafro, A., Fox, S. H., et al. (2007). Factors associated with dopaminergic drug–related pathological gambling in Parkinson disease. *Archives of Neurology, 64*(2), 212–216.

Weinberger, M., Hutchison, W. D., Lozano, A. M., Hodaie, M., & Dostrovsky, J. O. (2009). Increased gamma oscillatory activity in the subthalamic nucleus during tremor in Parkinson's disease patients. *Journal of Neurophysiology, 101*(2), 789–802.

Willshaw, D., & Li, Z. (2002). Subthalamic–pallidal interactions are critical in determining normal and abnormal functioning of the basal ganglia. *Proceedings of the Royal Society of London, Series B: Biological Sciences, 269*(1491), 545–551.

Printed by Printforce, the Netherlands